# CONTROL AND DYNAMIC SYSTEMS

*Advances in Theory and Application*

Volume 15

*CONTRIBUTORS TO THIS VOLUME*

RUTHERFORD ARIS
J. S. ARORA
JENS G. BALCHEN
LEONARD BECKER
LEONARD CHIN
MORTON M. DENN
JOSEPH J. DISTEFANO III
E. J. HAUG, JR.
PATRICK H. MAK
WILLIAM W-G. YEH

# CONTROL AND DYNAMIC SYSTEMS

## *ADVANCES IN THEORY AND APPLICATION*

Edited by
*C. T. LEONDES*

School of Engineering and Applied Science
University of California
Los Angeles, California

*VOLUME 15 1979*

ACADEMIC PRESS New York San Francisco London
A Subsidiary of Harcourt Brace Jovanovich, Publishers

ACADEMIC PRESS RAPID MANUSCRIPT REPRODUCTION



ACADEMIC PRESS, INC.
111 Fifth Avenue, New York, New York 10003

*United Kingdom Edition published by*
ACADEMIC PRESS, INC. (LONDON) LTD.
24/28 Oval Road, London NW1 7DX

LIBRARY OF CONGRESS CATALOG CARD NUMBER: 64-8027

ISBN 012-012715-6

PRINTED IN THE UNITED STATES OF AMERICA

79 80 81 82     9 8 7 6 5 4 3 2 1

# TABLE OF CONTENTS

# CONTRIBUTORS

Numbers in parentheses indicate the pages on which authors' contributions begin.

*Rutherford Aris* (41), Department of Chemical Engineering and Materials Science, University of Minnesota, 421 Washington Ave. S. E., Minneapolis, Minnesota

*J. S. Arora* (247), Division of Materials Engineering, College of Engineering, The University of Iowa, Iowa City, Iowa

*Jens G. Balchen* (99), Division of Engineering Cybernetics, The Norwegian Institute of Technology, University of Trondheim, Trondheim, Norway

*Leonard Becker* (195), Engineering Systems Department, University of California, Los Angeles, California

*Leonard Chin* (277), Communication Navigation Technology Directorate, Naval Air Development Center, Warminster, Pennsylvania

*Morton M. Denn* (147), Department of Chemical Engineering, University of Delaware, Newark, Delaware

*Joseph J. DiStefano III* (1), Departments of Engineering Systems and Medicine, University of California, Los Angeles, California

*E. J. Haug, Jr.* (247), Division of Materials Engineering, College of Engineering, The University of Iowa, Iowa City, Iowa

*Patrick H. Mak* (1), Biocybernetics Laboratory, University of California, Los Angeles, California

*William W-G. Yeh* (195), Engineering Systems Department, University of California, Los Angeles, California

# PREFACE

This volume continues the theme of Volume 14, models for complex and/or large-scale engineering systems. During the 1950s and 1960s techniques for the analysis and synthesis of systems were rather well developed. But even with a good, and continually improved, foundation for the analysis and synthesis of dynamic systems control, there remains the major issue of effective and requisite techniques for the modeling of such systems; and this, of course, is the starting point for the analysis and synthesis. The purpose of the two most recent volumes in this series is to establish through a sufficiently diverse array of complex systems models an adequately comprehensive base of techniques for approaching a wide variety of applied systems problems.

The first contribution in this volume, "Optimal control policies for the prescription of clinical drugs" by Patrick H. Mak and Joseph J. DiStefano III, exemplifies a significant trend in interdisciplinary efforts in areas with potential for great utility, in this case, biomedical engineering. Such an enormously complex and challenging effort requires individuals of dedication and uncommon interdisciplinary expertise. The present contribution by two such individuals should be a standard reference in this area.

Rutherford Aris is an internationally recognized pioneer in efforts to introduce modern control technology to chemical systems engineering. In the next contribution, "Methods in the modeling of chemical engineering systems," Professor Aris shares with us some most-notable results of his research during his tenure as a Fairchild Fellow as a visiting professor at the California Institute of Technology. This chapter too should prove to be a source reference for many years to come.

Advances in a variety of research efforts in agriculture have resulted in many important accomplishments. It is not at all unreasonable to expect such significant results in other areas as well. This is the motivation for the interesting contribution, "Modeling, prediction, and control of fish behavior," by the internationally recognized authority Jens Balchen. This contribution has potential significance well beyond this area, for the techniques Professor Balchen presents are broadly applicable.

Morton Denn, another pioneer of international stature, is well recognized for his modern systems engineering advances in process control engineering. In "Modeling for process control," we are indeed fortunate to have in the results of Professor Denn's research a contribution that will be an important reference for many years to come in the area of process control.

In a worldwide environment in which there is an increasing awareness and appreciation of the need for effective use of limited natural resources and therefore of the urgency for introducing practical system optimization techniques, the next contribution, "Water resource systems models" by William Yeh and Leonard Becker is extremely timely. Theirs is an excellent presentation of many issues of substantial complexity and importance having wide applicability to the utilization of a diverse array of limited natural resources now and in the future.

Until fairly recently structural systems have been approached as static systems, when, in fact, many such systems are more largely dynamic systems. One of the acknowledged international leaders in dynamic structural systems research and development is E. J. Haug. The next contribution, "Sensitivity analysis and optimization of large-scale structures" by J. S. Arora and E. J. Haug, presents some of the recent important and fundamental results in Professor Haug's continuing efforts in dynamic structural systems optimization.

Finally, for many issues of large-scale systems modeling, there is incomplete knowledge of dynamic systems parameters, particularly so in the case of stochastic systems. In the final contribution, "Advances in adaptive filtering" by Leonard Chin, an impressively comprehensive treatment of the many major issues in this very significant area is presented. This contribution will also undoubtedly become a basic reference for many years to come.

# CONTENTS OF PREVIOUS VOLUMES

Volume 5



Volume 6

## Volume 7



## Volume 8

# Optimal Control Policies for the Prescription of Clinical Drugs: Dynamics of Hormone Replacement for Endocrine Deficiency Disorders

PATRICK H. MAK[a]

*Biocybernetics Laboratory*
*University of California*
*Los Angeles, California*

and

JOSEPH J. DISTEFANO III

*Departments of Engineering Systems and Medicine*
*University of California*
*Los Angeles, California*



[a]*Currently with the Jet Propulsion Laboratory, California Institute of Technology, Pasadena, California.*




ISBN 0-12-012715-6

## I. INTRODUCTION

The goals of biological science and engineering are, for the most part, fundamentally different. Biology is primarily an analytic science, whereas engineering is concerned more with synthesis. On the other hand, medicine and engineering have much more in common. For example, the problems of clinical therapy and control engineering have many similar features. In fact, it is probably true that most if not all problems in clinical therapy can be formulated as problems in optimal control. We hasten to add, however, that mere formulation of such problems in the manner indicated is far from a guarantee for a practical solution.

The purpose of this chapter is first to illustrate how one such clinical problem can be recast as an optimal control problem and, second, to show in some detail how a very practical set of feasible solutions to this particular problem can be obtained. Many other example applications of optimal control theory to clinical problems can be found in the literature of the last decade. No attempt is made to review them here.

## II. CASE STUDY: OPTIMAL CONTROL POLICIES IN ENDOCRINE DISEASE

### *A. THE THYROID ENDOCRINE SYSTEM AND THE CLINICAL PROBLEM*

The specific endocrine system under consideration is the thyroid system. In this section we discuss, quite briefly, only those aspects of this process relevant to our problem. Readers interested in additional physiological or clinical details are referred to any recently published text in endocrinology as well as to the references at the end of this chapter.

Two thyroid hormones, namely, thyroxine and triiodothyronine (abbreviated $T_4$ and $T_3$, respectively), are secreted by the thyroid gland into the blood circulation. These "control signals" are distributed throughout the body and, generally speaking, are responsible for maintaining normal metabolic function. Disorders of the thyroid gland are among the most common diseases in clinical endocrinology, only second in importance to disorders of the pancreas such as diabetes mellitus. One of the most prevalent is primary hypothyroidism. This disorder is manifested by failure of the thyroid gland itself to secrete adequate amounts of $T_3$ and $T_4$, which results in abnormally low levels of these hormones in blood. A common clinical manifestation of this disease is goiter.

The classic method of treating hypothyroid patients is by supplementation of their inadequate hormone supplies with oral dosages of synthetic thyroid hormones or thyroid extract. Synthetic $T_4$ by itself appears particularly suitable in this regard. Patients treated with this compound alone have remarkably constant blood $T_3$ as well as $T_4$ levels throughout the day, simulating the relatively constant levels found in normal subjects [7]. It is important to note that the resulting normal $T_3$ levels observed in patients treated with $T_4$ alone are due to the conversion of $T_4$ to $T_3$ in the tissues. That is, some of the $T_4$ given orally is transformed chemically into $T_3$. It has recently been discovered that this conversion process can account for as much as 85% of all circulating $T_3$ in the body [1-6], in normal as well as treated individuals, indicating that the major source of circulating $T_3$ is this conversion process rather than the thyroid gland itself. In the context of hypothyroid replacement therapy, this fact is significant because the traditionally recommended average $T_4$ dose (300-400 μg/day) was based on the previously held belief that all $T_3$ is derived from the thyroid gland, and that a supernormal $T_4$ level is needed to compensate for the insufficient glandular $T_3$ secretion in hypothyroid patients. Since 1970, however, the question of what $T_4$ replacement dosage is "optimum" for hypothyroid patients has been reevaluated. It has been shown that 300 μg/day can result in metabolic alterations suggestive of subclinical *hyper*thyroidism [8] ($T_3$, $T_4$ levels too high); and another group of investigators showed that 90% of their primary hypothyroid patients could be restored to normal by much lower dosages, between 100-200 μg/day [9].

If both $T_4$ and $T_3$ are normally secreted by the thyroid gland, why not give both hormones to hypothyroid patients? It is interesting that the use of both $T_4$ and $T_3$ in replacement therapies has been discouraged by clinical thyroidologists. Their reluctance is based on the observation that patients maintained by the U.S. Food and Drug Administration (FDA) approved $T_3$:$T_4$ ratio of 1:4 showed a rapid transient rise in their plasma $T_3$ levels to as high as 2 to 3 times normal levels after ingestion of the combination drug, before gradually declining back to normal after about 24 hr [7,10]. Although no observable deleterious effects have been reported as a consequence, these $T_3$ fluctuations have led most clinicians to question the suitability of $T_3$ as a replacement agent, particularly considering that normal range blood $T_3$ levels can be restored by $T_4$ alone. Looking ahead, treatment with $T_4$ alone is strongly supported by our computations (discussed later), and that the optimal $T_3$:$T_4$ replacement ratio is almost 1:35, nearly 10 times smaller than the FDA approved ratio.

How does the clinician determine how much replacement hormone to prescribe? In clinical practice, whether $T_4$ alone or $T_3$ and $T_4$ combination therapy is utilized, the dosages usually are determined on a trial-and-error basis. A small trial dose is given initially. The dosage is then periodically adjusted in fixed amounts for up to several months until a final stable optimum condition is achieved. However, unless the patient is monitored carefully during this trial period, there is a good chance that the patient will become transiently hyperthyroid ($T_3$,$T_4$ levels too high) before the proper maintenance dosage can be determined. These fluctuations from hypo- to hyperthyroid, and vice versa, on occasion precipitate deleterious effects in some patients, a situation clearly inconsistent with sound clinical management, a feasible goal. The remainder of this chapter is concerned with an alternative optimal control theoretic solution to this problem. First, we consider an appropriate model of the pertinent portions of the process of interest.

*B. A MODEL OF THE DYNAMICS OF THYROID HORMONE METABOLISM IN HYPOTHYROID SUBJECTS*

The dynamics of the normal system have been studied extensively in the Biocybernetics Laboratory at UCLA [6,11-13]. The structure of a model describing overall $T_3$ and $T_4$ metabolism is depicted in Fig. 1. All nomenclature are given in Appendix A. The model has six pools, one each for $T_3$ and $T_4$ in plasma, *fast* and *slow* pools. The fast pools consist of all extravascular tissue spaces having rapid exchange dynamics with hormone in the plasma pool, mainly liver and kidney. The slow pools consist of all tissue spaces having relatively slow hormone exchange dynamics with the plasma pool; they consist mainly of skeletal muscle.

The *hypothyroid* patient model has four inputs: two endogenous secretion rates ($SR_3$ and $SR_4$) from the thyroid gland, both of which are zero or small fractions of the normal rates; and two exogenous absorption rates ($AR_3$ and $AR_4$) that result from the ingestion of oral replacement dosages. The functional relationships between the continuous absorption rates and the discrete oral dose rates are governed by the dynamics of gut absorption and are discussed in Appendix B. In this study, the optimal control inputs to be determined are the AR's; the secretion rates can be estimated from measurements. Once the optimal AR's have been determined, the oral dose rates can be calculated. Additional details about the model structure are given in references [6,11].

The model equations are (1)-(6) [6]. The state variables represent the $T_3$ and $T_4$ concentrations in each of the six pools.

$$\dot{x}_1 = [-(c_1+c_4)r_1 + c_2x_2 + c_5x_3 + u_1 + u_2]/c_{15} \quad , \qquad (1)$$

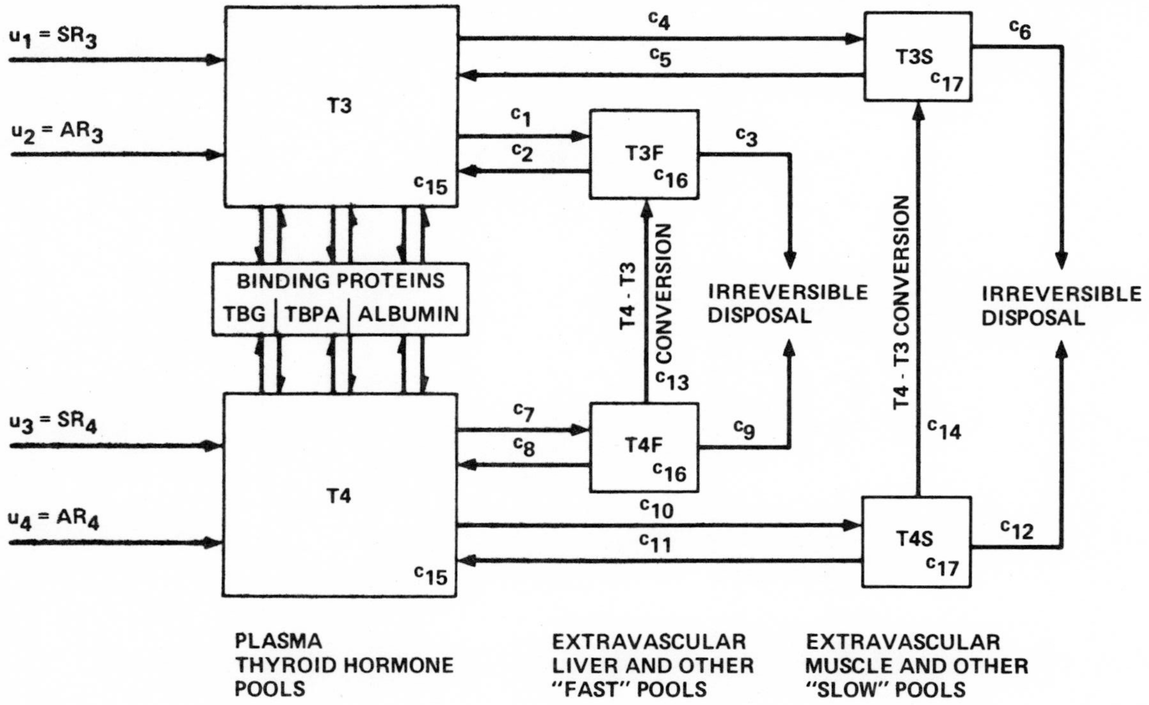


*FIG. 1. Structure and connectivity of a model of thyroid hormone metabolism in the human, adapted from DiStefano* et al. *[6]. All symbols are defined in Appendix A.*

$$\dot{x}_2 = [c_1 r_1 - (c_2+c_3)x_2 + c_{13}x_5]/c_{16} \quad , \tag{2}$$

$$\dot{x}_3 = [c_4 r_1 - (c_5+c_6)x_3 + c_{14}x_6]/c_{17} \quad , \tag{3}$$

$$\dot{x}_4 = [-(c_7+c_{10})r_2 + c_8 x_5 + c_{11}x_6 + u_3 + u_4]/c_{15} \quad , \tag{4}$$

$$\dot{x}_5 = [c_7 r_2 - (c_8+c_9+c_{13})x_5]/c_{16} \quad , \tag{5}$$

$$\dot{x}_6 = [c_{10} r_2 - (c_{11}+c_{12}+c_{14})x_6]/c_{17} \quad . \tag{6}$$

The symbols $r_1$ and $r_2$ in these equations are nonlinear functions of $x_1$ and $x_4$. For an average hypothyroid patient [19] the relationships are

$$r_1 \doteq x_1(0.0026 + 0.0055\ x_4 + 0.0028\ x_4^2 - 0.008\ x_4^3) \quad , \tag{7}$$

$$r_2 = x_4(0.0002 + 0.00053\ x_4 + 0.00082\ x_4^2 - 0.0012\ x_4^3) \quad . \tag{8}$$

Only two state variables are directly measurable: $x_1$ and $x_4$, the plasma $T_3$ and $T_4$ concentrations. We will have occasion to use the following vector notation in the sequel:

$$\underset{\sim}{x} = [x_1\ x_2\ x_3\ x_4\ x_5\ x_6]^T = [T_3\ T_{3F}\ T_{3S}\ T_4\ T_{4F}\ T_{4S}]^T , \tag{9}$$

$$\underset{\sim}{u} = [u_1\ u_2\ u_3\ u_4]^T = [SR_3\ AR_3\ SR_4\ AR_4]^T \quad , \tag{10}$$

$$\underset{\sim}{c} = [c_1\ c_2\ \ldots\ c_{17}]^T \quad . \tag{11}$$

Model identification: The initial conditions $x_1(0)$ and $x_4(0)$ are measurable directly. The remaining ones are obtained in terms of these two by setting Eqs. (2), (3), (5), and (6) to zero and solving the resulting algebraic equations; the system is assumed to be initially in steady state. The system is observable, and $\underset{\sim}{u}(0) = [SR_3\ 0\ SR_4\ 0]^T$ is determined from Eqs. (1) and (4). $SR_3$ and $SR_4$ are fractions of the respective euthyroid levels, $SR_{3N}$ and $SR_{4N}$, and their values depend on the severity of the hypothyroid condition. For example, the case when both $SR_3$ and $SR_4$ are zero corresponds to an athyreotic patient (without a thyroid gland) with no residual thyroid function.

The model is characterized by 14 clearance rates and three volume parameters, a total of 17 unknowns, $c_1, \ldots, c_{17}$. Using measurable $T_3$ and $T_4$ impulse responses as our data base, only 12 independent algebraic combinations of these 17 unknown parameters are identifiable and were estimated [6]. To quantify the model completely, five parameters were independently estimated: $c_1$, $c_{10}$, $c_{14}$, $c_{16}$, and $c_{17}$. The fast pool volume $c_{16}$ was estimated as 2.5 liters; liver volume is about 2 liters, kidney volume is about 0.5 liter [15,16]. The slow pool volume $c_{17}$ was approximated by the volume of skeletal muscle, which is about one-third of the body weight. Based on these two selections and the physical requirement that all $c$s must be nonnegative, the maximum and minimum bounds for $c_1$, $c_{10}$, and $c_{14}$ were calculated from the 12 computable combinations. It turned out that the bounds on $c_1$ and $c_{10}$ were very narrow, differing by less than 10%. Hence, their midranges were selected. Bounds for $c_{14}$, however, were not so tight, varying between zero (no $T_4$ to $T_3$ conversion in muscle) and 0.0789. Therefore, three values of $c_{14}$ (0.001, midrange, and 90% of $c_{14max}$) were selected within the range, and simulations were performed to study the possible effect on the optimal solution. The remaining $c$s were calculated from the values of these five parameters and the 12 computable combinations. The three sets of values, one set for each value of $c_{14}$, are shown in Table 1.

### C. *COMPUTATION OF STEADY-STATE MAINTENANCE DOSAGES*

Steady-state maintenance dosages, which represent the amounts of oral hormone necessary for maintaining a patient's euthyroid condition following the transient period of therapy,

TABLE 1

*The Three Sets of Parameter Values of the Model Based on Three Different Selections of $c_{14}$ (See Text)*

| Parameter | Parameter values | | | Units |
|---|---|---|---|---|
| | $c_{14} = 0.001$ | $c_{14}$ = midrange | $c_{14} = 0.9\ c_{14max}$ | |
| $c_1$ | 22.33 | 22.33 | 22.33 | liter/hr |
| $c_2$ | 5.297 | 5.297 | 5.297 | liter/hr |
| $c_3$ | 0.128 | 0.128 | 0.128 | liter/hr |
| $c_4$ | 1.88 | 1.88 | 1.88 | liter/hr |
| $c_5$ | 0.972 | 0.972 | 0.972 | liter/hr |
| $c_6$ | 0.378 | 0.378 | 0.378 | liter/hr |
| $c_7$ | 4.112 | 4.112 | 4.112 | liter/hr |
| $c_8$ | 2.576 | 2.576 | 2.576 | liter/hr |
| $c_9$ | 0.00629 | 0.0115 | 0.01578 | liter/hr |
| $c_{10}$ | 0.4265 | 0.4265 | 0.4265 | liter/hr |
| $c_{11}$ | 1.37 | 1.37 | 1.37 | liter/hr |
| $c_{12}$ | 0.0906 | 0.052 | 0.0206 | liter/hr |
| $c_{13}$ | 0.0106 | 0.00535 | 0.00107 | liter/hr |
| $c_{14}$ | 0.001 | 0.03945 | 0.071 | liter/hr |
| $c_{15}$ | 3.05 | 3.05 | 3.05 | liters |
| $c_{16}$ | 2.5 | 2.5 | 2.5 | liters |
| $c_{17}$ | 25.0 | 25.0 | 25.0 | liters |

can be determined in a relatively simple manner, without an elaborate optimization procedure. These results thus provide a check on the dynamic optimization solution.

*1. Maintenance with both $T_3$ and $T_4$.* In primary hypothyroidism, both $T_3$ and $T_4$ secretion rates are fractions of normal and these fractions are generally different because the thyroid gland preferentially secretes $T_3$ in this state. Let $z_3$ and $z_4$ be the respective fractions of residual function. Then, in the steady state, one has to replace $1 - z_3$ and $1 - z_4$ of the euthyroid secretion rates, respectively. Also, if the dosage is incompletely absorbed, the percentage absorption must be taken into consideration. Equations (12) and (13) represent the resulting relationships:

$$MD_3 = [(1-z_3)\ SR_{3N}]/AB_3 \quad , \tag{12}$$

$$MD_4 = [(1-z_4)\ SR_{4N}]/AB_4 \quad , \tag{13}$$

where, for $i = 3, 4$, $MD_i$ = daily $T_i$ maintenance dose rate; $AB_i$ = fraction absorbed for an oral $T_i$ dose. Consider an athyreotic patient for example. In this case, both $z_3 = z_4 = 0$, as there is no residual thyroid function. Therefore, in the steady state, full replacement must be made with respect to the euthyroid secretion rates, i.e.,

$$MD_3 = SR_{3N}/AB_3 \quad , \tag{14}$$

$$MD_4 = SR_{4N}/AB_4 \quad , \tag{15}$$

For an average human, the euthyroid (normal) endogenous secretion rates for $T_3$ and $T_4$ have been estimated to be 0.366 and 5.41 nmoles/hr, respectively [6], whereas $AB_3 = 1$ (100%) and $AB_4 = 0.5$ (50%) [17,18]. Therefore, the steady-state daily maintenance dosages for an average athyreotic patient are

$$MD_3 = 0.366 \text{ nmole/hr} = 5.73\ \mu\text{g } T_3/\text{day} \quad ,$$

$$MD_4 = 5.41/0.5 \text{ nmoles/hr} = 201.8 \ \mu\text{g } T_4/\text{day}$$

The average $T_3:T_4$ maintenance dose is therefore approximately 1:35.

2. *Maintenance with $T_4$ Alone.* The advantage of using $T_4$ alone as the replacement agent has been discussed. In this case, the daily maintenance dosage can be calculated as follows. Let $CV_{4-3}$ be the fraction of $T_4$ secreted by the thyroid that is converted to $T_3$. This fraction has been estimated to be approximately 0.36 or 36% [12]. One mole of $T_4$ yields one mole of $T_3$ upon conversion. Therefore the dose rate of $T_4$ necessary to replace $(1-z_3)SR_{3N}$ of $T_3$ [Eq. (12)] is $(1-z_3)SR_{3N}/CV_{4-3}$. Together with the amount necessary to maintain the $T_4$ level, the total $T_4$ maintenance dose is therefore

$$MD_4 = [(1-z_4)SR_{4N} + (1-z_3)SR_{3N}/CV_{4-3}]/AB_4 \tag{16}$$

where incomplete gut absorption of $T_4$ is again accounted for. For the average athyreotic patient discussed, Eq. (16) reduces to

$$MD_4 = [SR_{4N} + SR_{3N}/CV_{4-3}]/AB_4 \quad . \tag{17}$$

Substituting the values from the previous section, the average daily $T_4$ alone maintenance dosage is

$$MD_4 = [5.41 + 0.366/0.36]/0.5 \text{ nmoles/hr}$$
$$= 240 \ \mu\text{g } T_4/\text{day} \quad .$$

This value is in agreement with recent clinical results [7,9].

## D. *CRITERION FOR THERAPEUTIC OPTIMALITY*

1. *The Criterion Function.* Most physicians have preferred recovery patterns in mind when they are treating patients. For example, in elderly patients, the preferred pattern is gradual recovery, allowing ample time for patients

to adapt to the oral hormone(s). In practice, plasma $T_3$ and $T_4$ levels (among other factors) may be monitored periodically to see if they are approaching the normal condition in accordance with this criterion and any deviations from it usually call for either an increase or decrease in the dosage prescription. In this section, we quantify this trial-and-error approach of patient treatment, but with somewhat greater generality.

The first step is to select a suitable therapeutic optimization criterion, one which provides a quantitative measure of the relative goodness of different dosage regimen control policies. In this regard, two important questions of interest are (1) how much time should it take to achieve the euthyroid status (time for the transient response); and (2) how should the euthyroid condition be achieved (the dynamics of the transient response)? If the normal condition is restored too rapidly, this may pose a potential hazard to patients with other ailments, such as the possible danger of cardiac complications in patients with heart disease. If they are approached too slowly, this would unnecessarily prolong the patient's hypothyroid condition, which also entails potential dangers. Clearly, neither situation is desirable. The specific solution to this problem depends on the individual patient's condition. For these reasons, a rather general form of criterion is chosen.

Let $x_{1N}(t)$ and $x_{4N}(t)$ be the desired recovery patterns of $x_1(t)$ and $x_4(t)$, respectively. The deviation of $x_1$ and $x_4$ from $x_{1N}$ and $x_{4N}$ at any time is denoted by $\underset{\sim}{e}(t)$, i.e.,

$$\underset{\sim}{e}(t) = \begin{bmatrix} x_1(t) - x_{1N}(t) \\ x_4(t) - x_{4N}(t) \end{bmatrix} . \tag{18}$$

The following quadratic form has been chosen as the criterion for therapeutic optimality:

$$J(\underset{\sim}{u}) = \frac{1}{2}\, \underset{\sim}{e}^{T}(t_f) B\, \underset{\sim}{e}(t_f) + \frac{1}{2}\int_0^{t_f} \underset{\sim}{e}^{T}(t) Q\, \underset{\sim}{e}(t)\, dt \quad , \qquad (19)$$

where $t_f$ is the fixed final time, and $B$ and $Q$ are positive semidefinite diagonal matrices. Thus, the desired length of the treatment period $t_f$, can be selected, along with the manner in which plasma $T_3$ and $T_4$ concentrations are to reach euthyroid levels for $t \in [0, t_f]$. The latter is done by simply specifying $x_{1N}(t)$ and $x_{4N}(t)$.

2. *Transient Recovery Patterns*. Two different recovery patterns are considered as practical examples. Other forms may be chosen for other clinical objectives or patient conditions. We first consider a *sigmoidal pattern*, which has the desirable property of minimizing the rates of change of hormone levels during early therapy, thus providing sufficient time for the patient to adjust to the situation. As mentioned earlier, this type of gradual recovery is especially useful with elderly patients or patients who are particularly sensitive to thyroid medication. The two sigmoid patterns $x_{1N}(t)$ and $x_{4N}(t)$ are specified using hyperbolic tangent functions:

$$x_{1N}(t) = \frac{1}{2}\{[x_{1ss} + x_1(0)] + [x_{1ss} - x_1(0)] \tanh[\frac{t}{Ta} - 3]\}, \qquad (20)$$

$$x_{4N}(t) = \frac{1}{2}\{[x_{4ss} + x_4(0)] + [x_{4ss} - x_4(0)] \tanh[\frac{t}{Ta} - 3]\}, \qquad (21)$$

where $t$ is expressed in hours, and $x_{1ss}$ and $x_{4ss}$ are the euthyroid steady-state levels of $x_1$ and $x_4$, respectively. These two steady-state values were either measured or calculated in the same manner as were the initial conditions with

$\underset{\sim}{u} = [SR_{3N} \quad 0 \quad SR_{4N} \quad 0]^T$. We have chosen to restore our patient to the euthyroid condition in four weeks and have set $Ta = 112$ in Eqs. (20), (21) to achieve this.

Second, we consider a variation of a time-optimal recovery pattern. Ideally we would like to have

$$x_{1N}(t) = x_{1ss} \tag{22}$$

$$x_{4N}(t) = x_{4ss} \tag{23}$$

for all $t$ and, with no constraints on the control, the classic solution to this problem for a *linearized* version of our model includes a physically unrealizable impulse input function. This is also expected to be at least approximately true for the nonlinear model. To avoid this problem, a slightly sub-optimal control is sought. This is done by imposing a less restrictive terminating condition (Section E) so that the minimum $J(u)$ obtained will always be greater than zero.

## *E. MATHEMATICAL FORMULATION OF THE DYNAMIC OPTIMIZATION PROBLEM*

To solve the optimal control problem posed in the previous section, we use Pontryagin's minimum principle [19]. The Hamiltonian function for this problem is defined as

$$H(\underset{\sim}{x},\underset{\sim}{u},\underset{\sim}{p}) = \frac{1}{2}\, \underset{\sim}{e}^T(t)\; Q\; \underset{\sim}{e}(t) + \underset{\sim}{p}^T(t) \underset{\sim}{f}(\underset{\sim}{x},\underset{\sim}{u};\underset{\sim}{c}) \quad , \tag{24}$$

where $\underset{\sim}{p}^T(t) = [p_1(t) \; \ldots \; p_6(t)]$ are the costates, and $\underset{\sim}{f}(\underset{\sim}{x},\underset{\sim}{u};\underset{\sim}{c})$ are the vector functions associated with the state equations (1)-(6). Applying the conditions of optimality [19], we arrive at the following 12th order, nonlinear, two-point boundary value problem. The first six equations are state equations, (1)-(6). The other six are the costate equations $p(t)$, given by

$$\dot{\underset{\sim}{p}} = \frac{\partial H}{\partial \underset{\sim}{x}}$$

or

$$\dot{p}_1 = \left[\frac{c_1 + c_4}{c_{15}}\right] \frac{\partial r_1}{\partial x_1} p_1 - \left[\frac{c_1}{c_{16}}\right] \frac{\partial r_1}{\partial x_1} p_2 - \left[\frac{c_4}{c_{17}}\right] \frac{\partial r_1}{\partial x_1} p_3 - q_1(x_1 - x_{1N}) \tag{25}$$

$$\dot{p}_2 = - \left[\frac{c_2}{c_{15}}\right] p_1 + \left[\frac{c_2 + c_3}{c_{16}}\right] p_2 \tag{26}$$

$$\dot{p}_3 = - \left[\frac{c_5}{c_{15}}\right] p_1 + \left[\frac{c_5 + c_6}{c_{17}}\right] p_3 \tag{27}$$

$$\dot{p}_4 = \left[\frac{c_1 + c_4}{c_{15}}\right] \frac{\partial r_1}{\partial x_4} p_1 - \left[\frac{c_1}{c_{16}}\right] \frac{\partial r_1}{\partial x_4} p_2 - \left[\frac{c_4}{c_{17}}\right] \frac{\partial r_1}{\partial x_4} p_3 + \left[\frac{c_7 + c_{10}}{c_{15}}\right] \frac{\partial r_2}{\partial x_4} p_4 - \left[\frac{c_7}{c_{16}}\right] \frac{\partial r_2}{\partial x_4} p_5 - \left[\frac{c_{10}}{c_{17}}\right] \frac{\partial r_2}{\partial x_4} p_6 - q_2(x_4 - x_{4N}) \tag{28}$$

$$\dot{p}_5 = \left[\frac{c_{13}}{c_{16}}\right] p_2 - \left[\frac{c_8}{c_{15}}\right] p_4 + \left[\frac{c_8 + c_9 + c_{13}}{c_{16}}\right] p_5 \tag{29}$$

$$\dot{p}_6 = \left[\frac{c_{14}}{c_{17}}\right] p_3 - \left[\frac{c_{11}}{c_{15}}\right] p_4 + \left[\frac{c_{11} + c_{12} + c_{14}}{c_{17}}\right] p_6 \tag{30}$$

Twelve boundary conditions are required. Six are specified at $t = 0$: the initial conditions $\underset{\sim}{x}(0)$. The remaining six are specified on the costates at the final time as the transversality conditions (free end point, fixed final time):

$$p_1(t_f) = b_1[x_1(t_f) - x_{1N}(t_f)] \tag{31}$$

$$p_2(t_f) = 0 \tag{32}$$

$$p_3(t_f) = 0 \tag{33}$$

$$p_4(t_f) = b_2[x_4(t_f) - x_{4N}(t_f)] \tag{34}$$

$$p_5(t_f) = 0 \tag{35}$$

$$p_6(t_f) = 0 \tag{36}$$

The condition that the optimal control must minimize the Hamiltonian function, i.e., $\partial H/\partial \underset{\sim}{u} = \underset{\sim}{0}$, is used in the numerical algorithm for updating the control during the search for the optimal solution. Note that since $u_1$ and $u_3$ are constant, the Hamiltonian is minimized only with respect to $u_2$ and $u_4$.

### F. *COMPUTATION OF THE OPEN-LOOP OPTIMAL CONTROLS*

The optimal control problem was solved by numerical means for the two recovery patterns selected. In both cases, we assumed that our patient was athyreotic or with minimal thyroid function. The residual (initial) $T_3$ and $T_4$ concentrations in each of the pools were assumed to be a few percent of their respective normal levels.

1. *Case Study 1.* The patient's plasma $T_3$ and $T_4$ levels were required to follow the sigmoid shaped recovery pattern over a 28-day transient period, maintaining euthyroid levels thereafter. A first-order gradient search method was used to find the optimal solution, and the search was continued until the coefficient of variation of the error $\underset{\sim}{e}(t)$ was less than 5% for 30 days. Integrations were performed with a 4th order Runga-Kutta method, with integration step size 0.1 hr. The weighting matrices $B$ and $Q$ of $J(\underset{\sim}{u})$ were selected (after some trial runs) as

$$B = Q = \begin{bmatrix} 1/8 & 0 \\ 0 & 1/100 \end{bmatrix}. \tag{37}$$

This case study was repeated three times, for each of the three sets of parameter values (Table 1).

2. *Case Study 2.* This was a 10-day replacement study in which plasma $T_3$ and $T_4$ levels were required to reach normal values as soon as possible. The recovery patterns $x_{1N}$ and $x_{4N}$ were specified as the euthyroid steady-state values [Eqs. (22)

and (23)]. A transient stage of replacement was desired during the first seven days, with maintenance at the new steady state thereafter. The matrices $B$ and $Q$ were chosen as follows:

$$B = \begin{bmatrix} 1 & 0 \\ 0 & 1 \end{bmatrix}, \qquad Q = \begin{bmatrix} 1/4 & 0 \\ 0 & 1/50 \end{bmatrix}. \tag{38}$$

The same first-order gradient algorithm, integration method, and step size were used. However, the stopping condition was slightly different. This was necessary to avoid the possibility of an impulse control as the optimal solution as explained earlier. The coefficient of variation of error was less than 1% during the last two days of therapy; no restriction was imposed during the transient period. The controls obtaired under this terminating condition are therefore suboptimal. $c_{14}$ was given only its midrange value.

### *G. OPTIMAL ABSORPTION RATE RESULTS*

*1. Case Study 1.* The optimal absorption rates, $u_2 = AR_3$ and $u_4 = AR_4$, are shown in Fig. 2. Not only did the two plasma trajectories follow their respective sigmoidal recovery patterns very well, but the four $T_3$ and $T_4$ fast and slow pool concentrations also were sigmoidal. This was true regardless of the value of $c_{14}$. Because of the similarities in the response curves only those results with $c_{14}$ = midrange are shown.

Little change was observed in the six responses during the first few days. Most of the increase occurred between the second and the third week. At the end of the third week, practically all the trajectories had already reached their respective steady-state levels and were maintained at these levels throughout the remaining period. The two optimal absorption rates $u_2$ and $u_4$ showed similar characteristics. There was an

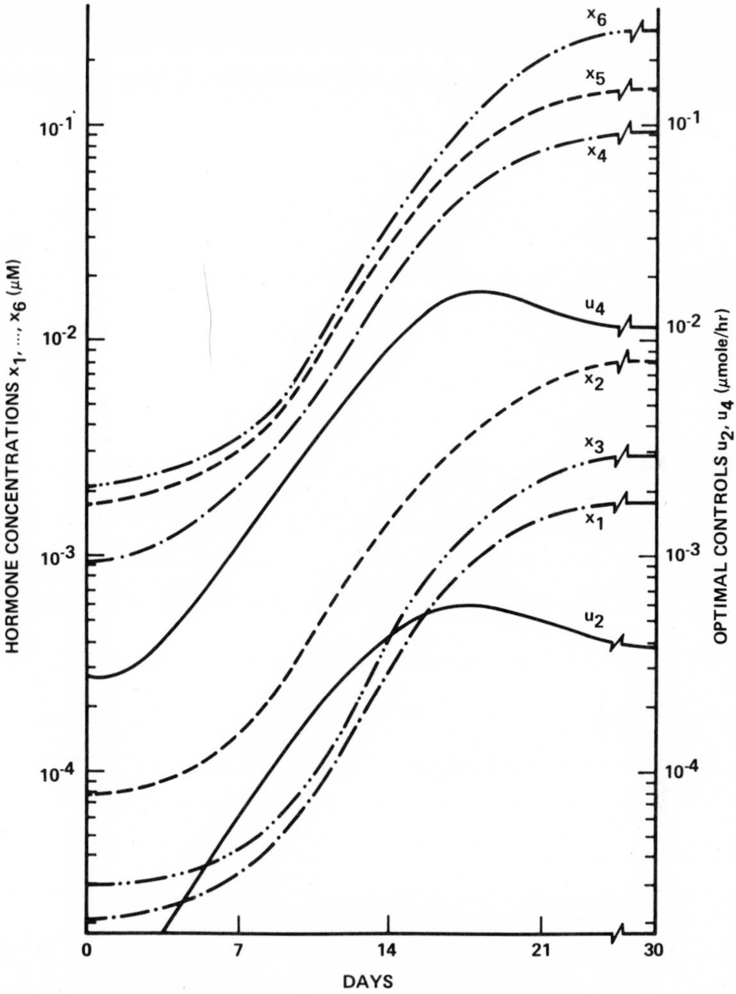


*FIG. 2. Optimal solutions for case study 1.*

initial transient decrease during the first day; then both increased very rapidly (about 50-fold) for the next three weeks, peaking around the midperiod. They then fell gradually and finally leveled off at around $u_2 = 0.00033$ and $u_4 = 0.0048$ μmole/hr in all cases.

Variations in $c_{14}$ had little or no effect on the $T_4$ responses. However, this was not the case for $T_3$. When $c_{14}$ was decreased to 0.001 from its midrange value, a corresponding decrease of about 25% in the steady-state $T_3$ concentration in the slow pool was observed. This was to be expected as $c_{14}$ represents the $T_4$-$T_3$ conversion parameter in the slow pool. A slight increase of about 2% was observed in the fast pool, while there was no change in the plasma $T_3$ concentration. When $c_{14}$ was increased to 90% of its maximum value, no change was observed in plasma $T_3$, but a 2% decrease in $x_2$ and an increase of about 20% in $x_3$ were observed in the steady state. For $u_2$, there was an average 12% increase during the transient stage when $c_{14} = 0.9\ c_{14max}$, but a 20% drop during the same period when $c_{14}$ = midrange. The steady-state values in all three cases were approximately the same. This was because the optimization was performed only with respect to the plasma levels. Had the other pool concentrations been incorporated into the criterion, the results undoubtedly would be different. No change in $u_4$ was observed for any of the three cases.

2. *Case Study 2*. All six hormone concentrations approached steady-state levels asymptotically, but at different rates (Fig 3). The fastest rates were observed in $x_1$ and $x_3$, where the steady-state levels were reached after only about 15 hr. $x_4$ took about two days, while the remaining three hormone levels required about four days. In all cases, the respective steady-state levels were maintained once they had been reached. This was true also in the $T_3$ and $T_4$ fast and slow pools, even though the optimization criterion was minimized only with respect to the plasma trajectories $x_1$ and $x_4$. The steady-state values achieved in these four extravascular pools were the same as in case study 1 when $c_{14}$ was its midrange value. The suboptimal absorption rates $u_2$ and $u_4$ were both rapidly decaying exponentials, but leveled off after about five days at 0.00032 and

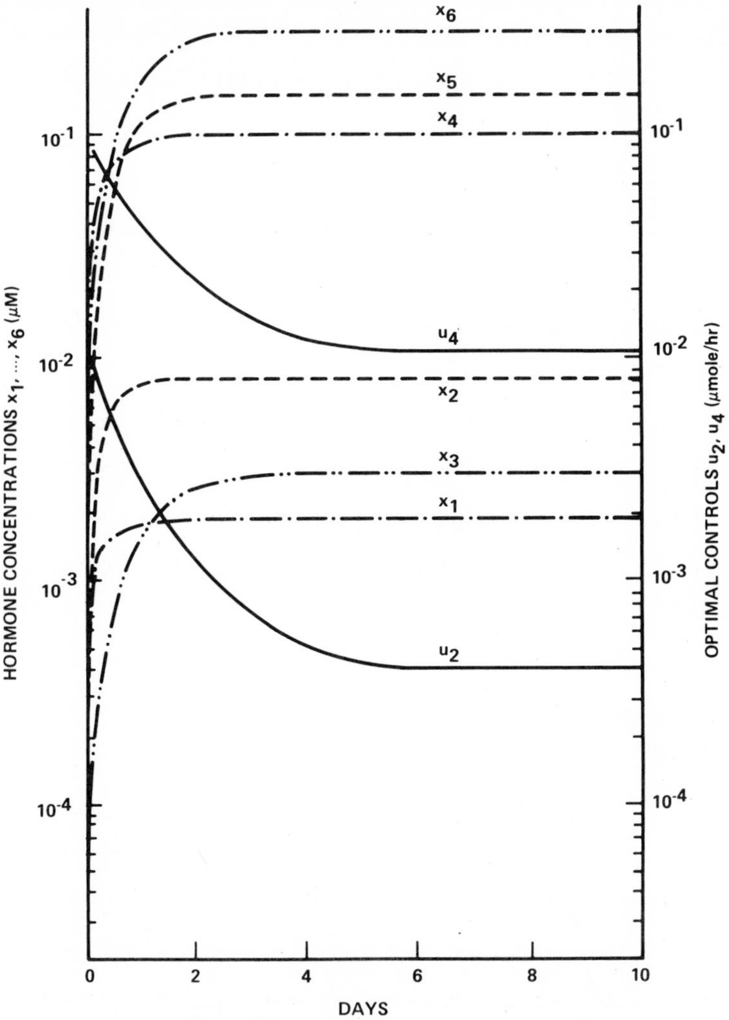


*FIG. 3. Optimal solutions for case study 2.*

0.0048 μmole/hr, respectively, though a slight undershoot was observed around the 6th and 7th days in both cases.

### *H. OPTIMAL DAILY DOSAGES*

In principle, if $T_3$ and $T_4$ were introduced into the bloodstream at the same rates as the optimal $u_2 = AR_3$ and $u_4 = AR_4$, our simulated hypothyroid patient would achieve the euthyroid state in the manner governed by the curves in Figs. 2 and 3.

In other words, the optimal $AR_3$ and $AR_4$ correspond to the optimal rates at which the hormones should arrive in blood, which could be achieved by (say) intravenous infusion of the hormones. However, practical considerations eliminate the intravenous route; hormone replacement is a chronic (usually lifetime) therapy. The only practical route of administration of this hormone therapy is the oral one. We have calculated the oral dosages in the following manner.

*1. $T_3$ and $T_4$ Combination Dosages.* Since the optimal $AR_3$ and $AR_4$ are rates, then integrating them for each 24 hr will give the amounts of $T_3$ and $T_4$ that must be absorbed within each 24-hr period. But since the whole oral dose for either $T_3$ or $T_4$ is not completely absorbed, adjustments must be made accordingly, i.e., for the jth day.

$$OD_i^j = \int_{24j}^{24(j+1)} AR_i(t)\,dt \times MW_i/AB_i \ , \tag{39}$$

where $i = 3,4$; $t$ is expressed in hours; $MW_i$ is the molecular weight of $T_i$ (651 for $T_3$ and 777 for $T_4$); and $AB_i$ is the fraction of $T_i$ absorbed (Section C). The results for both case studies are presented in Tables 2 and 3. The dosages represent the optimal $T_3$ and $T_4$ combination daily oral replacement dosages. It is readily seen that the optimal maintenance dosages in all cases were about the same as those determined from steady state consideration in Section C. Also, the optimal replacement dose ratio for both the transient and the maintenance periods varied from about 1:30 to about 1:50.

*2. $T_4$ Alone Daily Dosage.* The daily $T_4$ dosages were determined in two ways. First, based on the dosages summarized in Tables 2 and 3, we directly calculated the $T_4$ dosage using

TABLE 2

*Optimal $T_3$ and $T_4$ Combination and $T_4$ Alone Daily Dosages for Case Study 1 when $c_{14}$ = midrange*[a]

| Day | $T_3$ and $T_4$ combination ($c_{14}$ = midrange) $T_3$ (μg) | $T_4$ (μg) | $T_3$:$T_4$ ratio | $T_4$ dosages in μg $c_{14}$ = midrange |
|---|---|---|---|---|
| 1 | 0.15 | 5.6 | 1:38 | 6.6 |
| 2 | 0.14 | 5.3 | 1:39 | 6.2 |
| 3 | 0.17 | 6.3 | 1:37 | 7.4 |
| 4 | 0.25 | 8.6 | 1:35 | 10.3 |
| 5 | 0.36 | 12.3 | 1:34 | 14.7 |
| 6 | 0.53 | 18.0 | 1:34 | 21.5 |
| 7 | 0.78 | 26.8 | 1:34 | 31.9 |
| 8 | 1.12 | 39.8 | 1:36 | 47.2 |
| 9 | 1.58 | 58.5 | 1:37 | 69.3 |
| 10 | 2.20 | 84.1 | 1:38 | 98.7 |
| 11 | 3.02 | 117.3 | 1:39 | 137.3 |
| 12 | 4.05 | 157.5 | 1:39 | 184.3 |
| 13 | 5.25 | 202.1 | 1:38 | 236.9 |
| 14 | 6.54 | 246.4 | 1:38 | 289.7 |
| 15 | 7.74 | 284.8 | 1:37 | 336:1 |
| 16 | 8.66 | 312.1 | 1:36 | 369.5 |
| 17 | 9.15 | 325.6 | 1:36 | 386.2 |
| 18 | 9.17 | 325.6 | 1:36 | 386.4 |
| 19 | 8.78 | 315.3 | 1:36 | 373.5 |
| 20 | 8.12 | 299.0 | 1:36 | 352.8 |
| 21 | 7.34 | 280.8 | 1:37 | 329.5 |
| 22 | 6.85 | 263.9 | 1:38 | 309.3 |
| 23 | 6.52 | 250.0 | 1:38 | 293.2 |
| 24 | 6.15 | 239.3 | 1:39 | 280.0 |
| 25 | 5.92 | 231.6 | 1:39 | 270.8 |
| 26 | 5.81 | 225.1 | 1:39 | 263.6 |
| 27 | 5.78 | 216.6 | 1:38 | 254.9 |
| 28 | 5.75 | 207.2 | 1:36 | 245.3 |
| 29 | 5.72 | 201.9 | 1:35 | 239.8 |
| 30 | 5.72 | 201.8 | 1:35 | 239.7 |

[a]*These dosages were obtained by integrating the respective optimal absorption rates (Fig. 2) and adjusting for incomplete absorption (see text).*

*TABLE 3*

*Optimal $T_3$ and $T_4$ Combination and $T_4$ Alone Daily Dosages for Case Study 2*[a]

| | $T_3$ and $T_4$ combination therapy | | | $T_4$ alone therapy |
|---|---|---|---|---|
| Day | $T_3$ (μg) | $T_4$ (μg) | $T_3$:$T_4$ ratio | $T_4$ dosage (μg) |
| 1 | 58.74 | 1540.2 | 1:26 | 1929.7 |
| 2 | 24.63 | 738.0 | 1:30 | 901.3 |
| 3 | 12.61 | 408.6 | 1:32 | 492.2 |
| 4 | 7.35 | 260.8 | 1:35 | 309.5 |
| 5 | 5.42 | 200.7 | 1:37 | 236.6 |
| 6 | 5.12 | 182.9 | 1:37 | 216.8 |
| 7 | 5.37 | 192.5 | 1:36 | 228.1 |
| 8 | 5.53 | 196.6 | 1:36 | 233.2 |
| 9 | 5.72 | 200.2 | 1:35 | 238.1 |
| 10 | 5.72 | 200.9 | 1:35 | 238.8 |

[a]*These dosages were obtained by integrating the respective optimal absorption rates (Fig. 3) and adjusting for incomplete absorption (see text).*

the same approach as in Section C:

$$ODT_4^{\,j} = \left[\frac{OD_3^j}{CV_{4-3}}\right]\left[\frac{MW_4}{MW_3}\right]\left[\frac{AB_3}{AB_4}\right] + OD_4^j \quad , \tag{40}$$

where $ODT_4^{\,j}$ represents the daily $T_4$ alone replacement oral dose rate for the jth day. Second, we solved the optimization problem again with $u_2 = 0$ and determined the new optimal control $u_4$, and subsequently the daily dosages. The two methods yielded essentially the same results, and are presented in the last column of Tables 2 and 3 for the two cases studied.

3. *Weekly Dosage Regimens.* On examining the various daily dosage regimens, it is evident that requiring a patient to take different dosages every day is again a somewhat impractical matter. Therefore, further approximation was made to arrive at a reasonable compromise between practical administration and optimal dosages.

The daily $T_3$ and $T_4$ combination dosages were first divided into weekly segments, and the average dosage was computed for each week. The results are shown in Table 4. Therefore, rather than taking different disages every day as demanded by the optimal results, the daily dosages in this case differed only from week to week until the euthyroid state was reached. Beyond that point, the steady-state maintenance dosage was the same every day. Obviously, this weekly regimen is not optimal with respect to the optimization criterion chosen. However, it does represent a practical administrative dosage schedule derived from the optimal study. The $T_3$:$T_4$ replacement ratio also ranged from about 1:30 to 1:50.

Similar calculations were made for the $T_4$ alone daily dosages, and the weekly regimens are also summarized in Table 4.

4. *Optimal Regimens from Commercial Preparations.* The weekly dosages determined are not the same as those available commercially, nor are they exact multiples of those available.

*TABLE 4*

*$T_3$ and $T_4$ Combination and $T_4$ Alone Weekly Replacement Dosages for Case Studies 1 and 2*[a]

| | Case study 1 | | | | Case study 2 | | | |
|---|---|---|---|---|---|---|---|---|
| | $T_3$ and $T_4$ combination therapy | | | | $T_3$ and $T_4$ combination therapy | | | |
| Week | $T_3$ | $T_4$ | $T_3$:$T_4$ | $T_4$ alone | $T_3$ | $T_4$ | $T_3$:$T_4$ | $T_4$ alone |
| 1 | 0.4 | 12 | 1:30 | 15 | 17.0 | 500 | 1:30 | 610 |
| 2 | 3.5 | 130 | 1:37 | 150 | 5.7 | 200 | 1:35 | 240 |
| 3 | 8.5 | 305 | 1:36 | 360 | | | | |
| 4 | 6.0 | 230 | 1:38 | 270 | | | | |
| 5 | 5.7 | 200 | 1:35 | 240 | | | | |

[a] *The dosages were obtained by averaging the optimal daily dosages in Tables 2 and 3.*

TABLE 5

*Combinations of Available Pharmaceutical Preparations which Yield the Computed Weekly $T_3$ and $T_4$ Combination Dosages for Case Study 1*

| | Pharmaceutical preparations | | | | | | | | | | |
|---|---|---|---|---|---|---|---|---|---|---|---|
| | | | | | | | | Total $T_3$ and $T_4$ content | | Desired $T_3$ and $T_4$ content | |
| Week | Thyrolar -1 | Synthroid (mg) 0.025 | 0.05 | 0.1 | 0.15 | 0.2 | 0.3 | $T_3$ (μg) | $T_4$ (μg) | $T_3$ (μg) | $T_4$ (μg) |
| 1 | 0 | 1/2 | | | | | | 0 | 12 | 0.4 | 12 |
| 2 | 1/4 | | | 1 | | | | 3.2 | 125 | 3.5 | 130 |
| 3 | 3/4 | | 1 | | | 1 | | 9.0 | 290 | 8.5 | 305 |
| 4 | 1/2 | | | | | 1 | | 6.2 | 225 | 6.0 | 230 |
| 5 | 1/2 | 1 | 1 | 1 | | | | 6.2 | 200 | 5.7 | 200 |
| | Cytomel 0.005 mg | Synthroid (mg) 0.025 | 0.05 | 0.1 | 0.15 | 0.2 | 0.3 | | | | |
| 1 | 0 | 1/2 | | | | | | 0 | 12 | 0.4 | 12 |
| 2 | 1/2 | 1 | | 1 | | | | 2.5 | 125 | 3.5 | 130 |
| 3 | 1-1/2 | | | | | | 1 | 7.5 | 300 | 8.5 | 305 |
| 4 | 1 | 1 | | | | 1 | | 5.0 | 225 | 6.0 | 230 |
| 5 | 1 | | | | | 1 | | 5.0 | 200 | 5.7 | 200 |

TABLE 6

*Combinations of Available Pharmaceutical Preparations which Yield the Computed Weekly Dosages of $T_4$ for Case Study 1*

| Week | Synthroid (mg) .025 | .05 | .1 | .15 | .2 | .3 | Total $T_4$ content (μg) | Desired $T_4$ content (μg) |
|---|---|---|---|---|---|---|---|---|
| 1 | 1/2 | | | | | | 12.5 | 15 |
| 2 | | 1 | 1 | | | | 150 | 150 |
| 3 | | 1 | | | | 1 | 350 | 360 |
| 4 | | 1 | | | 1 | | 250 | 270 |
| 5 | | 1 | | | 1 | | 250 | 240 |

Thus, further simplifications were made to approximate the weekly dosage if the currently available pharmaceutical tablets are to be used. Tables 5 and 6 summarize the final results for $T_3$ and $T_4$ combination and $T_4$ alone dosages for case study 1 with $c_{14}$ = midrange. For combination therapy, the dosage can be approximated by either combining Cytomel tablets with Synthroid tablets, or by combining Thyrolar tablets with Synthroid tablets. For $T_4$ therapy, the dosages can be approximated by Synthroid tablets as shown.

5. *"Optimal" Frequency of Dosage Administration.* All dosages computed in the previous sections represent total daily intake. How often during the day should our 'patient' take these dosages? Should the patient take the dosage all at once? Does a once-a-day dosage regimen work just as well as two to three times a day, etc.?

$T_3$ and $T_4$ combination therapy. Figure 4 summarizes the highlights of plasma $T_3$ and $T_4$ responses as a result of the $T_3$ and $T_4$ combination therapy for 40 days. These responses were obtained by simulating the system with the weekly $T_3$ and $T_4$ replacement dosages indicated in Table 5. The (discrete) inputs to the model were the oral $T_3$ and $T_4$ dosages. The absorption dynamics were governed by the gut absorption model (Appendix B). The absorption rates of $T_3$ and $T_4$ were thus direct inputs to the plasma pool, and the two hormones were subsequently metabolized in accordance with Eqs. (1) - (6). Therefore, this study simulates the responses in the actual metabolic pathways following ingestion of the hormone. Because absorption is not instantaneous, the $T_3$ and $T_4$ plasma levels do not increase smoothly and sigmoidally to the euthyroid state (the optimal way). Rather, some initial

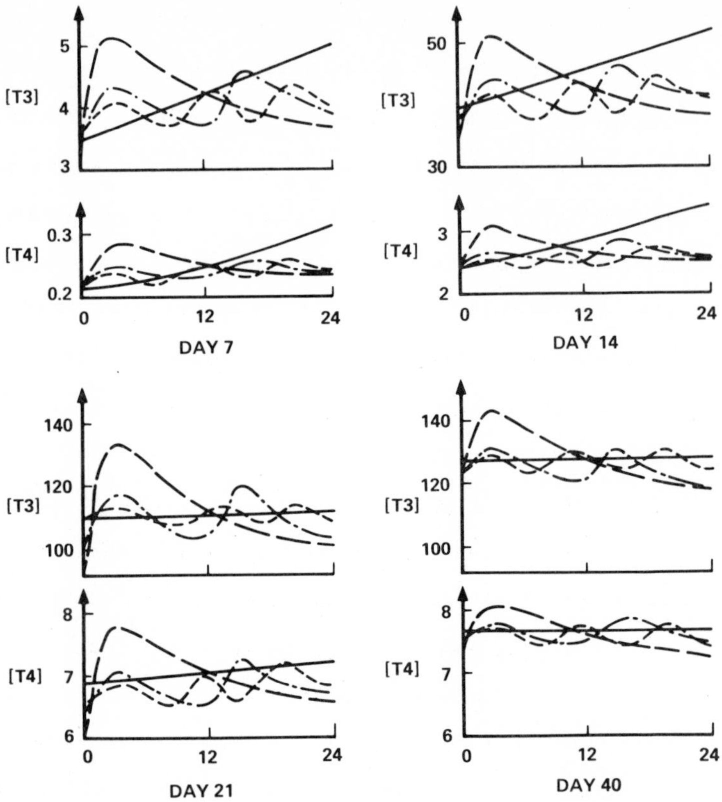


*FIG. 4. Plasma $T_3$ and $T_4$ responses for $T_3 + T_4$ combination therapy during four days of a 40-day replacement period for different frequencies of dosage intake. $T_3$ concentrations ≡ [$T_3$] in ng/100 ml; $T_4$ concentrations ≡ [$T_4$] in μg/100 ml. — — — 1-a-day, -·-·- 2-a-day, ----- 3-a-day; ——— desired sigmoid curve.*

transient fluctuations are observed during the absorption phase. The objective here is to minimize these fluctuations and at the same time maintain a practical schedule. Since it is impossible to show simultaneously and clearly on one graph the overall 40-day response and daily hormonal fluctuation, four representative daily responses were selected for illustration and discussion: days 7, 14, 21, and 40. Also, for comparison purposes, dosage regimes were simulated for three different frequencies, once a day, twice a day, and three times a day. In all cases, the total daily intake was the same.

The overall $T_3$ and $T_4$ levels tracked the respective sigmoidal pattern differently for each case, as illustrated in Fig. 4. Consider the $T_3$ response first. For the one-a-day dosage, the maximum daily fluctuation in the plasma $T_3$ concentration was about 40% from the preingestion level during the transient state; this decreased to about 20% during the maintenance period. When the dosage was given twice a day, there was a significant reduction in the daily fluctuation in the $T_3$ concentration. During the transient state, it dropped from 40 to 20%; and during the maintenance period it dropped from 20 to 10%. That is, a 50% reduction in the maximum daily fluctuation in the $T_3$ level was achieved when the dosage was given twice a day instead of once a day. When the dosage was given three times a day, there was a further decrease in the daily fluctuation in plasma $T_3$ level, as expected. However, the drop was not as significant as before. During the transient period, the percent maximum fluctuation dropped to 14% during the transient stage and to about 7% during the maintenance period.

From these results it can be seen that if we require plasma $T_3$ and $T_4$ levels to follow the optimum sigmoid curves as closely as possible, the three times a day regimen is obviously the choice. This regimen produced the smallest percent maximum daily fluctuation and therefore a smaller least-square error when measured with respect to the best curves. For the same reason, the once-a-day regimen is the least desirable. However, judging from the same results, both $T_3$ and $T_4$ responses to the twice-a-day dosage did not differ significantly from the results of the thrice-a-day regimen. Since this is a chronic therapy, and from the viewpoint of convenience to the patient, the twice-a-day regimen appears to be the most reasonable compromise for this $T_3$ and $T_4$ combination therapy.

$\underline{T_4 \text{ alone therapy}}$. The same simulation procedure was applied. Figure 5 summarizes the results for the same representative days. For plasma $T_4$, the percent maximum daily fluctuations were about the same as the combination therapy. However, this is not true for $T_3$. The plasma $T_3$ level increased very smoothly in a sigmoidal fashion all the way to the euthyroid state during the entire 40-day period. This was true whether $T_4$ was administered once a day, twice a day,

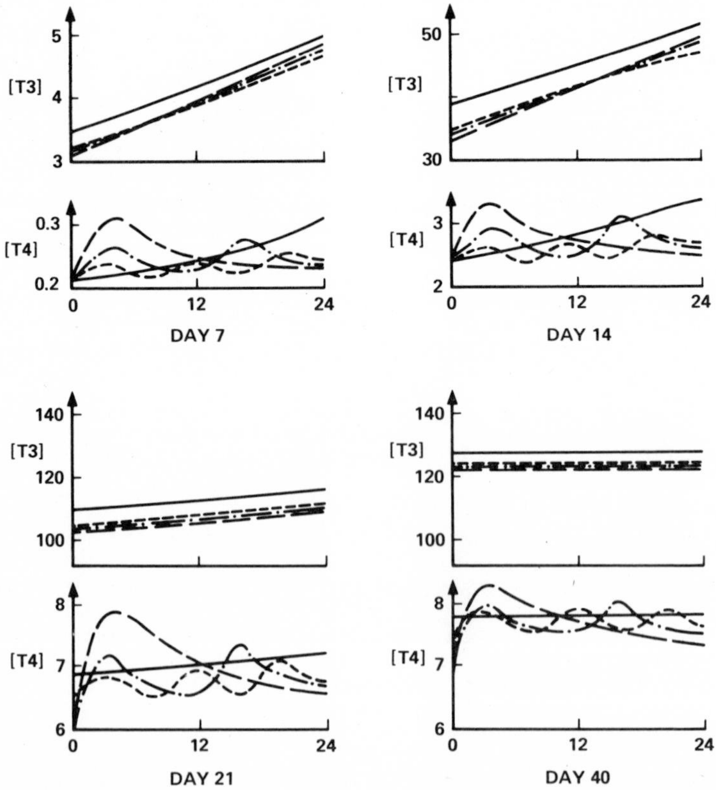


*FIG. 5. Plasma $T_3$ and $T_4$ responses for $T_4$ alone therapy during four days of a 40-day replacement period, for different frequencies of dosage intake. $T_3$ concentrations ≡ [$T_3$] in ng/100 ml; $T_4$ concentrations ≡ [$T_4$] in μg/100 ml. — — — 1-a-day, -.-.- 2-a-day, ----- 3-a-day; ——— desired sigmoid curve.*

or three times a day. There was no observable daily fluctuation in the plasma $T_3$ level. These model simulation results are in excellent agreement with clinical observations [7,9,10].

From these results, it can be seen that replacement with $T_4$ alone achieves essentially the same goals as $T_3$ and $T_4$ combination therapy and does it without causing any significant fluctuations in the $T_3$ concentration. These results, together with those of the previous section, suggest that a twice-a-day regimen is the most desirable one, in terms of convenience and minimum daily fluctuation in $T_4$ levels.

## *I. CLINICAL IMPLEMENTATION, VARIANTS, AND OTHER APPLICATIONS*

We have shown that optimal control methods can be applied successfully to determine the optimal daily dosages of thyroid hormone necessary for treatment of primary hypothyroid patients. In the first test case studied, the $T_3$ and $T_4$ plasma trajectories were sigmoid in shape when the optimal condition was achieved, as desired. The other four hormone trajectories were also sigmoidal. These results have two important clinical implications. First, they indicate that the two plasma hormone concentrations can be forced to follow prespecified recovery patterns, at least sigmoidal ones. Thus, the physician may be able to select a pattern appropriate for each individual patient, depending on their specific clinical history. As such, this approach can easily be adapted to different patients, or even the same patient whose condition may have changed during therapy. Second, assuming the model is at least approximately correct, the sigmoidal manner in which steady-state hormone concentrations were achieved in all extravascular as well as vascular pools is advantageous, particularly for "high-risk" hypothyroid patients, i.e., those usually very sensitive to thyroid hormone. Thyroid hormone effects are manifested at the cellular level.

We note that the daily replacement dosages would be smaller and increase at an even slower rate if the transient therapeutic period were increased to several months in the optimization computations. More generally, the initial and final values of $\underset{\sim}{x}(t)$ for the optimization computation could be fixed at the hormone levels at the time of any patient visit and a desired level at the next visit, respectively. In this manner, very close control over hormonal variations could be maintained from one visit to another, if this were required.

The optimal control approach to drug therapy described in this chapter is directly applicable to systems other than the thyroid hormone regulation system. For example, it has been applied successfully to the problem of controlling blood clotting abnormalities in patients with thromboembolitic disease; optimal dosages of the anticoagulant Warfarin were computed employing a similar criterion as well as a model of pertinent aspects of hemostatic mechanism [20].

## APPENDIX A. NOMENCLATURE

| | |
|---|---|
| $AB_3$ | fraction absorbed for an oral $T_3$ dose |
| $AB_4$ | fraction absorbed for an oral $T_4$ dose |
| $AR_3 = u_2$ | $T_3$ absorption rate in μmoles/hr |
| $AR_4 = u_4$ | $T_4$ absorption rate in μmoles/hr |
| $B$ | semidefinite diagonal weighting matrix |
| $b_1$ | diagonal element of $B$ |

| | |
|---|---|
| $b_2$ | diagonal element of $B$ |
| $c_1$ | $T_3$ plasma to fast pool clearance parameter in liter/hr |
| $c_2$ | $T_3$ fast pool to plasma clearance parameter in liter/hr |
| $c_3$ | $T_3$ fast pool irreversible loss parameter in liter/hr |
| $c_4$ | $T_3$ plasma to slow pool clearance parameter in liter/hr |
| $c_5$ | $T_3$ slow pool to plasma clearance parameter in liter/hr |
| $c_6$ | $T_3$ slow pool irreversible loss clearance parameter in liter/hr |
| $c_7$ | $T_4$ plasma to fast pool clearance parameter in liter/hr |
| $c_8$ | $T_4$ fast pool to plasma clearance parameter in liter/hr |
| $c_9$ | $T_4$ fast pool irreversible loss clearance parameter in liter/hr |
| $c_{10}$ | $T_4$ plasma to slow pool clearance parameter in liter/hr |
| $c_{11}$ | $T_4$ slow pool to plasma clearance parameter in liter/hr |
| $c_{12}$ | $T_4$ slow pool irreversible loss clearance parameter in liter/hr |
| $c_{13}$ | $T_4$ to $T_3$ conversion clearance parameter in fast pool in liter/hr |
| $c_{14}$ | $T_4$ to $T_3$ conversion clearance parameter in slow pool in liter/hr |
| $c_{15}$ | plasma volume in liters |

| | |
|---|---|
| $c_{16}$ | fast pool volume in liters |
| $c_{17}$ | slow pool volume in liters |
| $CV_{4-3}$ | fraction of $T_4$ secreted by the thyroid that is converted to $T_3$ |
| $\underset{\sim}{e}(t)$ | error vector |
| $FT_3 = r_1$ | plasma free $T_3$ |
| $FT_4 = r_2$ | plasma free $T_4$ |
| $J(u)$ | criterion for therapeutic optimality |
| $k_{1i}$ | gut dissolution rate constant in time$^{-1}$ for $T_i$, $i$ = 3 or 4 |
| $k_{2i}$ | gut elimination rate constant in time$^{-1}$ for $T_i$, $i$ = 3 or 4 |
| $k_{3i}$ | gut absorption rate constant in time$^{-1}$ for $T_i$, $i$ = 3 or 4 |
| $MD_3$ | daily $T_3$ maintenance dose rate in μg/day |
| $MD_4$ | daily $T_4$ maintenance dose rate in μg/day |
| $m_{Di}$ | strength of an oral $T_i$ dose, $i$ = 3 or 4, in μg |
| $MW_3$ | molecular weight of $T_3$ |
| $MW_4$ | molecular weight of $T_4$ |
| $OD_3^j$ | total amount of oral $T_3$ intake on the $j$th day for $T_3$ and $T_4$ combination therapy in μg |
| $OD_4^j$ | total amount of oral $T_4$ intake on the $j$th day for $T_3$ and $T_4$ combination therapy in μg |
| $ODT_4^j$ | total amount of oral $T_4$ intake on the $j$th day for $T_4$ alone replacement therapy in μg |
| $OR_3$ | oral $T_3$ dose rate in μg/day |
| $OR_4$ | oral $T_4$ dose rate in μg/day |
| $\underset{\sim}{p}(t)$ | costate vector |
| $Q$ | semidefinite diagonal weighting matrix |
| $q_1$ | diagonal element of $Q$ |

| | |
|---|---|
| $q_2$ | diagonal element of $Q$ |
| $SR_3 = u_1$ | hypothyroid $T_3$ secretion rate in μmole/hr |
| $SR_{3N}$ | euthyroid $T_3$ secretion rate in μmole/hr |
| $SR_4 = u_3$ | hypothyroid $T_4$ secretion rate in μmole/hr |
| $SR_{4N}$ | euthyroid $T_4$ secretion rate in μmole/hr |
| $T_3 = x_1$ | plasma total $T_3$ in μmole/liter |
| $T_{3F} = x_2$ | fast pool total $T_3$ in μmole/liter |
| $T_{3S} = x_3$ | slow pool total $T_3$ in μmole/liter |
| $T_4 = x_4$ | plasma total $T_4$ in μmole/liter |
| $T_{4F} = x_5$ | fast pool total $T_4$ in μmole/liter |
| $T_{4S} = x_6$ | slow pool total $T_4$ in μmole/liter |
| $Ta$ | scaling factor |
| $t_f$ | desired length of the treatment period in hr |
| $x_{1N}$ | prespecified nominal curve (concentration) of $x_1$ in μmole/liter |
| $x_{1ss}$ | euthyroid steady-state level of $x_1$ in μmole/liter |
| $x_{4N}$ | prespecified nominal curve (concentration) of $x_4$ in μmole/liter |
| $x_{4ss}$ | euthyroid steady-state level of $x_4$ in μmole/liter |
| $z_3$ | residual fraction of thyroid $T_3$ secretory function |
| $z_4$ | residual fraction of thyroid $T_4$ secretory function |

## APPENDIX B. THE GUT ABSORPTION MODEL

From a functional point of view, a drug in pill form exists in two physical states after ingestion: a solid state that represents the undissolved portion of the tablet, and an aqueous state that represents the dissolved portion. In most cases dissolution of a drug is usually complete and direct excretion in stool involves the dissolved portion that does not get absorbed into the blood. If one selects this overall view, the gut can be compartmentalized into two functional pools, one for each state of the drug. Consequently, the complex processes of dissolution and absorption are simplified as first-order kinetics depending on the two physical states. With respect to the thyroid hormones, a two-compartment model of gut absorption dynamics results, one for each of $T_3$ and $T_4$, as illustrated in Fig. 6. In the figure, $i = 3, 4$ and $m_{Di}$ is the discrete oral dose input of $T_i$ in microgram (assumed to be an impulse input of strength $m_{Di}$); $m_{Si}$ the amount of the oral $T_i$ dose in undissolved form in microgram at any time $t$; $m_{Li}$ the amount of the oral $T_i$ dose in dissolved form in microgram at any time $t$; $k_{1i}$ the dissolution rate constant in time$^{-1}$ units for $T_i$; $k_{2i}$ the direct gut excretion rate constant of the dissolved dose in time$^{-1}$ units for $T_i$; and $k_{3i}$ the absorption rate constant in time$^{-1}$ units for $T_i$.

The rate at which an oral dose is dissolved can be represented by $k_{1i}m_{Si}(t)$. The amount of the dissolved dose in the gut at any time, $m_{Li}(t)$, is a function of drug dissolution, absorption into the plasma, and excretion from the gut. Assuming first-order kinetics, the dynamics of dissolution and absorption can be represented by

$$\dot{m}_{Si} = - k_{1i}m_{Si} + m_{Di}\ \delta(t) \quad , \qquad \text{(B-1)}$$

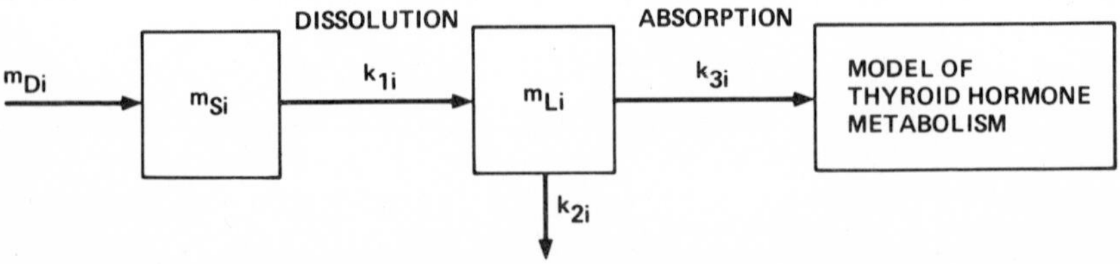


*FIG. 6. A two-pool model for $T_3$ or $T_4$ absorption in the gut.*

$$\dot{m}_{Li} = k_{1i}m_{Si} - (k_{2i}+k_{3i})m_{Li} \quad . \tag{B-2}$$

The initial conditions are

$$m_{Si}(0) = 0 \tag{B-3}$$

$$m_{Li}(0) = 0 \quad . \tag{B-4}$$

Solving Eqs. (B-1) and (B-2), subject to these initial conditions and the impulse input $m_{Di}\ \delta(t)$, we get the basic equations of the gut model:

$$m_{Si}(t) = m_{Di} \exp(-k_{1i}t) \quad , \tag{B-5}$$

$$m_{Li}(t) = \frac{m_{Di}k_{1i}}{k_{1i}-(k_{2i}+k_{3i})} (\exp[-(k_{2i}+k_{3i})t] - \exp[-k_{1i}t]) \quad . \tag{B-6}$$

The values of the parameters $k_{1i}$, $k_{2i}$, and $k_{3i}$ were obtained by fitting this model to available clinical observations [7] and the results are summarized in Table 7. In addition, the rate of $T_3$ or $T_4$ absorption from the gut to the plasma pool at any time $t \geq 0$ is

$$AR_i(t) = k_{3i}m_{Li}(t) \tag{B-7}$$

and the fraction absorption ($AB_i$) can be calculated as

$$AB_i \quad \frac{\int_0^\infty AR_i(t)\,dt}{m_{Di}} = \frac{k_{3i}}{k_{2i}+k_{3i}} \tag{B-8}$$

*TABLE 7*

*Parameter Values of the Gut Absorption Model*

| | $T_3$ | $T_4$ |
|---|---|---|
| $k_1$ | 1.1 | 1.3 |
| $k_2$ | 0 | 0.6 |
| $k_3$ | 0.9 | 0.6 |

*REFERENCES*

1. L. E. BRAVERMAN, S. H. INGBAR, and K. STERLING, "Conversion of Thyroxine (T4) to Triiodothyronine (T3) in Athyreotic Human Subjects," *J. Clin. Invest.*, *49*, 855-864 (1970).
2. K. STERLING, M. A. BRENNER, and E. S. NEWMAN, "Conversion of Thyroxine to Triiodothyronine in Normal Human Subjects," *Science*, *196*, 1099-1100 (1970).
3. C. S. PITTMAN, J. B. CHAMBERS, Jr., and V. H. READ, "The Extrathyroidal Conversion Rate of Thyroxine to Triiodothyronine in Normal Man," *J. Clin. Invest.*, *50*, 1187-1196 (1971).
4. H. L. SCHWARTZ, M. I. SURKS, and J. H. OPPENHEIMER, "Quantitation of Extrathyroidal Conversion of L-Thyroxine to 3,5,3'-triiodo-L-thyronine in the Rat," *J. Clin. Invest.*, *50*, 1124-1130 (1971).
5. M. I. SURKS, A. R. SCHADLOW, J. M. STOCK, and J. H. OPPENHEIMER, "Determination of Iodothyronine Absorption and Conversion of L-Thyroxine (T4) to L-Triiodothyronine (T3) Using Turnover Rate Techniques," *J. Clin. Invest.*, *52*, 805-811 (1973).
6. J. J. DiSTEFANO III, K. C. WILSON, M. JANG, AND P. H. MAK, "Identification of the Dynamics of Thyroid Hormone Metabolism," *Automatica*, *11*, 149-159 (1975).
7. M. SABERI and R. D. UTIGER, "Serum Thyroid Hormone and Thyrotropin Concentrations during Thyroxine and Triiodothyronine Therapy," *J. Clin. Endocrin. Metab.*, *39*, 923-927 (1974).
8. L. E. BRAVERMAN, A. VAGENAKIS, P. DOWNS, A. FOSTER, K. STERLING, and S. H. INGBAR, "Effects of Replacement Doses of Sodium L-Thyroxine on the Peripheral Metabolism of Thyroxine and Triiodothyronine in Man," *J. Clin. Invest.*, *52*, 1010-1017 (1973).
9. J. STOCK, M. I. SURKS, and J. H. OPPENHEIMER, "Replacement Dosage of T4 in Hypothyroidism--A Reevaluation," *New Engl. J. Med.*, *209*, 529-533 (1974).
10. M. I. SURKS, A. R. SCHADLOW, and J. H. OPPENHEIMER, "A New Radioimmunoassay for Plasma L-Triiodothyronine: Measurements in Thyroid Disease and Patients Maintained on Hormonal Replacement," *J. Clin. Invest.*, *51*, 3104-3113 (1972).
11. J. J. DiSTEFANO III and R. F. CHANG, "Computer Simulation of Thyroid Hormone Binding, Distribution and Disposal Dynamics in Man," *Amer. J. Physiol.*, *221*, 1529-1544 (1971).
12. J. J. DiSTEFANO III, K. WILSON, M. JANG, and D. A. FISHER, "Estimation of Thyroid Hormone Secretion, Transport & Disposal Rates and the T4-T3 Conversion Rate in Man," Prog. Endocrine Society Mtg., Supple. *Endocrinology*, *92*, A178 (1973).
13. K. WILSON, D. A. FISHER, J. SACK, and J. J. DiSTEFANO III, "System Analysis and Estimation of Key Parameters of Thyroid Hormone Metabolism in Sheep," *Ann. Biomed. Eng.*, *5*, No. 1 (1977).
14. P. H. MAK, J. J. DiSTEFANO III, and D. A. FISHER, "Thyroid Hormone-Binding Protein Relationships in Blood," *Progr. 55th Annual Mtg.*, The Endocrine Society, Chicago, A-391 (1973).

15. R. R. CAVALIERI, C. MOSER, P. MARTIN, and V. PEREZ-MENDEZ, "Kinetics of Renal Uptake of L-Triiodothyronine (T3) in Man," *Proc. Int. Conf. Thyroid Hormone Metabolism*, Glasgow (1974).
16. R. R. CAVALIERI, M. STEINBERG, and G. L. SEARLE, "The Distribution Kinetics of Triiodothyronine Studies of Euthyroid Subjects with Decreased Plasma Thyroxine-binding Globulin and Patients with Graves' Disease," *J. Clin. Invest.*, *49*, 1041-1050 (1970).
17. M. T. HAYS, "Absorption of Oral Thyroxine in Man," *J. Clin. Endocrin. Metab.*, *28*, 749-756 (1968).
18. M. T. HAYS, "Absorption of Triiodothyronine in Man," *J. Clin. Endocrin. Metab.*, *30*, 675-677 (1970).
19. D. E. KIRK, *Optimal Control Theory: An Introduction*, Prentice-Hall, New Jersey, 1970.
20. P. H. MAK, "Optimal Control of Thyroid Hormone Replacement Therapy and Anticoagulant Therapy," Ph.D. Dissertation, School of Engineering and Applied Science, UCLA, 1976.

# Method in the Modeling of Chemical Engineering Systems

RUTHERFORD ARIS

*Department of Chemical Engineering and Materials Science*
*University of Minnesota*
*Minneapolis, Minnesota*






ISBN 0-12-012715-6

This discussion of the principles of mathematical modeling will be illustrated by examples from chemical engineering. Most of them are slight illustrations, given as they arise and not fully developed, but it seemed also useful to have one major example that would illustrate most aspects of the craft in one context. This is a kind of fold-out map, though it cannot be presented as such, and is given in the appendix, its equation numbers being prefixed with an E. As with a map, our references to it will not necessarily follow its own order and the reader may like to begin by getting an overview of this master example.

## I. THE NOTION OF A MODEL

The word "model," as its derivation from "modus" implies, involves the idea of a change of scale and this was the sense that we first met in childhood. But scale need not be understood only in a physical sense, we can also understand it to mean that we have a representation in a different scale of abstraction. Thus "mathematicals models" of physical processes are commonly regarded as more abstract, and by some engineers as less "real," than their prototypes. But this need not always be the case; in logic, for example, the theory of models deals with the concrete realization of an abstract axiom system. Indeed there is a certain degree of reciprocity in the relation of model to prototype and there is a sense in which the plant itself is a model of the equations. However, though its uses are many and varied, the word model or mathematical model in this essay will mean the system of equations $\Sigma$, which are held to be a representation of a physical system $S$. Though specific examples from a chemical engineering context will be given, I shall not attempt to treat of the substance of chemical engineering models or the details of the techniques used in their analysis, a matter left in better hands than mine, as Chapter 4 of this volume will show, but rather try to elucidate some of the principles of model building. There is a danger of sounding rather platitudinous in attempting this for the reader will have picked up the craft of model building for himself and may not be as thrilled as M. Jourdain to discover he has been speaking prose all the time. Nevertheless I believe the effort is worthwhile, if only for the fact that it is occasionally useful to stop and ask what one is doing, even though, as Whitehead remarks, we progress by increasing the number of things we can do without thinking.

The intention of model building is to gain some insight into a physical plant and, specifically in this volume, to understand its dynamics and control. It falls (as it must if his claims are just) into the framework of cognitional theory that Lonergan has elucidated in his great work on "Insight" [1]. This structure is a dynamic one grounded in experience, seeking understanding and calling for judgment. Moreover there is a constant interplay of these aspects so that judgment, for example, is called for not only in the final assessment of the value of a model but in excluding irrelevant detail from experience and in selecting appropriate paths to understanding.

It seems important to recognize at the outset that models do not come in isolation. There is no such thing as a unique model of a process, though, of course, once a model has been isolated it can be studied in its own right and for the sake of its own properties. But in general we recognize that there is a hierarchy of models of varying degrees of sophistication each adapted to a particular set of conditions and to a use for which it is intended. Thus a comparatively simple model of algebraic equations may serve the purpose of steady-state design, but a set of ordinary differential equations may be needed for control. On the other hand the same purpose, say control, might require a partial differential equation under some conditions but an ordinary equation would suffice under others.

To illustrate this consider the master example. Here the stirred tank reactor is represented by five different systems of equations, two appropriate to the steady state, $\Sigma_2$ and $\Sigma_3$, and the remainder to the transient. In the case of the steady state, $\Sigma_2$ consists of an elliptic partial differential equation with boundary conditions coupled to several nondifferential equations (i.e., equations with algebraic or transcendental functions). We might label this structure PDE/NDE. If the walls are thin, the PDE becomes an ordinary differential equation (ODE) whose solution can be obtained rather easily and allows the system to be reduced to NDE form entirely. Of the transient models $\Sigma_1$ is the most sophisticated, being a parabolic partial differential equation connected through its boundary conditions to ODEs. It applies to thick walls under all conditions but can be specialized to $\Sigma_5$, which is ODE only, if the conductivity of the wall is high even though this wall is thick. On the other hand if the wall is thin or the conductivity low, a model of similar sophistication but different equations $\Sigma_4$ can be used. Thus we might describe the conditions rather loosely as thick or thin and well or badly conducting walls and the purposes as design and control. Then the relationship of the several models is shown in Fig. 1;

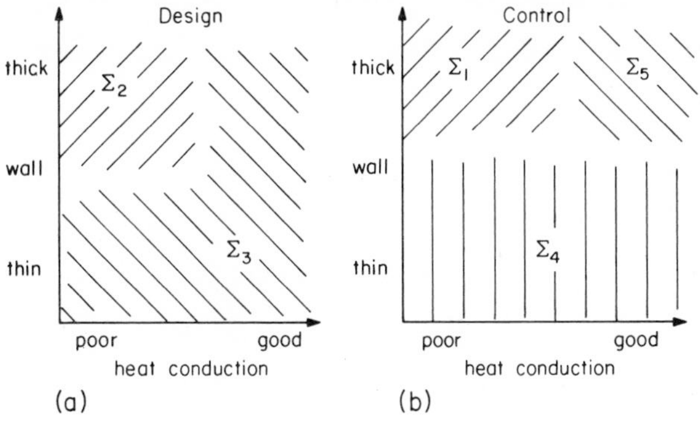


*FIG. 1. Relationships of models in the space of the conditions when the purpose is (a) design (b) control.*

shown are maps of constant purpose in the two-dimensional space of conditions. The areas are drawn with deliberate vagueness for there are no sharp divisions indicated here. Figure 2 attempts to show the levels of sophistication that are involved.

Gill's graphical summary of the regions of applicability of solutions to the equations of laminar dispersion in capillaries can also be interpreted in this light [2]. The problem is that of diffusion in parabolic flow of mean velocity $U$ through a cylindrical tube of radius $a$ and the full equations for the

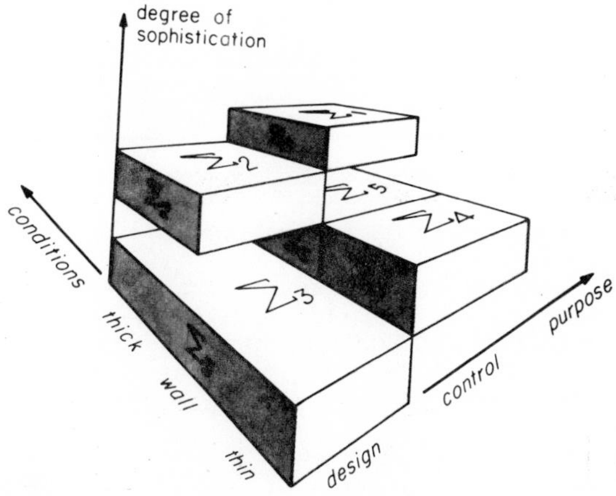


*FIG. 2. Relationships of models in the conditions-purpose space.*

concentration of matter being dispersed are

$$\frac{\partial c}{\partial t} + 2U\left(1 - \frac{r^2}{a^2}\right)\frac{\partial c}{\partial x} = D\frac{\partial^2 c}{\partial x^2} + D\frac{1}{r}\frac{\partial}{\partial r}\left(r\frac{\partial c}{\partial r}\right); \tag{1}$$

$$c(0, x, r) = 0, \quad x \geq 0, \quad 0 \leq r \leq a; \tag{2}$$

$$c(t, 0, r) = 1, \quad r > 0; \tag{3}$$

$$c(t, x, r) \to 0, \quad x \to \infty; \tag{4}$$

$$\frac{\partial c}{\partial r} = 0 \quad \text{at} \quad r = 0 \quad \text{and} \quad r = a. \tag{5}$$

These equations have to be solved numerically to obtain even the average concentration with great accuracy but, if the purpose to which the model is put permits, certain approximations can be made. Thus $\bar{c}(x, t) = \int_0^a 2(r/a)c(t, x, r)dr/a$ develops from a step function at time $t = 0$ to a curve of gradually decreasing slope (complementary error function type of curve) as time goes on (see Fig. 3). This can be approximated by

$$\bar{c} = \frac{1}{2}\operatorname{erfc}\,[(x - Ut)/2kt], \tag{6}$$

where $k$ is a dispersion coefficient. But this is a solution of a simpler model:

$$\Sigma_2 \begin{cases} \dfrac{\partial \bar{c}}{\partial t} + U\dfrac{\partial \bar{c}}{\partial x} = k\dfrac{\partial^2 \bar{c}}{\partial x^2}; & (7) \\ \bar{c}(t, 0) = 1; & (8) \\ \bar{c}(0, x) = 0; & (9) \\ \bar{c}(t, x) \to 0 \quad \text{as} \quad x \to \infty. & (10) \end{cases}$$

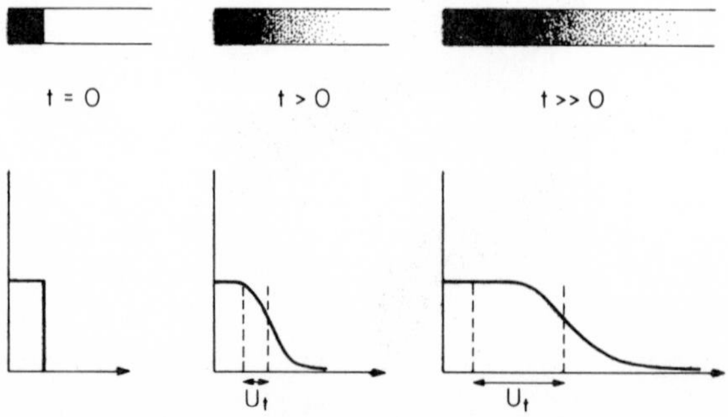


*FIG. 3. Softening of a sharp front due to dispersion.*

In essence, what we are saying is that a model of plug flow with the same mean velocity and an apparent longitudinal dispersion coefficient of $k$ can be used in place of $\Sigma_1$, a much more complicated model with convection and transverse diffusion as well. Taylor [3] showed that in some cases

$$\Sigma_2', \quad k = \frac{a^2 U^2}{48D}, \tag{11}$$

a result of striking simplicity and elegance obtained with that beautiful economy of argument that was the hallmark of his work. Aris [4] interpreted this result in terms of the moments and showed that

$$\frac{d}{dt}\int_0^\infty x\bar{c}(t, x)\,dx \sim U, \quad \frac{d}{dx}\int_0^\infty (x - Ut)^2\,\bar{c}(t, x)\,dx \sim 2k$$

at $t \to \infty$, provided that,

$$\Sigma_2'', \quad k = D + \frac{a^2 U^2}{48D}. \tag{12}$$

There is also the pure diffusion model applicable when the flow is very slow and here

$$\Sigma_2''', \quad k = D, \tag{13}$$

in Eq. (6). Finally there is the case of pure convection when diffusion plays no role and

$$\Sigma_3 \begin{cases} \bar{c} = 1 - \dfrac{x}{2Ut}, & x < 2Ut; \\ \quad = 0, & x > 2Ut. \end{cases} \tag{14}$$

But Gill *et al.* found that in certain cases a very good approximation was given by

$$\bar{c} = \frac{1}{2}\,\mathrm{erfc}\left[\frac{x - Ut}{2\sqrt{Dt(1 + P^2Q)}}\right] + \frac{1}{2}\exp\left[\frac{xU}{D(1 + P^2Q)}\right] \times \mathrm{erfc}\left[\frac{x + Ut}{2\sqrt{Dt(1 + P^2Q)}}\right] \tag{15}$$

with

$$P = aU/D$$

and

$$\Sigma_4': \quad Q = 0.028(Dt/a^2)^{0.55}, \quad Dt < 0.6a^2; \tag{16}$$

$$\Sigma_4'': \quad Q = 1/48, \quad Dt > 0.6a^2. \tag{17}$$

This last is really a model of a model since it represents an empirical, though informed, fitting in of constants. Such a map as is shown in Fig. 4 is of great value when it can be drawn, as Gill did, with some accuracy. It is a guide to the economical application of the whole corpus of work that surrounds a given subject.

Apostel [5] in a formal study of models has taken account of the idea of purpose in designating the modeling relationship as R(S, P, M, T) wherein the subject S with purpose P takes the entity M as a model for the prototype T. After a lengthy discussion he concludes that there is much to be done

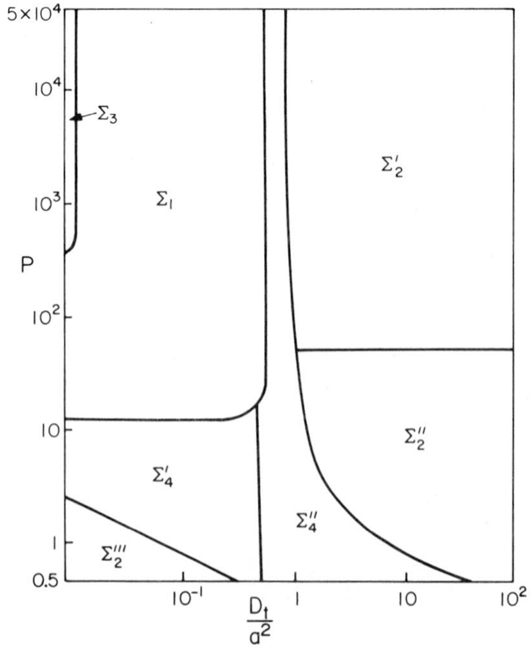


*FIG. 4. Domains of applicability of the models of dispersion.*

before a unified general theory can be attained but gives a hint of how a general definition of a model might be framed. Any subject using a system $\Sigma$ neither directly nor indirectly interacting with the system S to obtain information about S, is using $\Sigma$ as a model for S.

To revert briefly to the master example, it is clear that the relationship between $\Sigma$ and S depends on the hypotheses that have been invoked in the formulation and that these determine the relations between the models themselves. We shall have more to say about this in discussing the formulation of the model. The involvement of the different hypotheses can be diagrammed as in Fig. 5, where in passing radially from S to $\Sigma_i$ one crosses the zones representing the hypothesis involved.

## II. THE TYPES OF MODEL

Models in general can be classified in many ways. There are verbal, physical, and graphical models as well as the mathematical ones we are primarily concerned with. It is not possible to separate these aspects entirely and the rheologist,

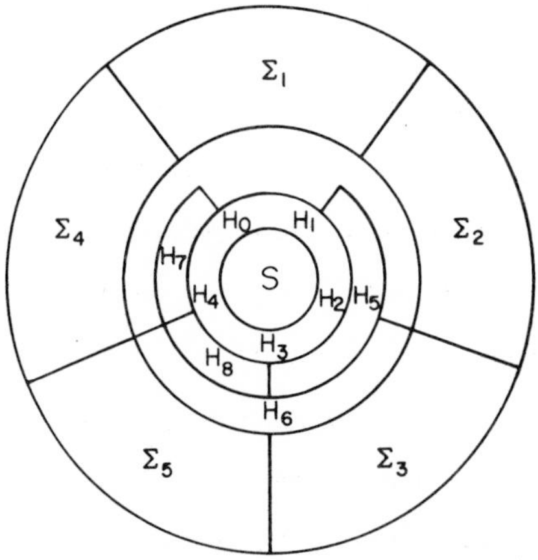


*FIG. 5. Use of the various hypotheses in the different models.*

for example, when he writes down a stress-rate of strain relation for a viscoelastic fluid has been known to have a model of springs and dashpots in mind as well as derivatives on paper. There is a famous passage in Duhem [6] contrasting the French and English in their approach to the physics of his day: "The employment of similar mechanical models...is a regular feature of the English treatises on physics. Here is a book intended to expound the modern theories of electricity ...in it there are nothing but strings which move around pulleys, which roll around drums, which go through pearl beads, which carry weights; and tubes which pump water while others swell and contract; toothed wheels which are geared to one another and engage hooks. We thought we were entering the tranquil and neatly ordered abode of reason, but we find ourselves in a factory." But what should the scientists of that "nation of shopkeepers" turn out to be if not a bunch of mechanics?

Even if we confine attention to mathematical models as sets of equations, ignoring any background images, we have some basic dichotomies. Thus there are probabilistic and deterministic models, empirical and mechanistic, discrete and continuous, lumped and distributed. A discrete model is one in which local physical structure is absent or is ignored; in a lumped model, it has been suppressed. Of the tools available for the analysis of discrete models, we have graph theory, game theory, linear, nonlinear, and dynamic programming, difference equations, and the theory of Markov chains to mention but a few. In a continuous model the local structure of time or space or both is taken into account and the methods of analysis, rather than of algebra, come to the fore. However the distinctions are not absolute nor clean-cut, either in substance or in method, and may manifest themselves at one time as ordinary versus partial differential equations, at another in linear algebraic versus integral equations. Moreover when it comes to computation on a digital computer, the continuous necessarily becomes discrete. The terms lumped parameter and distributed parameter systems seem misguided for it is variables not parameters that are lumped (discrete) or distributed (continuous).

The word lumping, in spite of its ungainly overtones, is useful in describing the process by which a number of things are put together in one. This may result in replacing a continuous system by a discrete one. An example of this is the network thermodynamics of Oster *et al.* [7], where problems of flow and transport in biological systems are treated by the ideas of electrical network theory. This converts parabolic equations into ordinary differential equations and elliptic into algebraic. An important discussion of lumping in the context of monomolecular reactions has been made by Wei and Kuo [8]. This reduces a large system of species into a continuous of species and in this sense lumping them together. Lumping of this sort does replace a large number of equations by a single equation, but this is often an integro-differential equation rather than an ordinary differential equation. In fact the sense of the word has been turned around and what is being done is the distribution of a large number of discrete variables into a continuous variable, vigorously stirring out the lumps. These two processes are worth considering further as they illustrate some of the subtleties of the relations between continuous and discrete models.

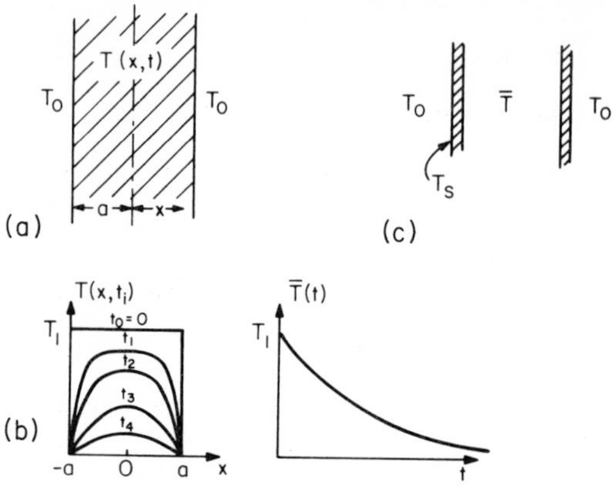


*FIG. 6. Loss of heat from a wall as a lumped model. (a) The geometry. (b) The distributed solution. (c) The lumped equivalent.*

Consider first the loss of heat of a wall to its environment, a thermal analog of Oster's case of diffusion through a membrane. Figure 6a shows the physical picture with $T(x, t)$, $-a \le x \le a$, the temperature at any point of the wall. Thus (incorporating the symmetry) the continuous model would be

$$\rho c_p \frac{\partial T}{\partial t} = k \frac{\partial^2 T}{\partial x^2}, \quad 0 \le x < z; \tag{15}$$

$$\frac{\partial T}{\partial t} = 0, \quad x = 0; \tag{16}$$

$$k \frac{\partial T}{\partial x} = h(T_0 - T); \tag{17}$$

$$T(x, 0) = T_1 . \tag{18}$$

The kind of solution that we expect for these equations is sketched in Fig. 6b. Let us make the equations dimensionless by writing

$$\xi = x/a, \ \tau = kt/\rho c_p a^2, \ u = (T - T_0)(T_1 - T_0), \tag{19}$$
$$\mu = ha/k .$$

Thus

$$u_\tau = u_{\xi\xi}, \quad 0 \le \xi \le 1; \tag{20}$$

$$u_\xi = 0, \quad \xi = 0; \tag{21}$$

$$u_\xi + \mu u = 0, \quad \xi = 1; \tag{22}$$

$$u(\xi, 0) = 1 . \tag{23}$$

The complete solution of this equation is elementary enough, being

$$u(\xi, \tau) = \sum_1^\infty \frac{2\mu}{\lambda_n^2(\sec \lambda_n + \sin \lambda_n)} \exp(-\lambda_n^2\tau) \cos \lambda_n\xi, \tag{24}$$

where

$$\lambda_n \tan \lambda_n = \mu, \quad n = 1, 2, \ldots \tag{25}$$

In particular the average temperature is

$$\underset{\sim}{u}(\tau) = \sum_1^\infty \frac{2\mu^2}{\lambda_n^2(\lambda_n^2 + \mu\lambda_n + \mu^2)} \exp(-\lambda_n^2\tau) . \tag{26}$$

The sequence of eigenvalues $\lambda_1$, $\lambda_2$ increases rapidly; for example, with large $\mu$, $\lambda_n = (2n-1)\pi/2$. Thus all the terms after the first quickly become negligible and

$$\frac{d}{d\tau} \underset{\sim}{u}(\tau) = -\lambda_1^2 \underset{\sim}{u}(\tau) . \tag{27}$$

This makes it look like a first-order system with a time constant of $\lambda_1^{-2}$. In terms of real time this is

$$a^2/\lambda_1^2 D, \tag{28}$$

or

$$4a^2/\pi^2 D \text{ as } \mu \to \infty. \tag{29}$$

The network analog consists in lumping the resistances and examining the driving potentials. A wall of unit area, thickness $2a$ and conductivity $k$ will conduct heat at a rate $Q = k\Delta T/2a$ and might therefore be regarded as having a resistance of $\Delta T/Q = 2a/k$. Let this resistance, which is in fact distributed over the thickness, be divided into two and lumped at the surfaces of the slab as shown in Fig. 6c. The temperature within the slab will also have to be lumped and since the resistance has been separated to the walls, the natural lumping is $\underset{\sim}{T}$, the mean temperature. The flow of heat out of the slab at each face will therefore be $(k/a)(T - \underset{\sim}{T}_s)$ where $T_s$ is the temperature at the outside surface of the slab. This matches the flow through the external film $h(T_s - T_0)$, so we have

$$q = \frac{k}{a}\,(\underset{\sim}{T} - T_s) = h(T_s - T_0) = \frac{k}{a}\,\left(1 + \frac{1}{\mu}\right)^{-1}\,(\underset{\sim}{T} - T_0)\ .$$

Now the thermal capacity of the slab is clearly $2a\rho c_p$, so

$$2a\rho c_p \frac{d\underset{\sim}{T}}{dt} = 2\frac{k}{a}\,\left(1 + \frac{1}{\mu}\right)^{-1}\,(T_0 - \underset{\sim}{T})$$

or

$$\frac{d\underset{\sim}{u}}{d\tau} = -\left(1 + \frac{1}{\mu}\right)^{-1}\,\underset{\sim}{u}\ . \qquad (30)$$

This gives a dimensionless time constant of $(1 + 1/\mu)$ or in dimensional terms

$$a^2(1 + \mu)/D\mu \qquad (31)$$

and in the limit $\mu \to \infty$,

$$a^2/D\ . \qquad (32)$$

Comparing (32) and (29), we see that we are off by a factor of 2.5. This suggests that agreement could be improved by assuming that the "dynamic" capacity of the wall is only 2/5 of its "static" capacity $2a\rho c_p$. For the more general case of external resistance, the ratio of the time constants of the lumped and distributed systems is

$$\frac{t_L}{t_D} = \frac{\mu}{(1 + \mu)\lambda_1^2} = \frac{\tan\lambda_1}{\lambda_1(1 + \lambda_1 \tan\lambda_1)}\ .$$

As $\mu$ decreases from infinity to zero, $\lambda_1$ decreases from $\pi/2$ to 0 and the ratio increases from $(4/\pi^2)$ to 1. It is not surprising that the two time constants should agree in the limit

$\mu \to 0$, or $k/a >>> h$, for if the resistance of the outside films dominates so completely, then the system is truly lumped.

This comparison illustrates some of the difficulties in going from the discrete to the continuous. A certain amount of accuracy can be recovered by the use of a pseudocapacity much as a virtual mass can be used in other cases, but it is not altogether satisfactory to have the wall capacity depend on the heat transfer coefficient. An analogous method of reducing the partial differential equations of a catalyst particle to ordinary differential equations was used by Hlaváček [9]. Reference to this is given in [10] and in [11] the one-point collocation method of Stewart and Viladsen [12] is also described.

To illustrate the kind of "lumping" that is really distribution consider Luss and Hutchinson's treatment of many parallel first order reactions [13]. In many situations it is not possible to describe a mixture of chemical species that boil between, say 350° and 500° and this might be taken as a largish lump. On the other hand if we talk about the number of moles $n(T)dT$ that boil in the range $(T, T + dT)$, we have really made a continuum, i.e., an infinity of species, out of a system that is necessarily discrete. In the case of species that can all undergo a reaction $A_i \to B_i$ with rate constant $k_i$, we may devise a continuum and talk about the "species" $A(k)dk$ as all that reacts with rate constant in the range $(k, k + dk)$. If $c(t, k)dk$ is the concentration of this material at time $t$ and if the reactions are all parallel first order,

$$c(t, k)dk = c(0, k)dk\ e^{-kt} . \tag{33}$$

Now in many cases we may only be interested in the total amount

$C(t) = \int_0^\infty c(t, k)dk$ and we see that

$$C(t) = \int_0^\infty e^{-kt} c(0, k)dk \tag{34}$$

is the Laplace transform of the initial distribution with time, for a change, playing the role of the transform variable.

It is interesting to enquire if there is an apparent rate law

$$\dot{C} = \frac{dC}{dt} = -f(C)$$

but although

$$\dot{C} = -\int_0^\infty k\, e^{-kt} c(0, k)dk , \tag{35}$$

it is seldom possible to invert (34) and so eliminate $t$. A notable exception is

$$c(0, k) = C(0)k^\alpha e^{-\beta k} \frac{\beta^{\alpha+1}}{\Gamma(\alpha + 1)} , \quad \alpha, \beta > 0 .$$

Let us make $c$ dimensionless by dividing by $C(0)$, i.e.,

$$u(t, k) = c(t, k)/\int_0^\infty c(0, k)dk \tag{36}$$

so that for this distribution,

$$u(0, k) = \beta^{\alpha+1} k^\alpha e^{-\beta k}/\Gamma(\alpha+1) . \tag{37}$$

Then, by Eq. (34)

$$U(t) = \frac{C(t)}{C(0)} = (1 + \frac{t}{\beta})^{-(\alpha+1)} \tag{38}$$

and

$$\dot{U}(t) = \frac{\alpha+1}{\beta} (1 + \frac{t}{\beta})^{-(\alpha+2)} = -\frac{\alpha+1}{\beta} U^\gamma , \tag{39}$$

where

$$\gamma = (\alpha+2)/(\alpha+1) . \tag{40}$$

Thus the lump appears to decay as a $\gamma$th order reactant with $\gamma$ depending on the parameter $\alpha$ in the initial distribution. It is not surprising that as $\alpha \to \infty$, $\gamma \to 1$ for the variance $\sigma^2$ of the initial distribution is $(\alpha+1)/\beta^2$ and the mean, $\mu$ is $(\alpha+1)/\beta$. Thus $\sigma^2/\mu^2 = 1/(\alpha+1) \to 0$ as $\alpha \to \infty$, the distribution becomes narrower and therefore appears to decay in first-order fashion. It is remarkable that for all $\alpha$ and $\beta$ the rate constant in Eq. (39) is $\mu = (\alpha+1)/\beta$, the mean value of $k$ in the initial distribution. If $\alpha = 0$, the apparent order is $\gamma = 2$ and it is noteworthy that second-order reaction rates have been used to correlate hydrocarbon cracking for some time.

Though it is seldom possible to get complete results of this kind, Luss exploited the convexity of the exponential to show that

$$e^{-\mu t} \leq U(t) \leq (\sigma^2 + \mu^2 e^{-\nu t})/(\sigma^2 + \mu^2) ,$$
$$\nu = (\sigma^2 + \mu^2)/\mu . \tag{41}$$

Such a result is extremely useful (and incidentally an excellent illustration of the value of the theory of inequalities) as it gives bounds on an observable in terms of certain calculable functionals, in this case the mean rate constant and their variance in the initial distribution. For an extended discussion of continuous mixtures see [14].

## III. THE FORMULATION OF MODELS

The basic origin of the equations in a mathematical model is the expression of a conservation principle, whether of mass, momentum, or energy. Considerations of relativity, under which mass and energy would be jointly rather than severally conserved, seldom arise in the chemical engineering context. Such balances are made on a discrete part of the system, as for example a plate in a distillation column, or on a typical volume element, either fixed or moving, in the continuum.

In a discrete element we can let $F$ be the net flux of the entity into the element, $G$ its rate of generation there, and $H$ the total amount present. Then $F$, $G$, and $H$ are functions of time and satisfy

$$F + G = \frac{dH}{dt} . \tag{42}$$

If we are dealing with a continuum, then these quantities must be defined as densities. Thus we let the vector $\underset{\sim}{f}$ denote a flux which is defined such that the flux across an element of area $dS$ in the direction of its normal $\underset{\sim}{n}$ is $\underset{\sim}{f}\cdot\underset{\sim}{n}\, dS$. Similarly, the generation must be defined as a rate per unit volume, so that in a volume element it is $g\, dV$, and $H$ becomes a concentration $h$. Then if $\Omega$ is an arbitrary, simply connected region of the continuum with a piecewise smooth surface $\partial\Omega$ whose outward normal is denoted by $\underset{\sim}{n}$, we have

$$- \int\int_{\partial\Omega} \underset{\sim}{f}\cdot\underset{\sim}{n}\, dS + \int\int_{\Omega}\int g\, dV = \frac{\partial}{\partial t}\int\int_{\Omega}\int h\, dV . \tag{43}$$

In this equation we use the fact that $\Omega$ is fixed to interchange the order of integration and differentiation and use Green's theorem on the surface integral. Then all terms can be brought to one side of the equation and we have

$$\iiint_{\Omega} \left[\frac{\partial h}{\partial t} + \nabla\cdot\underset{\sim}{f} - g\right] dV = 0 .$$

We now must make the hypothesis that $\underset{\sim}{f}$, $g$, and $h$ are sufficiently continuous that the integrand is continuous and then, since the region $\Omega$ is completely arbitrary,

$$- \nabla\cdot\underset{\sim}{f} + g = \frac{\partial h}{\partial t} . \tag{44}$$

If a volume is a material volume $\Omega(t)$ moving in a continuum where the velocity field is $\underset{\sim}{v} = \underset{\sim}{v}(\underset{\sim}{x}, t)$, then we need Reynolds' theorem for the interchange of differentiation with respect to time and integration. This is

$$\frac{d}{dt} \iiint_{\Omega(t)} h \, dV = \iiint_{\Omega(t)} \left( \frac{\partial h}{\partial t} + \underset{\sim}{v} \cdot \nabla h \right) dV \ . \qquad (45)$$

The fact that the flux through a surface element can always be expressed as $\underset{\sim}{f} \cdot \underset{\sim}{n} \, dS$ is the conclusion of an interesting type of argument that is sometimes useful in other contexts. Figure 7a shows a particular form of element, namely, at tetrahedron of volume $dV$ and with three sides perpendicular to the axes $0n_1$, $0n_2$, $0n_3$ and having $dS$ as the area of its slanting face. Then by definition of the direction cosine the face perpendicular to $0n_1$ has area $n_1 \, dS$. Let $f_i$ be the flux in the direction $0n_i$ and $\hat{f}$ the flux over the slant face. Then a balance can be struck over the tetrahedron for

$$F = (f_1 n_1 + f_2 n_2 + f_3 n_3 - \hat{f}) dS, \quad G = g \, dV, \quad H = h \, dV$$

and by Eq. (42), $F + G = H$. But if the volume is allowed to shrink in size, while keeping its proportions, $dS$ will decrease as the square of the size but $dV$ as the cube. It follows that

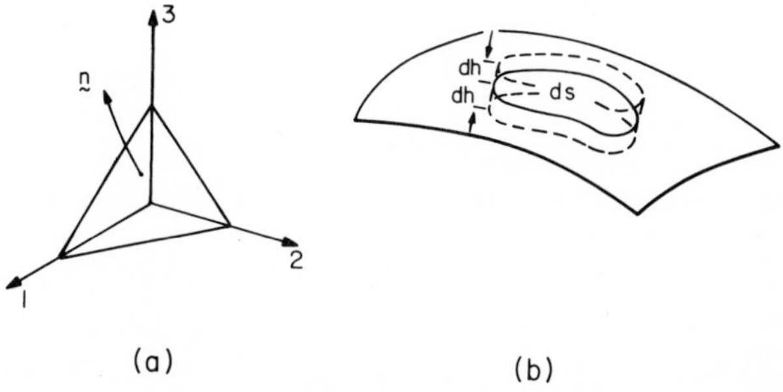


*FIG. 7. (a) The tetrahedral element. (b) The surface box.*

$G$ and $H$ become negligible in comparison with $F$ and hence in the limit $F = 0$. Thus

$$\hat{f} = f_1 n_1 + f_2 n_2 + f_3 n_3 = \underset{\sim}{f} \cdot \underset{\sim}{n} ,$$

if $\underset{\sim}{f}$ is the vector with components $(f_1, f_2, f_3)$.

A similar argument is used in the formulation of boundary conditions if the element over which the balance is made is an element of surface extended by a distance $dh$ on either side (Fig. 7b). Then letting $dh \to 0$ first reduces the volume to zero. It is another way of saying that in three dimensions a surface has no volume and hence no capacity for a quantity defined per unit volume. Thus in the boundary conditions (E4) and (E5), there is no term with a time derivative since the natural capacity of the surface is zero. If however a quantity is defined per unit area and is therefore a surface concentration, it may well show up in a boundary condition.

It should be evident that the general conservation principles are universally applicable and that nothing particular has been said so far that bears on the construction of the system. This comes in when we relate the flux or generation terms to the concentration. Referring to the master example we have an illustration of this in Eq. (E3). If $\underset{\sim}{f}$ is the flux of heat, $h = \rho c_p T_w$ can be regarded as its concentration (i.e., heat content per unit volume) in the wall and the generation is zero, $g = 0$. But if we assert Fourier's law of heat conduction, we are specifying the constitution of the wall as far as its thermal behavior is concerned. This would give $\underset{\sim}{f} = -k \nabla T_w$ for constant $k$. Equation (E1) is an illustration of Eq. (42) with $F = q_j c_{jf} - q c_j$, $G = \alpha_j V\, r(c_1, \ldots, c_s, T)$, $H = V c_j$. Again the constitutive relation is implied by expressing the reaction rate in terms of the concentrations and temperature.

Under certain circumstances it may be well to make explicit the distinction between general principles and specific constitutive relations. Thus the set of underlying hypotheses $H_0$ might be divided into two sets $H_0'$ and $H_0''$, the general and specific, respectively.

It is an aphorism of N. R. Amundson, the doyen of chemical engineering model builders, that "all boundary conditions arise from nature." By this he implies that the equations themselves contain more of the modeler's imagination, the hypotheses that he has imposed in demanding a particular type of model, than do the boundary conditions which are the ligamen between the constructed model and the natural world. There is a good deal in this though it must be remembered that hypotheses are used also in formulating the boundary conditions. In particular, (E4) and (E5) are the direct assertion of $H_3$, which, as the so-called Newton's law of cooling, might even have been regarded as a constitutive relation like Fourier's law of conduction.

## IV. THE JUST MOULDING OF MODELS

When a model has been constructed it is almost always a mistake to jump in and start an extensive series of computations with it. It needs to be considered carefully from different angles, moulded into different forms, examined in special cases and, in the soon-faded jargon of yesteryear, "massaged" until a thorough appreciation of its possibilities is attained. Indeed this might be called the foreplay of modeling for it calls for a like delicacy and tact. Unfortunately, except for Lin and Segel's excellent book [15], there are virtually no manuals for the art. Yet, though its higher forms may demand the touch of the poet, there is a level of competence that any of us may aspire to, and to which some training and practice may lead [16]. In this connection there is a lot to be leared from the enormously vital writting of Hammersley on the strategy and tactics of research in mathematics [17 - ]9]. His maxims for manipulations are worth quoting [17] since they give the flavor of his thinking as well as several points we shall return to later. They are (i) clean up the notation, (ii) choose suitable units, (iii) reduce the number of variables wherever possible, (iv) draw rough sketches and examine particular cases, (v) avoid rigor like the plague, it only leads to rigor mortis at the manipulative stage, and (vi) have about an equal amount of stuff on each side of the equation. He illustrates these with a maximization problem of deceptively innocent aspect. I will not repeat the longer list of maxims given in [16]; as Hammersley suggests, it is not always the best golfer who has the most clubs. We shall look at several aspects of the craft using the master example as proving ground for our efforts. Segel's paper [20] is worth studying too, though it is largely contained in Lin and Segel [15].

### A. *CHOICE OF VARIABLES AND NOTATION*

Hammersley's first maxim has more to it than meets the eye, for the right choice of notation, though in some respects a matter of personal taste, is not to be dismissed as trivial. It is easy enough to make mistakes under the best of circumstances but to burden oneself with a cumbersome and unsuitable notation is downright stupid. Except where some overriding convention has a prerogative, the symbols for the basic physical quantities should be generally as simple and mnemonic as possible. It is not always possible to avoid suffixes but again it is important that they should be natural. Thus, by calling the species $A_1, A_2, \ldots, A_s$, it is natural to denote their concentrations by $c_j$, $j = 1, 2, \ldots, s$. Similarly at the early stage it is better to call the volume of the coolant $V_c$ than to preempt yet another letter. How sympathetic one person's notation may appear to another is a matter of context and personal taste. For example chemical engineering texts often use $\Theta$ for the residence or holding time of the stirred tank $V/q$, while biochemical engineers use $D$ for its reciprocal, calling it the dilution rate; yet, since $D$ plays the role of a death rate, many would feel that $\Theta$ itself would be preferable. In choosing symbols for the dimensionless variables, it is seldom possible to make a consistent translation between Latin and Greek letters, though it is an admirable tradition to follow. Thus coordinates $(x, y, z)$ can often become $(\xi, \eta, \zeta)$ or time $t$ become dimensionless time $\tau$. Though both alphabets run out all too quickly, it is better to have recourse to Ps. 119 in the King James than to succumb to such symbolic solecisms as $V$ for a diffusion coefficient or $x$ for a mass. In general single letters should be used for a single quantity, but tradition has sanctified the union of a capital and lower case for the notable dimensionless groups such as the Reynolds number Re. This seldom causes confusion and is less unsightly than the convention $N_{Re}$. Chemical engineers seem to have led the field in coining names for dimensionless groups and hence immortalizing one another, much as the naturalists of the eighteenth century did with the taxonomy of Linnaeus, though it may be questioned whether it helps to perpetuate Damköhler's memory in four numbers with Roman numeral suffixes. In the master example the concentrations and temperatures $c_j$, $T$, $T_{\mathrm{w}}$, and $T_{\mathrm{c}}$ become $u_j$, $v$, $w$, and $\Theta$ in dimensionless form; I am not altogether happy with the last choice, but avoided $v_{\mathrm{c}}$ as I was aiming for ultimate freedom from suffixes for the dependent variables.

If the matter of notation is incidental, being governed by the canons of taste and common sense rather than high principle, the matter of making the variables dimensionless is of the essence. A physical magnitude has meaning with respect to an arbitrary set of standards and two quantities of the same dimenseions measured in the same units can be compared. But even the world-wide adoption of SI units cannot give it any intrinsic meaning and the current effort in metrization, while useful to trade and engineering, has less to contribute to science than would accrue from the revival of Latin as a lingua franca. It is only when quantities are made dimensionless that their magnitudes acquire an intrinsic meaning in the context of the model.

The principle of making the equations dimensionless is simple enough. Each variable, dependent or independent, is expressed as a product of some characteristic quantity of the same dimensions and a dimensionless variable. The equations are then rearranged until a suitable set of dimensionless parameters appears. Two principles govern this process: (i) constant quantities should be used as characteristic quantities; (ii) the dimensionless parameters should bear the burden of showing the comparative importance of the various terms in the equation.

To illustrate this consider Eq. (E1). The hypothesis $H_1$ assures us that $V$ and the $q_j$ are constant so that $q = \Sigma q_j$ is also constant. Any of the ratios $V/q_j$ defines a time that could be used as the characteristic time, but there is no basis for picking a particular $j$ and it is better to use the total flow rate $q$. Then $\tau = qt/V$ becomes a natural scale of dimensionless time. For example, if it were purely a matter of mixing with no reaction, Eq. (E1) would be

$$\frac{V}{q}\frac{dc_j}{dt} = \frac{dc_j}{d\tau} = \frac{q_j c_{jf}}{q} - c_j = \bar{c}_{jf} - c_j$$

or

$$c_j(\tau) = c_j(0) + \{\bar{c}_{jf} - c_j(0)\}\, e^{-\tau}$$

for constant $c_{jf}$, showing that $\tau$ is the natural time in the tank. If $q$ were not constant, then it would not be suitable. If it is going to fluctuate about a representative value, say $\bar{q}$, then one would take $\tau = \bar{q}t/V$ and let $q/\bar{q}$ be the dimensionless flowrate which would then fluctuate about 1. If the behavior of the system for all $q$ is to be studied, it would be inappropriate to take $\tau = \bar{q}t/V$ and we should look elsewhere for a characteristic time. For example, if $c$ is the concentration of an organism in a chemostat or fermentor (the biochemical engineer's name for a stirred tank) where it grows on a nutrient of concentration n that is fed to the chemostat at a rate $qn_f$, we have

$$V\frac{dc}{dt} = -\,qc + V\mu(n)c\ ,$$

$$V\frac{dn}{dt} = q(n_f - n) - \frac{V\mu(n)c}{Y}\ ,$$

where $Y$ is called the yield coefficient and $\mu$ the specific growth rate. Now $\mu$ has the dimensions of reciprocal time so that we would look to it, rather than $q/V$, for a characteristic time. For example, if

$$\mu(n) = \frac{kn}{K + n},$$

we could take $\tau = kt$ and let $\kappa = q/kV$ giving

$$\frac{dc}{d\tau} = -\kappa + \frac{nc}{K + n},$$

$$\frac{dn}{d\tau} = \kappa(n_f - n) - \frac{1}{Y}\frac{nc}{K + n},$$

and $\kappa$ becomes the dimensionless form of the flowrate we wish to vary from 0 to $\infty$. Again suppose that the full range of $n_f$ is to be explored then it is not good to take $n/n_f$ as the dimensionless nutrient concentration. Rather do we look to the other concentration, the Michaelis constant $K$, for the characteristic quantity. If $\gamma = c/YK$, $\nu = n/K$, and $N = n_f/K$, then

$$\frac{d\gamma}{d\tau} = -\kappa\gamma + \frac{\nu\gamma}{1 + \nu},$$

$$\frac{d\nu}{d\tau} = \kappa(N - \nu) - \frac{\nu\gamma}{1 + \nu}.$$

In these equations $\kappa$ and $N$ are two parameters whose influences we wish to explore and, because the system is so simple, all others have been eliminated. When the process of rendering the equations dimensionless is done correctly, we have the smallest number of parameters in the equations. For example, if we had taken $\tau = qt/V$, $u = c/Yn_f$, $v = n/n_f$, $\alpha = kV/q$, $\beta = K/n_f$, an alternative and perfectly correct choice, then the resulting equations

$$\frac{du}{d\tau} = -u + \frac{\alpha uv}{\beta + v},$$

$$\frac{dv}{d\tau} = 1 - v - \frac{\alpha uv}{\beta + v}$$

would still have two parameters $\alpha$ and $\beta$. If, as we should, we include the initial conditions $c = c_0$, $n = n_0$ we would have two further dimensionless constants $\gamma_0 = c_0/YK$, $\nu_0 = n_0/K$ or $u_0 = c_0/Yn_f$, $v_0 = n_0/n_f$. Had we used $c_0$ and $n_0$ to make the variables dimensionless, say $x = c/c_0$, $y = n/n_0$, $\tau = qt/V$, then we would have initial conditions $x = y = 1$, but four parameters ($Y$, $n_f/n_0$, $K/n_0$, and $Vk/q$) in the equations.

Returning to the master example and Eq. (E1), we realize that there is no particular merit in taking a different characteristic concentration, such as $c_j = q_j c_{jf}/q$, for each $c_j$. For one thing, some of the $c_j$ may be zero. It is better to take an average concentration $c^* = \Sigma q_j c_{jf}/q$. We might then expect, if the stoichiometric coefficients $\alpha_j$ are not too extreme, and the initial conditions $u_{j0}$ of reasonable magnitude, that the dimensionless concentrations would be of reasonable magnitude. What does "reasonable magnitude" mean? It is obviously intended to be a vague term and it would be a mistake to make it too precise until we have lived with the equations a little longer. We have chosen to make the reaction rate dimensionless by dividing it by the rate $qc/V$. This means that $q$, $c^*$, $V$ are taken in to the kinetic constants of the reaction rate expression. For example, if $r(c_1, T) = A \exp(-E/RT)c_1$, then $R(u_1, v) = (VA/qc^*)[\exp-(E/RT^*)/v]u_1$ with two dimensionless groups, a Damköhler number $(VA/qc^*)$ and an Arrhenius number $E/RT^*$. This choice might not in fact be the best as $VA/qc^*$ might be objectionably large when the preexponential factor $A$ is used. An alternative is to use the reaction rate at feed conditions to define

$$\tilde{R}(u_1,\ldots,v) = \frac{r(c^*u_1,\ldots,T^*v)}{r(c^*\gamma_1,\ldots,T^*v_f)} \tag{46}$$

Then $\tilde{R}$ is 1 for inlet conditions and is probably of reasonable magnitude in the reactor. The Damköhler number

$$\mathrm{Da} = \frac{Vr(c_{1f},\ldots,T_f)}{qc^*} \tag{47}$$

is now a better representation of the intensity of reaction. Thus (E15) might well be modified to

$$\frac{du_j}{d\tau} = \gamma_j - u_j + \alpha_j \,\mathrm{Da}\, \tilde{R}(u_1,\ldots,v) \;. \tag{48}$$

Similarly, the choice of $T^* = (-\Delta H)c^*/C_\mathrm{p}$ has the merit of being physically recognizable as of the order of adiabatic temperature rise at complete reaction, but for highly exothermic reactions may not be the most appropriate since it would make $v$ unreasonably small. If $T^*$ is chosen to be of the order of magnitude of $T_f$, we would have another number $\tilde{\omega} = (-\Delta H)c^*/C_\mathrm{p}T^*$ which is a dimensionless form of the adiabatic temperature rise. (In another context this is the Prater number, hence $\tilde{\omega}$ since Pr has been preempted by Prandtl.) Equation (E16) would later read

$$\frac{dv}{d\tau} = v_f - v + \tilde{\omega}\,\mathrm{Da}\,\tilde{R}\,(u_1,\ldots,v) \tag{49}$$

and all other equations would remain the same.

We have belabored this point a little, but it is important not to become too attached to one's first choice of dimensionless variables, or, for that matter, of notation, if an alternative will show better how things stand. For, to return to the second point, it is when $u_j$ and $v$ are of reasonable magnitude that the parameters Da, $\tilde{\omega}$, etc., show the importance of various terms. If, for example, Da of Eq. (46) is of the order of $10^{10}$, it would appear that the reaction is exceedingly fast and must in fact always be at equilibrium.

Two further points should be made before going further.

The first is that it is not always desirable to amalgamate products of parameters. Thus $\tilde{\omega}$ Da as a product of two parameters shows the physical meaning much better than a new parameter, $(-\Delta H)Vr_f/qC_pT^*$. Secondly, characteristic quantities should not be introduced unnaturally. For example, if the model of a plug flow tubular reactor envisages a semiinfinite tube, an actual length should not be introduced gratuitously. Thus the equations might be

$$D\frac{d^2c}{dz^2} - v\frac{dc}{dz} - kc = 0, \quad z > 0,$$

$$-D\frac{dc}{dz} + vc = vc_f, \quad z = 0,$$

$$c \text{ finite as } z \to \infty .$$

It is generally not desirable to introduce a characteristic length $L$ even though the solution may be wanted at $z = L$ or $2L$ or some set of fixed points. Rather $D/v$ or $v/k$ can be used as characteristic lengths to give $\zeta = vz/D$ or $kz/v$ and then $z = L$ will correspond to $\zeta = Z = vL/D$. On the other hand if the model of the reactor is of finite length $L$, there is a natural characteristic length and $\zeta = z/L$, with boundary conditions at $\zeta = 0$ and 1, is an appropriate choice.

### B. *REDUCTION IN NUMBER OF EQUATIONS*

It might be suspected that, since there is only one reaction, there would be some redundancy in the $S$ mass balances. In fact it is clear from (E15) or (47) that the reaction rate term is totally eliminated if the equation for $u_j$ is divided by $\alpha_j$ and subtracted from any other equation treated in the same manner. Thus

$$\frac{d}{d\tau}\left(\frac{u_j}{\alpha_j} - \frac{u_k}{\alpha_k}\right) = \left(\frac{\gamma_j}{\alpha_h} - \frac{\gamma_k}{\alpha_k}\right) - \left(\frac{u_j}{\alpha_j} - \frac{u_k}{\alpha_k}\right)$$

and $(u_j/\alpha_j) - (u_k/\alpha_k) \to (\gamma_j/\alpha_j) - (\gamma_k/\alpha_k)$ as $\tau \to \infty$ if $\gamma_j$ is asymptotically constant. This suggests that, for constant $\gamma_j$, we can write

$$u_j(\tau) = \gamma_j + \alpha_j\, u(\tau)$$

and obtain a single equation

$$\frac{du}{d\tau} = -\,u + R\,(\gamma_1 + \alpha_1 u, \ldots, v) = -\,u + P(u, v) \quad . \tag{50}$$

We have not, of course, gotten rid of $\gamma_j$ and $\alpha_j$, but they have been folded into the reaction rate expression, which is a very good place for them. However, unless $u_{j0} = \gamma_j + \alpha_j u_0$ for all $j$ and some $u_0$, the initial conditions will not be satisfied. In this case let

$$u_j(\tau) = \gamma_j + \alpha_j\, u(\tau) + (u_{j0} - \gamma_j)\, y(\tau) \quad , \tag{51}$$

then substituting and rearranging gives

$$\left(\frac{du}{d\tau} + u - R\right) + \frac{u_{j0} - \gamma_j}{\alpha_j}\left(\frac{dy}{d\tau} + y\right) = 0 \quad .$$

Now not all the ratios $(u_{j0} - \gamma_j)/\alpha_j$ are the same, otherwise $u_0$ could be their common value, hence taking two values of $j$ with different ratios and subtracting the common ratio, we see that the two brackets in the last equation are separately zero. But since $y(0) = 1$, $y(\tau) = e^{-\tau}$ and the effect of the incompatibility dies away quickly. Thus, although this makes $R$ a function of $u$, $v$, and $\tau$, its dependence on $\tau$ quickly disappears. We lose nothing in our study of the problem if we use Eq. (50) in place of (E15) or

$$\frac{du}{d\tau} = -\,u + \mathrm{Da}\;\tilde{P}(u, v) \tag{52}$$

in place of (48).

Notice that this is not a change of model, but a reduction in the number of equations within the model; in fact in this case within all models. In the steady-state models we do not even have to take account of $y$, since its steady-state value is zero. As a general rule the number of mass balances can be reduced from $S$ to $R$, the number of linearly independent reactions, though the further equation for $y$ may have to be considered and if there are persistent fluctuations in the feed rates this may push the number of equations needed back up to $S$ [21].

In chemical reactor problems it is often useful to look for linear combinations of concentrations which satisfy nonreactive equations that can either be solved or show a notably simpler behavior. For example, if the reaction is again $\Sigma\alpha_j A_j = 0$ and it takes place at a rate or per unit volume in a porous pellet $\Omega$, through which the reactants diffuse with effective Knudsen diffusion coefficients $D_j$, we have the equations

$$D_j \nabla^2 c_j + \alpha_j r(c_1, \ldots, T) = 0 \quad \text{in} \quad \Omega \tag{53}$$

with

$$c_j = c_{js} \quad \text{on} \quad \partial\Omega \; .$$

Then

$$\nabla^2 \{(D_j c_j/\alpha_j) - (D_k c_k/\alpha_k)\} = 0 \quad \text{in} \quad \Omega$$

and is constant over $\partial\Omega$. But this implies that it is constant everywhere and hence we can substitute

$$c_j = c_{js} + (\alpha_j/D_j) w$$

and have

$$\nabla^2 w + r(c_{1s} + (\alpha_1/D_1) w, \ldots, T) = 0 \quad \text{in} \quad \Omega \; ,$$

with

$$w = 0 \quad \text{on} \quad \partial\Omega \quad .$$

The energy balance gives

$$k_e \nabla^2 T = (\Delta H) r(c_1, \ldots, T) \quad \text{in} \quad \Omega$$

$$T = T_s \quad \text{on} \quad \partial\Omega$$

and this suggests substituting

$$T = T_s + (-\Delta H/k_e) w \quad .$$

Thus all the equations collapse into one, namely,

$$\nabla^2 w = - R(w) = - r[c_{1s} + (\alpha_1/D_1)w, \ldots, T_s$$

$$+ (- \Delta H/k_e) w] \quad \text{in} \quad \Omega$$

with

$$w = 0 \quad \text{on} \quad \partial\Omega \quad .$$

Notice, however, that this is of restricted application. Except for symmetric regions, such as a sphere, it does not apply with the more general boundary conditions

$$D_j \frac{\partial c_j}{\partial \nu} = k_j (c_{jf} - c_j) \quad \text{on} \quad \partial\Omega \quad .$$

Nor can it be extended to transients except in the special case of $(D_j \rho c_p / k_e) = 1$ for all $j$. In this case we write the transient equations as

$$\frac{\partial c_j}{\partial t} = D_j \nabla^2 c_j + \alpha_j r \quad ,$$

$$\rho c_p \frac{\partial T}{\partial t} = k_e \nabla^2 T + (-\Delta H) r$$

and assume the initial values are uniform and $c_{j0}$, $T_0$, respectively. Then putting

$$c_j = c_{js} + (\alpha_j/D_j) w + (c_{j0} - c_{js}) v \quad ,$$

$$T = T_s + (-\Delta H/k_e) w + (T_0 - T_s) v \quad ,$$

and letting

$$D = D_j = k_e/\rho c_p$$

gives

$$\left(\frac{\partial w}{\partial t} - D\nabla^2 w - R\right) + \frac{(c_{j0} - c_{js})D_j}{\alpha_j}\left(\frac{\partial v}{\partial t} - D\nabla^2 v\right) = 0 \quad .$$

Also $w = v = 0$ on $\partial\Omega$ and $w = 0$, $v = 1$ initially. If all the quantities $(q_0 - q_s)D_j/\alpha_j$ are the same, then we can fix $w_0$ and ignore $v$. But if they are different, then the same argument as before shows that the two brackets are equal to zero. Thus $v$ is the distribution of temperature in $\Omega$ which is initially uniform and has zero boundary values and, as we know, subsides to zero everywhere. Thus, as with the stirred tank, the full story of its stability properties is told by the study of one equation. However, even if the $D_j$ were all equal but their common value not $k_e/\rho c_p$, the system would be much more complicated and its stability properties critically dependent on $k_e/\rho c_p D$.

## C. *THE EXAMINATION OF SPECIAL CASES*

In gaining experience with a model, not even the most drastic of simplifications should be overlooked. For example, if we claimed that the whole stirred tank system was held at a fixed temperature, this would reduce all models very much. It could be attained as the limit of infinite $\beta_0$, $\beta_c$, $\lambda$, and $\chi$ for then $v = w_i = w = w_0 = \Theta = \Theta_f$ and any incoming temperature fluctuations are damped out instantaneously. By the reduction of the last section we would have

$$\frac{du}{d\tau} = - u + \mathrm{Da}\ \tilde{P}(u)$$

as the governing equation. In particular the steady state is a solution of the equation

$$u = \text{Da}\ \tilde{P}(u) \quad .$$

Now $u = 0$ implies $u_j = \gamma_j$ and $\tilde{R} = 1$ by Eq. (46). Thus $\tilde{P}(0) = 1$. Also $u_j$, being a concentration must be positive so by Eq. (51) $u \leq u_m = \min\ (\gamma_j/-\alpha_j)$ for all for which this ratio is positive. In fact $u = u_m$ implies that some reactant species has been used up and, hence, that $P$ would be zero or, if the reaction were reversible [$u_m > u_e$, where $P(u_e) = 0$] $P$ would be negative. But $P$ is continuous and so must have a form analogous to one of those in Fig. 8. If it has a form such as a, b, or c it is clear that there is one and only one intersection with the line $u$/Da. As Da, the intensity of the reaction, increases from zero to infinity this intersection moves monotonically along the curve and $u$ increases from 0 to $u_e$. This is what we would expect. However with a curve of type $d$, for which $P(w)/u$ is not monotonic, there is a range of Da, say ($\text{Da}_*$, $\text{Da}^*$), in which the line of slope 1/Da can intersect the curve $P$ three times.

We have thus arrived, in a very special case, at the possibility of multiple steady states and if these things be done in the green tree! In particular if we take the steady-state

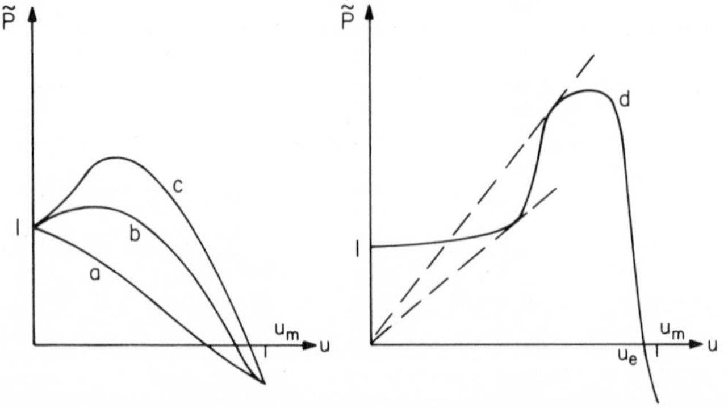


*FIG. 8. Forms of the rate function P(u).*

equations (E23) and (E24) and put $u_j = \gamma_j + \alpha_j u$, we have two simultaneous equations

$$u = P(u, v) = (1 + \hat{\beta})v - (v_f + \hat{\beta}\Theta_f)$$

whence

$$u = P\left(u, \frac{v_f + \hat{\beta}\Theta_f}{1 + \hat{\beta}} + \frac{u}{1 + \hat{\beta}}\right) = G(u) \quad .$$

Clearly a sufficiently complicated $P(u, v)$ can yield a truly droll $G(u)$.

Another special case is the adiabatic reactor given by Eqs. (E15) and (E16) with no transfer $\beta_i = 0$. We shall look at its dynamics in the next section; but here it may be remarked that again the functions can get complicated.

A limiting case worth considering is when $\beta^* \to \infty$ in $\Sigma_5$. Equation (E27) may be written

$$\frac{\delta_c}{\beta^*}\frac{d\Theta}{d\tau} = \frac{\chi}{\beta^*}(\Theta_f - \Theta) + v - \Theta \quad .$$

Now $\chi(\Theta_f - \Theta)$ is certainly bounded and we will neglect the first term on the right side. This leaves us with a singular perturbation problem

$$\varepsilon \frac{d\Theta}{d\tau} = v - \Theta \quad .$$

Physically this means that $\Theta$ follows $v$ with a very small time constant $\varepsilon$. Thus if $v$ varies in the time scale of $\Theta$, the departure of $\Theta$ from $v$ will be of the order of $e^{-\tau/\varepsilon}$, which is quite negligible. But if $\Theta = v$ as $\beta^* \to \infty$, we have by adding (E21) and (E27)

$$(1 + \delta_c)\frac{dv}{d\tau} = (v_f + \chi\Theta_f) - (1 + \chi)v + R(u_1, \ldots, v) \quad .$$

This equation shows how, in this limit, the static capacity $(1 + \delta_c)$ and the dynamic capacity $(1 + \chi)$ have both been increased by the presence of the cooling jacket.

We do not have space here to treat the methods of singular perturbation properly. Suffice it to say that Lin and Segel [15] have a most lucid introduction and Cole [22] a comprehensive treatment. For applications to the stirred tank reactor see Varma and Aris [23] and the papers cited there.

*D. APPROXIMATE METHODS AND GETTING A FEEL FOR THE SOLUTION*

As experience with a model and its cognates grows, it is useful to draw sketches of how the functions may look. Again it is important not to fix one's ideas too closely or tie them down to first impressions, but a certain wild approximation is helpful at an early stage. Thus the partial differential equation may have to be solved by a detailed numerical method eventually, but at first a lumped equivalent or one-point collocation approximation may be worth looking at. In this case the boundary conditions (E18) and (E19) that have to be applied to (E11) do not yield a simple quadratic polynomial that might approximate the profile in the wall. If the wall is treated as a plate of uniform thickness $d_w$, then the eigenvalue approximation gives a dimensionless time constant of $1/(\lambda\mu^2)$ where $\mu$ is the smallest root of $\mu \cos \mu = (\lambda^2\mu^2\delta_w\delta'_w/\beta_o\beta_i - 1)/\lambda\{\delta_w/\beta_i) + (\delta'_w/\beta_o)\}$. Certainly in the initial stages of an investigation it is worth treating partial differential equations by an approximating ordinary equation.

Ordinary differential equations themselves do not of course have obvious solutions, though when there are only two of them, the phase plane and method of isoclines give valuable insight. This was the type of argument used by Amundson and Bilous [24] in the first complete analysis of the stirred tank. Up to that time the analysis of stirred tank behavior had been based on the steady-state equations

$$-u + P(u, v) = 0 \quad , \tag{54}$$

$$(v_f + \hat{\beta}\Theta_f) - (1 + \hat{\beta})v + P(u, v) = 0 \quad , \tag{55}$$

i.e., Eqs. (E23) and (E24) with $u_j = \gamma_j + \alpha_j u$ and $P(u, v) = R(u_1, \ldots, v)$. Thus van Heerden [35] argued that if the first were solved for $u = U(v)$, the second could be written

$$(1 + \hat{\beta})v - (v_f + \hat{\beta}\Theta_f) = P[U(v), v) \tag{56}$$

and the right side would represent the heat generated, whereas the left side would be the heat removed, either by coolant or by heating up the feed. The curve $P[U(v), v]$ is shown in Fig. 9; the left-hand side of Eq. (56) is of course a straight line and with suitable $\hat{\beta}$ and $\bar{v}_f = (v_f + \hat{\beta}\Theta_f)/(1 + \hat{\beta})$, its intersections are as shown at A, B, and C. The intermediate state must be unstable since a slight departure to a higher temperature leads to an excess generation of heat and vice versa. But the reverse argument applied to A and C does not prove stability, for it is based on a steady-state picture and stability is a dynamic question. Liljenroth [26] had advanced similar arguments in 1918; he clearly saw how the instability would arise and did so without writing down even an algebraic equation! It so happens that the stability of A and C is

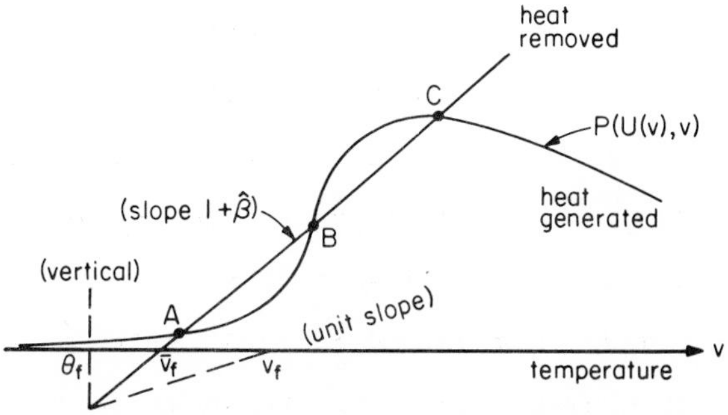


*FIG. 9. Heat generation and removal in the steady state.*

assured by the steeper slope of the straight line at those intersections in the adiabatic case, but it was Amundson and Bilous who first saw that two conditions would be necessary. These they obtained by linearizing

$$\frac{du}{d\tau} = -u + P(u, v) \quad , \tag{57}$$

$$\frac{dv}{d\tau} = (1 + \hat{\beta})(\bar{v}_f - v) + P(u, v) \tag{58}$$

about the steady-state $(u_s, v_s)$. They showed that the necessary and sufficient conditions for stability were

$$2 + \hat{\beta} - P_u(u_s, v_s) - P_v(u_s, v_s) > 0$$

and

$$(1 + \hat{\beta})[1 - P_u(u_s, v_s)] - P_v(u_s, v_s) > 0 \quad ;$$

it is the latter condition which corresponds to the disposition of curve and line at the intersections in Fig. 9.

But Amundson and Bilous went further than this by introducing the phase plane from nonlinear mechanics. If the Eqs. (57) and (58) are written

$$\frac{du}{d\tau} = F(u, v) \quad ,$$

$$\frac{dv}{d\tau} = G(u, v) \quad ,$$

then the loci $F = 0$ and $G = 0$ in the $(u, v)$ plane are also isoclines on which $du/dv$ is zero and infinite, respectively. Moreover $F < 0$ above the curve $F = 0$ and $F > 0$ below it, while $G < 0$ to the right of $G = 0$ and $G > 0$ to the left. We can therefore put short vertical and horizontal lines through the curves $G = 0$ and $F = 0$, respectively, and put an arrowhead on each to show the direction in which a trajectory would be going when crossing these curves. We can then fill in each of the areas of the diagram in Fig. 10a with an arrow directed into the appropriate quadrant. For example, in the lobe between

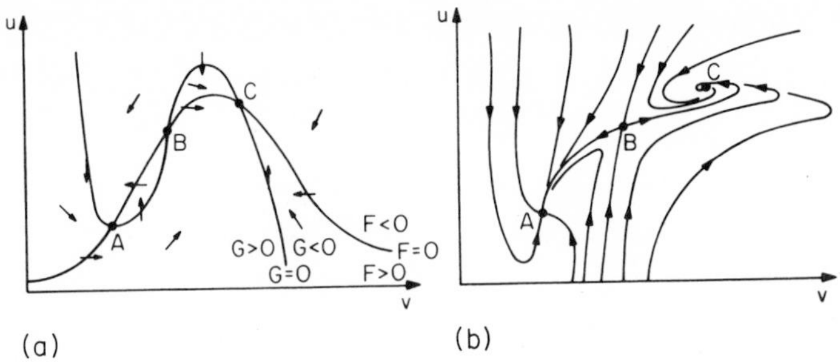


*FIG. 10. The phase plane with three steady states. (a) Isoclines. (b) Trajectories.*

the $F$ and $H$ curves between A and B, the arrow on the segment of $G = 0$ goes vertically upwards while that on $F = 0$ goes to the left; it follows that all trajectories passing through that lobe must go "northwesterly." It is clear from the general swirl of the arrows that a trajectory is going to have difficulty in getting to B and in fact the phase plane of Fig. 10b shows that only along the knife edge of a separatrix can one hope to reach B. This is of course impossible in practice since even rounding error would put the calculation on one side or the other. It would be possible to plot the isoclines, i.e., the curves in the $u,v$-plane along which $du/dv = F/G$ is constant, but this is not always necessary and can become too complicated to be worthwhile.

The subsequent study of the stirred tank has made extensive use of the phase plane as a method of presentation. Amundson and Aris [27] showed how the unstable steady state could be brought under control and how limit cycles might arise in the intermediate stages. Two-phase reactors and more complex reactions occupied Amundson and his colleagues for several years (see references in [10]) but the last word has been given comparatively recently in a pair of beautifully comprehensive and nicely argued papers by Uppal *et al.* [28]. The best general account of the stability problems of chemical engineering is Denn's book [29].

## V. THE EFFECTIVE PRESENTATION OF MODELS

One of the beauties of a model, or at any rate a sufficiently simple one, is that it can be studied comprehensively. With a little care one can often ensure that every typical case has been studied. The phase plane does this for the solutions of a two-dimensional system for, in such a diagram as Fig. 10b, the eye can very easily interpolate and see any solution. It would also be possible to put in a selection of isochrones (though their origins on each trajectory would necessarily be arbitrary), so that it would become clear where $u$ and $v$ were changing rapidly and where slowly, but there is a danger of cluttering up these figures beyond the point of greatest clarity. The limitation of phase portraits, as they are often called, is that they are not easy to draw in more than two dimensions and one has to make do with a number of two-dimensional projections. Certainly colored holography would allow four-dimensional presentation, but this is clearly out of common reach; stereoscopic pictures are possible but in their commonest form they demand a decoupling of the eyes that not everyone can manage [30].

The artistic standards of such a journal as Scientific American are a model of clarity in overcoming many of the difficulties of three dimensions, as Zeeman's brilliant survey of catastrophe theory shows [31]. A large part of the appeal of this theory, particularly in the soft sciences, is that it is capable of graphically presenting a whole rather than the several parts. It is interesting to note that chemical engineers have been dealing with mathematical catastrophes in chemical reactor theory for 20 yrs or more, and catastrophe, in the popular sense of the word, in reactor practice for much longer.

The steady state of the stirred tank as given by $v_s$, the solution of

$$(1 + \hat{\beta})(v - \underset{\sim}{v}_f) = P[U(v), v] = Q(v)$$

can be presented as the cusp catastrophe. Suppose $Q(v)$ has the form shown in Fig. 11 which the slope at the point of inflection is say $\sigma > 1$. When $\beta > \sigma - 1$, there can only be one intersection of the straight line with the curve and the larger $\hat{\beta}$ is, the closer will its abcissa $v_s$ be to $v_f$. If $\hat{\beta}$ is infinite $v_s = \underset{\sim}{v}_f$ and we might call the reactor isothermal with its coolant since $\underset{\sim}{v}_f$ equals $\Theta_f$. On the other hand, for $\hat{\beta} = 0$ the reactor is adiabatic and for $\underset{\sim}{v}_f$ between A and B, there will be three solutions. In fact, for all $\beta < \sigma - 1$, there will be a range of $\underset{\sim}{v}_f$ giving three solutions, but as $\hat{\beta}$ approaches $\sigma = 1$, this range narrows down to the point $\underset{\sim}{v}_{fc}$. The high part of the curve is where the reaction rate is large and we can speak of the reaction as "ignited." At high temperatures the reaction rate drops off since it is limited by "equilibrium,"

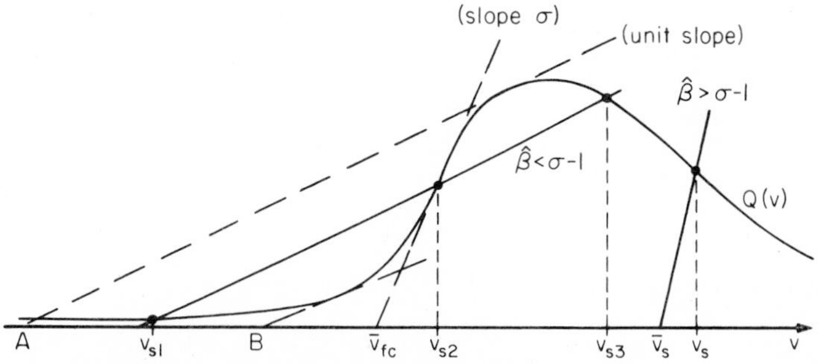


*FIG. 11. Critical points of the heat balance diagram.*

but at low temperatures it goes so slowly we would call it "quenched."

To present this catastropic form we plot $v_s$ the temperature at steady state as a surface over the plane of $v_f$ and $(1 + \hat{\beta})^{-1}$. The latter is chosen to make the picture look better for now $\hat{\beta} = \infty$ will be $(1 + \hat{\beta})^{-1} = 0$ on the axis at the back of the picture. When $(1 + \hat{\beta})^{-1} > \sigma^{-1}$, there is a range of $\underset{\sim}{v}_f$ for which there are three steady states and hence three values of the reaction rate. Such a diagram as Fig. 12 makes the instability of the intermediate steady state very clear, for moving about on the surface it is clearly unnatural to try to get up underneath the fold. What may happen naturally is that in moving about on the surface (i.e., changing $\underset{\sim}{v}_f$ and $\hat{\beta}$ very slowly), we come to the edges of the fold and have to jump from one branch of the surface to the other. Thus on the

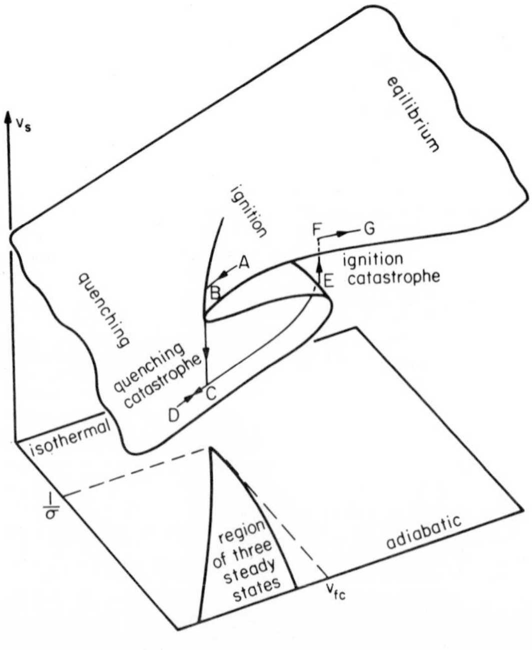


*FIG. 12. The steady state as a cusp catastrophe.*

path AB, for example, as $\underset{\sim}{v}_f$ is decreased beyond its value at B there is nowhere to go but the lower part of the surface: CD. This is the quenching catastrophe. On the other hand if $\underset{\sim}{v}_f$ is increased again on the path DCE, we can keep on the lower surface until at E we have to jump to the upper FG. This is the ignition catastrophe. We have perhaps belabored the point a little and the reader may find it mathematically flatulent, but of the dramatic forcefulness of this method of presentation there can be no question.

There is of course, more to catastrophe theory than the presentation of models. On the one hand there is the topology and Thom, its principal architect, has shown how catastrophes may be fully classified. On the other hand there are the applications, not merely to the social sciences, but to the whole question of morphogenesis in biology [32]. The idea is that in the space of observables there may be a closed subset $K$, the catastrophe set, such that the form (morphos) of the system remains essentially the same until its state encounters $K$. The study begins with the local structure of $K$ in the hope of building up to a global picture. As has been said, chemical engineers have been "talking prose" for years, but there is no doubt that interaction with catastrophists will be fruitful. At a meeting in Kiev, Y. A. Yoblonsky of the Institute of Catalysis in Novosibirsk pointed out that even some of the more complex reaction rate expressions had the form of the cusp catastrophe surface.

Comprehensiveness in presentation requires much skill when there are several parameters and the work of Uppal *et al.* [28] illustrates how well this can be done. They use equations which are essentially the same as (57) and (58) but with a dimensionless temperature $v' = E(T - T_f)/RT_f^2$, where $E$ is the activation energy of rate constant for a first-order irreversible reaction. Thus $P(u, v')$ takes the form of $\mathrm{Da}(1 - u) \exp[v'/(1 + v'/\gamma)]$, $\gamma = E/RT_f$. At a later stage they show that the value of $\gamma$ is not important with respect to structural changes and clearly it is a great simplification to make it infinite. It allows them so specify types of steady-state dependence in various regions of the space of two of the parameters, and for each of these types to show the dependence on a third. Then for each different region of the latter, a typical phase plane shows the dynamics of the system. When it is remembered that the term steady state has been used loosely in the above and that what they study are really invariant sets (i.e., include limit cycles), it is clear that a vein of rich ore in the mine of dynamical systems has been opened up, the wealth of which was scarcely dreamed of in the philosophy of 20 yrs ago.

## APPENDIX: AN ILLUSTRATIVE EXAMPLE

### *DESCRIPTION*

A stirred tank reactor consists of a cylindrical vessel of volume $V$ with incoming and outgoing pipes. The incoming pipes bring reactants $A_1, A_2, \ldots, A_r$, at volume flowrates $q_1, q_2, \ldots, q_r$, and the outgoing pipe takes of the mixture of products $A_{r+1}, \ldots, A_s$ and the remnants of the reactants, at a flowrate of $q = q_1 + q_2 + \ldots + q_r$. Thus the volume $V$ remains constant. The reaction can be written as $\Sigma\alpha_j A_j = 0$, where $\alpha_1, \ldots, \alpha_r$ are negative stoichiometric coefficients and $\alpha_{r+1}, \ldots, \alpha_s$ are positive. This cylinder is immersed in another cylinder of annular

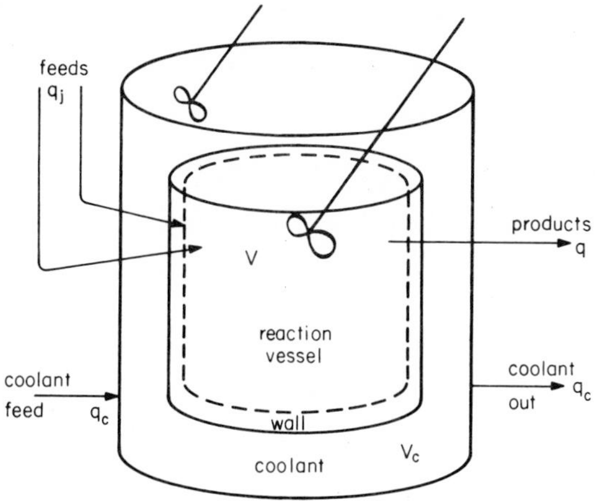


*FIG. 13. Schematic diagram of the stirred reactor.*

volume $V_c$, also perfectly stirred, which is fed cooling water of temperature $T_{cf}$ at a flowrate of $q_c$. Other details will be mentioned as we proceed. It should be mentioned that this description has already been deliberately simplified since the geometry of a real jacketed reactor would undoubtedly be more complicated than that of simple cylinders. However I have no desire to pile Pelion on Ossa.

*Hypotheses*

Let us lump together the applicability of all physical laws, such as the conservation of matter and energy or Fourier's law of heat conduction and call this the underlying general hypothesis $H_o$. The following hypotheses can be extracted from the description or be excogitated as relevant to the setting:

$H_1$: the mixing is perfect so that the concentrations $c_j$, the reaction temperature $T$ and $T_c$, the temperature of the coolant jacket, are all independent of position, though they may be functions of time. The volumes $V$ and $V_c$ are constant, as also are the flowrates $q_j$ and the feed temperatures $T_{jf}$. The work done by the stirrers may be ignored.

$H_2$: the reaction rate is a function $r(c_1, \ldots, c_s, T)$ such that the rate of change in the number of moles of $A_j$ by reaction alone is $\alpha_j r$ per unit volume.

$H_3$: the heat transfer to the inner and outer sides of the wall where the surface temperatures will be denoted by $T_i$ and $T_o$, respectively, can be described by transfer coefficients $h_i$ and $h_o$ such that the heat transferred per unit area is $h_i(T - T_i)$ and $h_o(T_o - T_c)$, respectively.

$H_4$: the heat capacity of the reaction mixture does not change significantly.

$H_5$: the system is in steady state.

$H_6$: the curvature of the wall is negligible and the sharp corners can be ignored.

$H_7$: the conductivity of the wall is extremely high.

$H_8$: the heat capacity of the wall is negligible.

*Derivation of the Most General Model* $\Sigma_1$

Using the principle of the conservation of matter from the underlying hypothesis $H_o$, we have the following balance for each species:

$$\begin{bmatrix}\text{rate of change of}\\ \text{number of moles of}\\ A_j \text{ in reactor}\end{bmatrix} = \begin{bmatrix}\text{rate of}\\ \text{feed of}\\ A_j\end{bmatrix} - \begin{bmatrix}\text{rate of}\\ \text{withdrawal}\\ \text{of } A_j\end{bmatrix} + \begin{bmatrix}\text{rate of formation}\\ \text{of } A_j \text{ by means}\\ \text{of the reaction}\end{bmatrix}$$

If $c_{jf}$ is the concentration (moles/volume) of $A_j$ in its feed stream, this translates immediately into the ordinary differential equation

$$V \frac{dc_j}{dt} = q_j c_{jf} - qc_j + \alpha_j Vr(c_1,\ldots,c_S,T) \quad . \tag{E1}$$

In obtaining this equation we have invoked $H_1$ and $H_2$.

If $h_j(c_1,\ldots,c_S,T)$ is the enthalpy per mole of $A_j$, and the work done by the stirrer is ignored, then conservation of energy implies

$$V \frac{d}{dt} \Sigma\, c_j h_j = \Sigma q_j c_f h_{jf} - q\Sigma c_j h_j - A_i h_i (T - \overline{T}_i) \quad .$$

The symbol $h_{jf}$ denotes the specific enthalpy of $A_j$ evaluated for its feed conditions. In the last term $A_i$ is the total internal wall area and since the heat transfer coefficient $h_i$ is independent of position we need only average the inner surface temperature of the wall. Thus $H_1$ and $H_3$ are used here. We now simplify this equation by subtracting from it the sum over $j$ of equations (E1) each multiplied by $h_j$. Thus

$$V\, \Sigma c_j \frac{dh_j}{dt} = q_j c_{jf}(h_{jf} - h_j) - (\Sigma\alpha_j h_j) V_r - A_i h_i (T - \overline{T}_i) \quad .$$

Next we observe that $\alpha_j h_j = \Delta H$ is the heat of reaction and that since $h_j$ is an intensive thermodynamic variable (i.e., $\Sigma c_j \frac{\partial h_j}{\partial c_k} = \Sigma c_j \frac{\partial h_k}{\partial c_j} = 0$)

$$\Sigma c_j \frac{dh_i}{dt} = \Sigma c_j c_{pj} \frac{dT}{dt} + \Sigma\Sigma\, c_j \frac{\partial h_j}{\partial c_k} \frac{dc_k}{dt} = C_p \frac{dT}{dt} \quad ,$$

where $c_{pj}$ is the heat capacity of $A$ per mole and $C_p$ is the heat capacity of the mixture per unit volume. We now invoke $H_4$ and write $q_j c_{jf}(h_{jf} - h_j) = qC_p(T_f - T)$ to give

$$VC_p \frac{dT}{dt} = qC_p(T_f - T) + (-\Delta H) Vr(c_1,\ldots,T) - A_i h_i (T - \overline{T}_i) \quad . \tag{E2}$$

This form allows us to check the commonsense of the equation for we can write it as

$$\begin{bmatrix}\text{rate of}\\ \text{change}\\ \text{of heat}\\ \text{content}\end{bmatrix} = \begin{bmatrix}\text{heat}\\ \text{brought}\\ \text{in with}\\ \text{feed}\end{bmatrix} - \begin{bmatrix}\text{heat}\\ \text{taken out}\\ \text{with}\\ \text{products}\end{bmatrix} + \begin{bmatrix}\text{heat}\\ \text{generated}\\ \text{by}\\ \text{reaction}\end{bmatrix} - \begin{bmatrix}\text{heat}\\ \text{removed by}\\ \text{cooling}\\ \text{wall}\end{bmatrix} .$$

The wall has been simplified to be a finite cylinder of internal area $A_i$. If we denote the region it occupies by $D$ and its inner and outer surfaces by $D_i$ and $D_o$, we apply conservation principles and Fourier's law of heat conduction to obtain for the wall temperature $T_w$,

$$\rho_w c_{pw} \frac{\partial T_w}{\partial t} = k_w \nabla^2 T_w \quad \text{in} \quad D \quad , \tag{E3}$$

where $\rho_w$, $c_{pw}$, and $k_w$ are the density, specific heat, and conductivity of the wall, respectively. To obtain boundary conditions we have to call on $H_3$,

$$k_w \frac{\partial T_w}{\partial n} = h_i(T - T_i) \quad \text{on} \quad \partial D_i \quad , \tag{E4}$$

$$k_w \frac{\partial T_w}{\partial n} = h_o(T_c - T_o) \quad \text{on} \quad \partial D_o \quad , \tag{E5}$$

where $\partial/\partial n$ is the normal derivative on the surface directed outward from $D$.

Before seeing how these equations simplify, let us write the heat balance equation for the coolant. This is

$$V_c C_{pc} \frac{dT_c}{dt} = q_c C_{pc}(T_{cf} - T_c) + A_o h_o(\overline{T}_o - T_c) \tag{E6}$$

where $C_{pc}$ is the heat capacity of the coolant per unit volume, $A_o$ is the area of $\partial D_o$ and $\overline{T}_o$ the average outer temperature. $H_1$ and $H_3$ have been involved in deriving this equation as well as the underlying $H_0$.

Equations (E1-6), together with suitable initial conditions, give $S+2$ ordinary and one partial differential equation with its boundary conditions and constitute $\Sigma_1$, the most detailed model we shall consider. In obtaining it the hypotheses used have been $H_0$, $H_1$, $H_2$, $H_3$, and $H_4$.

*Derivation of the Steady-State Models $\Sigma_2$ and $\Sigma_3$*

Now let us invoke $H_5$ and assume that the system is at steady state. To do this is to set all time derivatives equal to zero and it leads to a partial differential equation, Laplace's, for $T_w$ connected through its boundary conditions to a set of algebraic equations. (The term algebraic equation is applied to any equation that is not a differential equation even though transcendental functions may appear in it.) Let this model be $\Sigma_2$.

However when we recognize that $T_w$ is a potential function, we can use Green's theorem to give

$$0 = \iiint_D k_w \nabla^2 T_w dN = \iint_{\partial D_i + \partial D_o} k_w \left( \frac{\partial T_w}{\partial n} \right) dS$$

$$= A_i h_i (T - \overline{T}_i) - A_o h_o (\overline{T}_o - T_c) \quad . \tag{E7}$$

Combining this with (E6) we have three expressions for the rate of removal of heat

$$Q_c = q_c C_{pc} (T_c - T_{cf}) = A_o h_o (\overline{T}_o - T_c) = A_i h_i (T - \overline{T}_i) \quad .$$

This gives

$$T - T_{cf} = Q_c \left\{ \frac{1}{q_c C_{pc}} + \frac{1}{A_o h_o} + \frac{1}{A_i h_i} + \frac{\underset{\sim}{T}_i - \underset{\sim}{T}_o}{Q_c} \right\}$$

but the last term is a messy one and has to be evaluated from the full solution of the potential equation. If however we invoke $H_6$ and take $d_w$ to be the thickness of the wall, then we have the local flux of heat per unit area equal to $k_w(T_i - T_o)/d_w$. Also ignoring curvature makes $A_o = A_i = A$ so that $Q_c = k_w A(\overline{T}_i - \overline{T}_o)/d_w$. Then

$$Q_c = hA(T - T_{cf}), \text{ where } \frac{1}{h} = \left\{ \frac{A}{q_c C_{pc}} + \frac{1}{h_o} + \frac{1}{h_i} + \frac{d_w}{k_w} \right\} . \tag{E8}$$

Thus the model simplifies to $\Sigma_3$ consisting of the equations

$$q_j c_{jf} - qc_j + \alpha_j Vr(c_1,\ldots,c_s,T) = 0, \quad j = 1,\ldots,s, \tag{E9}$$

$$qC_p(T_f - T) + (-\Delta H)Vr(c_1,\ldots,c_s,T) - \hat{h}A(T - T_{cf}) = 0 . \tag{E10}$$

Notice that $H_8$ has no relevance at all to this model and that were we to invoke $H_7$ it would not change the model but only modify the value of $\hat{h}$.

*Simplified Transient Models $\Sigma_4$ and $\Sigma_5$*

Let us now return to the transient model, dropping $H_5$, and see the effect of $H_7$ and $H_8$. A little caution is needed here or terms can get lost. First suppose that $k_w \to \infty$ ($H_7$), we cannot conclude that $T_w$ is constant in time by observing that this limit, like the steady-state hypothesis, leads to Laplace's equation for $T_w$. Rather the form of the temperature profile through the wall is invariant in time, for it takes up a uniform temperature $T_w$ between $T_i$ and $T_o$ with virtually no delay. Thus, for very large $k_w$, the Laplacian of the temperature in the wall becomes very small and the product $k_w \nabla^2 T_w$ is finite.

Let $T_w$ denote the temperature of the wall, which is not only uniform across its thickness, but uniform everywhere on account of $k_w \to \infty$ and the fact that the wall is exposed to uniform temperatures on both sides. Then integrating (E3) throughout $D$, using Green's theorem and the boundary conditions (E4) and (E5), gives

$$V_w \rho_w C_{pw} \frac{dT_w}{dt} = A_i h_i T + A_o h_o T_c - (A_i h_i + A_o h_o) T_w \qquad \text{(E11)}$$

Thus we have a model consisting of $(s + 3)$ ordinary differential equations (E1), (E2), (E6), and (E11), where in (E2) and (E6) $T_i$ and $T_o$ have both been made equal to $T_w$. Let this be the model $\Sigma_4$; it invokes $H_o$-$H_4$ and $H_7$. Note that we did not need $H_8$.

Suppose now that we assert $H_8$, but drop $H_7$. Again $T_w$ is governed by Laplace's equation and we can arrive at (E7). However we are still left with Laplace's equation unless we invoke $H_6$ and note that

$$Q_c = Ah_o(\overline{T}_o - T_c) = Ak_w(\overline{T}_i - \overline{T}_o)/d_w = Ah_i(T - \overline{T}_i)$$
$$= Ah^*(T - T_c) \quad ,$$

where

$$\frac{1}{h^*} = \frac{1}{h_o} + \frac{d_w}{k_w} + \frac{1}{h_i}$$

We then have a model $\Sigma_5$, consisting of $(S+2)$ ordinary equations, the $S$ Eq. (E1) and

$$VC_p \frac{dT}{dt} = qC_p(T_f - T) + (-\Delta H) Vr(c_1, \ldots, T) - Ah^*(T - T_c) \qquad \text{(E12)}$$

$$V_c C_{pc} \frac{dT_c}{dt} = q_c C_{pc}(T_{cf} - T_c) + Ah^*(T - T_c) \qquad \text{(E13)}$$

*Initial Conditions*

Nothing has yet been said as to the initial conditions. These are certainly needed to solve equations, but may or may not be important in the physical problem. If the problem is one of optimal control during start-up, then initial conditions are obviously of the essence, but if it is one of response to long-term perturbations, they may not need to be considered. However they may be simply denoted by a suffix o and written

$$c_j = c_{jo}, \quad T = T_o', \quad T_w = T_{wo}(\underset{\sim}{x}), \quad T_c = T_{co} \quad \text{at } t = 0, \tag{E14}$$

where $\underset{\sim}{x}$ denotes the coordinates within the wall.

*The Dimensionless Equations*

Up to this point everything has been very dimensional and it is not clear what we meant by large and small values. There are various characteristic lengths, times, etc., in the problem and we want to pick the most judicious set. In particular, constants to which we are going to give some limiting value should not be used to render others dimensionless, nor should those whose variation we are going to study.

Let $V/q$ be the characteristic time and $\tau = qt/V$;

$c^*$, a characteristic concentration, say $\Sigma q_j c_{jf}/q$ and

$u_j = c_j/c^*$;

$T^*$, a characteristic temperature, say $(-\Delta H)c^*/C_p$, and

$v = T/T^*$, $w = T_w/T^*$, $\Theta = T_c/T^*$, $v_f = T_f/T^*$, $\Theta_f = T_{cf}/T^*$,

$w_i = T_i/T^*$, $w_o = T_o/T^*$;

$d_w$, a characteristic length such as a mean wall thickness with which the independent variables in the Laplacian are made to be dimensionless.

The other parameters will emerge with the equations. If (E1) is divided by $qc^*$, it becomes

$$\frac{du_j}{d\tau} = \gamma_j - u_j + \alpha_j R(u_1,\ldots,v) \quad , \tag{E15}$$

where $\gamma_j = q_j c_{jf}/\Sigma q_j c_{jf}$, i.e., $\Sigma\gamma_j = 1$, is the $j$th fraction of feed and $R = Vr/qc^*$. Similarly, let (E2) be divided by $qC_p/T^*$ to give

$$\frac{dv}{d\tau} = v_f - v + R(u_1,\ldots,v) - \beta_i(v - v_i) \quad , \tag{E16}$$

where $\beta_i = A_i h_i/qC_p$.

We will use the same symbol as before for the Laplacian with respect to the dimensionless variables, so that (E3) can be divided by $\rho_w c_{pw} q/VT^*$ to give

$$\frac{\partial w}{\partial \tau} = \lambda \nabla^2 w \tag{E17}$$

when $\lambda = k_w V/q\rho_w c_{pw} d_w^{\,2}$. If $\partial/\partial\nu$ denotes the normal derivative in the dimensionless variables,

$$\lambda \frac{\partial w}{\partial \nu} = \frac{\beta_i}{\delta_w}(v - w_i) \quad \text{on} \quad \partial D_i \quad , \tag{E18}$$

where $\delta_w = A_i d_w \rho_w c_{pw}/VC_p$ is the ratio of the heat capacity of the wall to that of the contents. Similarly,

$$\lambda \frac{\partial w}{\partial \nu} = \frac{\beta_o}{\delta'_w}(\Theta - w_o) \quad \text{on} \quad \partial D_o \quad , \tag{E19}$$

where $\beta_o = A_o h_o/qC_p$ and $\delta'_w = (A_o/A_i)\delta_w$. Finally (E5) becomes

$$\delta_c \frac{d\Theta}{d\tau} = \chi(\Theta_f - \Theta) + \beta_o(w_o - \Theta) \quad , \tag{E20}$$

where $\delta_c = V_c C_{pc}/VC_p$ and $\chi = q_c C_{pc}/qC_p$.

In the later models we have $\hat{h}$ and $h^*$ and we make them dimensionless with $qC_p$ to give

$$\hat{\beta} = A\hat{h}/qC_p = \left\{\frac{1}{\chi} + \frac{1}{\beta_o} + \frac{1}{\beta_i} + \frac{1}{\lambda\beta_w}\right\}^{-1} \quad , \tag{E21}$$

$$\beta^* = Ah^*/qC_p = \left\{ \frac{1}{\beta_o} + \frac{1}{\lambda\delta_w} + \frac{1}{\beta_i} \right\}^{-1} \tag{E22}$$

Thus we have for $\Sigma_3$ the $(S + 1)$ nondifferential equations

$$\gamma_j - u_j + \alpha_j R(u_1,\ldots,v) = 0 \quad , \tag{E23}$$

$$v_f - v + R(u_1,\ldots,v) - \hat{\beta}(v - \Theta_f) = 0 \quad . \tag{E24}$$

The model $\Sigma_4$ consists of Eqs. (E15), (E16), and (E20) with the dimensionless form of (E11), namely,

$$\delta_w \frac{dw}{d\tau} = \beta_i v + \beta_o \Theta - (\beta_o + \beta_i) w \quad . \tag{E25}$$

Finally, the model $\Sigma_5$ is Eq. (E15) and the two equations:

$$\frac{dv}{d\tau} = v_f - v + R(u_1,\ldots,v) - \beta^*(v - \Theta) \quad , \tag{E26}$$

$$\delta_c \frac{d\Theta}{d\tau} = \chi(\Theta_f - \Theta) + \beta^*(v - \Theta) \quad . \tag{E27}$$

The initial conditions, as needed, are

$$u_j = u_{jo}, \quad v = v_o, \quad w = w_o(\xi), \quad \Theta = \Theta_o, \quad \tau = 0, \quad \underset{\sim}{\xi} = \underset{\sim}{x}/d_w. \tag{E28}$$

*SUMMARY OF PARAMETERS*

Reaction : $\alpha_j$, stoichiometric coefficients

—, parameters of the rate law, e.g., $E/RT_f$

Feed : $\gamma_j$, fraction of $A_j$ in feed

$v_f$, feed temperature

$\Theta_f$, coolant feed temperature

Capacities: $\delta_c$, heat capacity ratio of coolant to reactants

$\delta_w$, ratio of heat capacity of wall to reactants

$\chi$, ratio of heat carrying capacities of coolant to reactants

Transfer : $\beta_o, \beta_i$, dimensionless heat transfer coefficients

$\hat{\beta}, \beta^*$, composite heat transfer-coefficients

$\lambda$, dimensionless wall conductivity

*SUMMARY OF MODELS*

| Model $\Sigma$ | Hypotheses H | Equations E | Dimensionless equations | Remarks |
|---|---|---|---|---|
| 1 | 0-4 | 1-6 | 15-20 | cf(47) and (48) with (E15) and (E16) |
| 2 | 0-5 | 1-6 | 15-20 | Set $\partial/\partial t$ or $\partial/\partial\tau = 0$ |
| 3 | 0-6 | 9,10 | 23,24 | |
| 4 | 0-4,7 | 1,2,6,11 | 15,16,20,26 | |
| 5 | 0-4,6,8 | 1,12,13 | 15,26,27 | |

*NOTATION*

Only the notation of the master example is given: The extra notation of the text is ephemeral defined in its context.

$A_j$, chemical species, $j = 1, S$; $j = 1,\ldots,r$ for reactants, $r + 1,\ldots,S$ for products

$A_i, A_o$, inner and outer areas of reactor wall

$C_p$, heat capacity per unit volume of reactant mixture

$C_{pc}$, heat capacity per unit volume of coolant

$c_j$, concentration of $A_j$

$c_{jo}$, initial concentration of $A_j$

$c_{jf}$, feed concentration of $A_j$

$c_{pw}$, specific heat of wall

$c^*$, reference concentration

$d_w$, thickness of reactor wall

$h_j$, enthalpy per mole of $A_j$

$h_{jf}$, enthalpy per mole of $A_j$ under feed conditions

$h_i, h_o$, heat transfer coefficient at inner and outer wall surface

$\hat{h}, h^*$, composite heat transfer coefficients

| | |
|---|---|
| $k_w$, | thermal conductivity of wall |
| $n$, | outward normal to wall in $\partial/\partial n$ |
| $Q_c$, | total rate of heat removal |
| $q$, | flowrate of reacting mixture |
| $q_c$, | coolant flowrate |
| $q_j$, | feed rate of $A_j$ |
| $R$, | dimensionless reaction rate $Vr/qc^*$ |
| $r$, | reaction rate per unit volume |
| $T$, | temperature |
| $T_c, T_{cf}, T_f, T_w(\underset{\sim}{x})$, | temperature of coolant, coolant feed, reactor feed, and wall, respectively |
| $T_i, T_o, \underset{\sim}{T}_i, \underset{\sim}{T}_o$, | inner and outer wall temperatures and their averages |
| $T'_o, T_{wo}, T_{co}$, | initial reactor, wall, and coolant temperatures |
| $T^*$, | reference temperature |
| $t$, | time |
| $V$, | volume of reactor |
| $V_c, V_w$, | volume of coolant, wall |
| $v$, | dimensionless temperature $T/T^*$ |
| $v_f$, | $T_f/R^*$ |
| $w(\xi)$, | $T_w/T^*$ |
| $w_i/w_o$, | $T_i/T^*$, $T_o/T^*$ |
| $\underset{\sim}{x}$, | coordinates within wall |
| $\alpha_j$, | stoichiometric coefficients |
| $\beta_i, \beta_o$, | $h_iA_i/qC_p$, $h_oA_o/qC_p$ |
| $\hat{\beta}, \beta^*$, | dimensionless composite heat transfer coefficients; (E21), (E22) |
| $\gamma_j$, | dimensionless feed rate of $A_j$ |
| $\delta_c, \delta_w, \delta'_w$, | $V_cC_{pc}/VC_p$, $A_id_w\rho_wc_{pw}/VC_p$, $A_od_w\rho_wc_{pw}/VC_p$ |
| $\Delta H$, | heat of reaction |
| $\Theta$, | dimensionless coolant temperature, $T_c/T^*$ |

$\Theta_o, \Theta_f$, $T_{co}/T^*$, $T_{cf}/T^*$

$\lambda$, $k_w V/q\rho_w c_{pw} d_w^2$

$\nu$, dimensionless normal in $\partial/\partial\nu$

$\underset{\sim}{\xi}$, $\underset{\sim}{x}/d_w$

$\rho_w$ density of wall

$\tau$, $qt/V$

$\chi$, $q_c C_{pc}/qC_p$

*ACKNOWLEDGMENT*

I am indebted to the generosity of the Fairchild Foundation and to the hospitality of the California Institute of Technology for a period there as a Sherman Fairchild Distinguished Scholar under conditions so ideal that this and other products of my stay must inevitably fall short in accomplishment of what their kindness afforded in opportunity. Mrs. Yolande Johnson deciphered my scrawl and turned it into an excellent typescript; I am most grateful to her.

*REFERENCES*

1. B. J. F. LONERGAN, "Insight." Philosophical Press, 19 .
2. W. N. GILL, V. ANANTHAKRISHNAN, and A. J. BARDUHN, *Am. Inst. Chem. Eng. J. 11*, 1063 (1965).
3. G. I. TAYLOR, *Proc. Roy. Sci. A219*, 186 (1953).
4. R. ARIS, *Proc. Roy. Sci. A219*, 67 (1956).
5. L. APOSTEL, *in* "The Concept and the Role of the Model in Mathematics and Natural and Social Sciences." (H. Freudenthal, ed.) Reidel Publishing Co., Dordrecht, 1961.
6. P. Duhem, "The Aim and Structure of Physical Theory," 1st ed. (P. P. Wiener, Trans.), p. 906, Princeton University Press, Princeton, 1954.
7. G. F. OSTER, A. S. PERELSON, and A. KATCHALSKY, *Q. Rev. Biophys. 6*, 1 (1973).
8. J. WEI and J. C. W. KUO, *Ind. Eng. Chem. Fundamentals, 8*, 114 (1969).
9. V. HLAVÁCEK, M. MAREK, and M. KUBICEK, *J. Catal. 15*, 17, 31 (1969).
10. R. ARIS, *Adv. Chem. 109*, 578 (1972).
11. R. ARIS, "The Mathematical Theory of Diffusion and Reaction in Permeable Catalysts." Clarendon Press, Oxford, 1975.

12. J. VILADSEN and W. E. STEWART, *Am. Inst. Chem. Eng. J. 15*, 28 (1969).
13. D. LUSS and P. HUTCHINSON, *Chem. Eng. J. 1*, 129 (1970).
14. G. GAVALAS and R. ARIS, *Phil. Trans. Roy. Sci. A260*, 351 (1966).
15. C. C. LIN and L. A. SEGEL, "Mathematics Applied to Deterministic Problems in the Natural Sciences." Macmillan, New York, 1975.
16. R. ARIS, *Chem. Eng. Educ. 10*, 114 (1976).
17. J. M. HAMMERSLEY, *Bull. Inst. Math. Appl. 9*, 276 (1973); *10*, 368 (1974).
18. J. M. HAMMERSLEY, *Bull. Inst. Math. Appl. 10*, 235 (1974).
19. J. M. HAMMERSLEY, *Bull. Inst. Math. Appl. 9*, 214 (1973).
20. L. A. SEGEL, *SIAM Rev. 14*, 547 (1972).
21. R. H. S. MAH and R. ARIS, *Ind. Eng. Chem. Fundamentals 2*, 90 (1963).
22. J. COLE, "Perturbation Methods in Applied Mathematics." Ginn/Blaisdell, Waltham, Mass., 1968.
23. A. VARMA and R. ARIS, Stirred pots and empty tubes, Ch. 2 of the "Wilhelm Memorial Volume" (N. R. Amundson and L. Lapidus, eds.) Prentice Hall, Englewood Cliffs, 1977.
24. N. R. AMUNDSON and O. BILOUS, *Am. Inst. Chem. Eng. J. 1*, 513 (1965); *2*, 117 (1956).
25. C. VAN HEERDEN, *Ind. Eng. Chem. 45*, 1242 (1953).
26. F. G. LILJENROTH, *Chem. Met. Eng. 19*, 287 (1918).
27. N. R. AMUNDSON and R. ARIS, *Chem. Eng. Sci. 7*, 121, 132, 148 (1958).
28. A. UPPAL, W. H. RAY, and A. POORE, *Chem. Eng. Sci. 29*, 967 (1974); *31*, 205 (1976).
29. M. M. DENN, "Stability of Reaction and Transport Processes." Prentice Hall, Engelwood Cliffs, 1974.
30. O. E. ROSSLER, *Z. Naturforsch. 31a*, 259 (1976).
31. E. C. ZEEMAN, *Sci. Am. 234*, No. 4, 65 (April 1876).
32. R. THOM, "Structural Stability and Morphogenesis." (D. A. Fowler, Trans.) Benjamin, Reading, 1975.

# Modeling, Prediction, and Control of Fish Behavior

JENS G. BALCHEN

*Division of Engineering Cybernetics*
*The Norwegian Institute of Technology*
*University of Trondheim*
*Trondheim, Norway*






ISBN 0-12-012715-6

## I. INTRODUCTION

General system theory and control theory have found applications during the last decade in a large variety of fields including technical, economical, social, and biological systems. Mathematical modeling of processes and systems in these fields has become a useful tool to understand and describe basic mechanisms as well as large-scale behavior and interactions. Furthermore, results in the theory of state and parameter estimation have been applied to automatic updating of models of dynamic systems against actual observations of system behavior, thereby making it possible to calculate dynamic phenomena that are not directly measurable. Finally, mathematical models of complex systems can play a role in the derivation of control strategies through the use of the well-developed theory of optimal control.

One kind of large-scale system that has so far received only limited attention in these respects is the marine ecological system in which the major species of fish play an important role. It is the intention of this chapter to develop some principles for modeling and control of an important sector of the marine ecosystem; namely, fish behavior.

It is strongly emphasized that the marine ecosystem consists of many highly interacting subsystems for which individual sets of models are being developed. Such subsystems are the subsystem of physical oceanography (geophysics) describing the ocean currents, turbulence, temperature, and density; the subsystem of chemical oceanography describing the dynamic distribution of minerals that form a basis for the growth of phytoplankton; the subsystem describing the growth and distribution of various kinds of zooplankton that feed on the phytoplankton and are themselves the most important sources of food for many species of fish and other animals; and the subsystem of fishes describing the different species and their interaction with the environment, preferably modeled by separate but interacting models of the biological state of the individuals, the population of the species, and the behavior of individuals and schools.

Extensive programs are in progress in many countries on building mathematical models of this hierarchy of interacting subsystems. Such models may be used for purposes of research directed toward a better understanding of the functioning of the many details and their interactions. Furthermore, the models may be used as part of estimation and prediction schemes in which current and future states of important system variables are computed for the purpose of planning the utilization of marine resources and for planning fisheries operations. Finally, models of the biological state and behavior of fish are useful in deriving strategies for the control of fish behavior.

## II. MATHEMATICAL MODELING OF FISH BEHAVIOR

The hierarchy of interacting subsystems constituting the total marine ecosystem is illustrated in Fig. 1. Mathematical models describing each of the blocks will be nonlinear partial differential equations which, in order to be implemented for numerical calculations, must be discretized in time and space. In building such models the concept of state space is useful. The total ecosystem state is a multidimensional vector which is a function of both time and space coordinates and which can be split into subvectors describing each of the blocks in Fig. 1. The state vector of one aggregate of a particular species

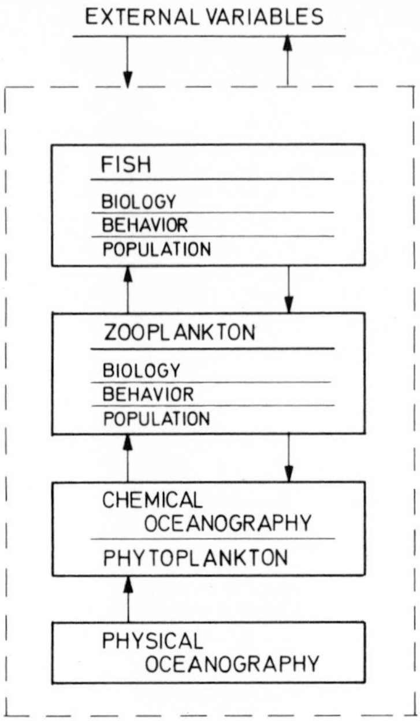


*FIG. 1. The hierarchy of interacting subsystems.*

may be divided into subvectors describing population, biological state, and behavior.

Since the purpose of this chapter is to study only modeling, prediction and control of fish behavior, we shall regard the states of the other subsystems of Fig. 1 as being results of measurements in the real system or through estimates produced by estimation schemes containing the models and using measurements in the real system. Thus all dynamic variables external to the fish shall be represented by the vector $\underset{\sim}{v}(t,\underset{\sim}{z})$ whose elements are functions of both time and space coordinates. In the following we shall study only one particular aggregate of fish and thus regard other aggregates as belonging to the environment described by the vector $\underset{\sim}{v}(\cdot)$.

An aggregate of fish is by definition a collection of individuals with identical behavior characteristics appearing together in such a way that their collective, rather than individual, behavior can be modeled. The relationship between the behavior of an individual and the behavior of an aggregate (school) in terms of movement is discussed in [1]. A major feature reported in [1] is that the behavior of the individual is suppressed by the behavior as a result of the internal forces of the school. Because of this, the variance of the movement of the school subject to a stochastic external stimulus is much less than that of an individual subject to the same stimulus.

It is a nearly impossible task to describe all the details of behavior for even the most primitive of animals. However, since behavior can be classified into at least two categories, small-scale and large-scale behaviors, one would hope, taking into account the stochastic nature of the problem, that it would be possible to model large-scale behavior with a reasonable degree of accuracy.

Parameterization of the average outcome of small-scale behavior (attacking prey, mating, hiding, etc.) is necessary in modeling large-scale behavior. Large-scale behavior is to a great extent the outcome of the small-scale behavior reflected upon the environment. An example of this is that the mechanism by which a certain species discovers and attacks its food (plankton) influences the nature and the reason for its migration. Since plankton quite often appear in patches, the dynamics and distribution of the patches must be modeled as a stochastic process in order to describe how the fish will move. Only large-scale movements will be considered here.

The behavior of an individual as well as an aggregate of fish in terms of large-scale movements is assumed to be a unique function of the "biological state" of the fish and of the state of the environment. A theory which describes how the biological state and the environmental state interact to produce the fish movement and which is based on the principle of "maximization of comfort" will be presented in Section II,C.

### A. *THE BIOLOGICAL STATE*

In order to understand and model fish behavior, it is first necessary to describe its biological state. In general, the biological state of a fish will certainly be a very complex description, but for the purpose of deriving a model of large-scale behavior, only a few variables dominate. We shall consider the biological state to consist of quantities relating to physiological storages of energy and quantities relating to hormonal influences. In addition to five energy storage states, three "hormonal" states will be assumed, but their model will not be specifically derived here.

A great deal of literature is available regarding the biochemical processes active in fish. Most of this literature, however, are descriptive and do not present a comprehensive, unified model of the interactions between the phenomena which control the biological process. Earlier attempts to model fish growth [2], [3] did not make use of differential equations, but instead they attempted to fit certain time functions to observations. Even though the models first derived are not directly useful in the context of this chapter, they provide considerable information that can be used for designing better models [4]-[32].

During the last few years some suggestions for mathematical models based on differential equations have been presented. Even though some of these models are rather simple, they represent a considerable improvement over those developed earlier [33], [34].

In modeling the biological state of an individual fish, as well as an aggregate of fish, the principle of state space is employed, which produces a set of first-order, coupled, ordinary differential equations. Each of these differential equations will, through integration, lead to one of the state variables. Since any biological system is extremely complex, it is necessary to define equivalent states that represent gross physical and biochemical phenomena with interconnections that resemble those in the real system. The choice of the structure of the approximate model is crucial to the success of the model. The first five state variables of the biological state, which describe the storages of energy, will be related through a dynamic energy balance [35]-[37]. This simply means that whatever is taken in by the fish in the form of food and oxygen is converted to produce either growth (reversible or irreversible), active energy (movement), or losses. The mathematical model will have the structure shown by Fig. 2. The five energy state variables are as follows: $x_1$, energy stored in stomach; $x_2$, energy stored in "blood"; $x_3$, energy stored in irreversible growth; $x_4$, energy stored in reversible storage except gonads; and $x_5$, energy stored in gonads.

Figure 2 indicates that there are a number of interconnections between these five energy storages. These are designated as $f_{ij}$ and represent the flow of energy from storage $j$ to $i$. The environment is designated by the subscript 0.

The energy flow $f_{10}(\cdot)$ is the intake of energy through feeding. This intake is certainly a function of many variables, such as hunger, the amount of food available, foraging

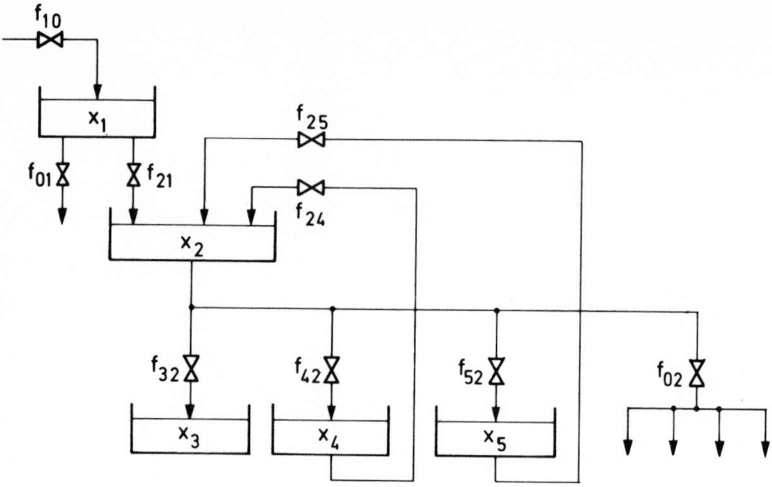


*FIG. 2. The structure of the mathematical model of energy storages.*

ability, etc. The detailed relationships determining $f_{10}(\cdot)$ will be discussed later.

The quantity $f_{01}(\cdot)$ is the energy lost in the feces. The gastric evacuation is a function of the volume of material in the stomach.

The rate of conversion of the energy in the stomach contents to "blood" is represented by $f_{21}(\cdot)$. "Blood" is the intermediate storage facility before the energy is converted into the four final forms. In gross terms $f_{21}(\cdot)$ can be regarded as digestion, which is the process of degradation of relatively large and complex nutrient molecules (carbohydrates, proteins, fat). The anabolic processes that lead to the building up of new structures and tissues in the fish are described by the functions $f_{32}(\cdot)$, $f_{42}(\cdot)$, and $f_{52}(\cdot)$.

The consumption of energy for external purposes is denoted $f_{02}(\cdot)$. This quantity is often referred to as "total respiration" and consists of various kinds of energy losses and active energy used for foraging and swimming.

The energy flow $f_{32}(\cdot)$ leads to irreversible growth (skeleton, skin, parts of tissue). The energy flow $f_{42}(\cdot)$ is the accumulation of biological materials that can be reconverted into usable energy via the blood. Fat in the tissue and the internal organs (liver, etc.) and much of the protein in the tissue are in this category. The energy storage in the gonads, however, is excluded from $x_4$ and modeled by a separate state $x_5$ since it serves a special purpose and varies differently. The net energy inflow to the reversible storage is actually $f_{42}(\cdot) - f_{24}(\cdot)$, where $f_{24}(\cdot)$ is the catabolism of stored energy via the blood to irreversible growth, gonads, and total respiration. Since $f_{42}(\cdot)$ and $f_{24}(\cdot)$ depend to some extent on different physiological mechanisms, they are shown separately, although they may well be regarded as one function encompassing both phenomena. Details of the functions $f_{32}(\cdot)$ and $f_{42}(\cdot)$ will be given later.

The growth and reduction of the energy storage in the gonads bear some similarity to those in the reversible storage, but different mechanisms reflecting the biological priority make the time pattern different. It is known, for instance, that the gonads may grow even though the fish receives no food. The energy supply to the blood under such circumstances comes from the reversible storage ($x_4$). Under severe starvation and demand for external energy, and when the reversible storage ($x_4$) has gone below a critical limit, energy may be drawn back to the blood from the gonads. Details of the function $f_{52}(\cdot)$ and $f_{25}(\cdot)$ will also be given later.

The energy flow for total respiration $f_{02}(\cdot)$ consists of four terms:

$f_{021}(\cdot)$ = standard metabolism, which is the energy consumption that sustains life without any physical work being done and without food being digested. The standard metabolism is a function that increases with both the environmental temperature and the weight of the fish.

$f_{022}(\cdot)$ = energy consumption during foraging. This is proportional to the standard metabolism $f_{021}(\cdot)$. It increases with the intensity of foraging ($p$), which in turn is a function of the level of hunger, physical vigor, a hormonal state, and the concentration of food.

$f_{023}(\cdot)$ = energy consumption due to active migration. This is proportional to the standard metabolism $f_{021}(\cdot)$ and increases with swimming speed.

$f_{024}(\cdot)$ = deaminization losses (the energy cost of digestion). It has been shown [22] that this loss is nearly proportional to the energy flow $f_{21}(\cdot)$ from the stomach to the blood.

The total metabolism is then

$$f_{02}(\cdot) = f_{021}(\cdot) + f_{022}(\cdot) + f_{023}(\cdot) + f_{024}(\cdot) \quad . \tag{1}$$

Using the notation introduced in Fig. 2, we can write the various equations describing the energy balance in the fish:

$$\dot{x}_1 = f_1(\cdot) = f_{10}(\cdot) - f_{01}(\cdot) - f_{21}(\cdot) \quad ; \tag{2}$$

$$\dot{x}_2 = f_2(\cdot) = f_{21}(\cdot) + f_{24}(\cdot) + f_{25}(\cdot) - f_{32}(\cdot) - f_{42}(\cdot) - f_{52}(\cdot) - f_{02}(\cdot) \quad ; \tag{3}$$

$$\dot{x}_3 = f_3(\cdot) = f_{32}(\cdot) \quad ; \tag{4}$$

$$\dot{x}_4 = f_4(\cdot) = f_{42}(\cdot) - f_{24}(\cdot) \quad ; \tag{5}$$

$$\dot{x}_5 = f_5(\cdot) = f_{52}(\cdot) - f_{25}(\cdot) \quad . \tag{6}$$

In addition to these five energy state variables we have, as already mentioned, three hormonal states described by

$\dot{x}_6 = f_6(\cdot)$, $x_6$ = conc of thyroid hormone (7)

$\dot{x}_7 = f_7(\cdot)$, $x_7$ = conc of adrenocortical steroids (8)

$\dot{x}_8 = f_8(\cdot)$, $x_8$ = conc of the sexual hormone (9)

These state variables will be used in the development of the energy state equations. Further description of the hormonal state state equations is found in [38].

*1. Detailed Analysis of the Functions $f_{ij}(\cdot)$*

The details of the rate functions $f_{ij}(\cdot)$ appearing in Eqs. (2) - (6) and in Fig. 2 can be derived to some extent from data available in the literature. Since the structure of the model employed here is somewhat more complex than those used in previous investigations, some of the relationships must be hypothesized and thus must eventually be experimentally verified. In any case, the hypotheses presented here are based on well-known physiological principles.

The space available does not permit a detailed discussion of the functions $f_{ij}(\cdot)$, but some of the most important features follow.

The flow of energy into the fish $f_{10}(\cdot)$ is

$$f_{10}(\cdot) = \alpha_{101} q_{10}(\cdot) \quad ,$$

where $\alpha_{101}$ is the specific energy content of the food and $q_{10}(\cdot)$ is the volumetric flow of the food. This volumetric flow is modeled as a product of three functions; namely, a function of the intensity of foraging ($p$), a function of the concentration of food ($v_2$), and a function of the efficiency of foraging which is assumed to be primarily related to light intensity ($v_3$). Thus it is proposed that the energy flow is described by

$$f_{10}(\cdot) = \alpha_{101} f_{101}(p) f_{102}(v_2) f_{103}(v_3) \quad . \tag{11}$$

Modeling this relationship as a product of three functions has a basis similar to that in statistics that gives the joint probability of three statistically independent events as the product of the probabilities of each event. The general shapes of the functions appearing in Eq. (11) are shown in Fig. 3. These functions can easily be given analytical representations that make them suitable for computer simulation.

The "intensity of foraging," $p$ in Eq. (11), must be functionally related to both the biological state variables of the fish and to the environmental state variables. It has been suggested in [37] that $p$ be approximated in a separable form by

$$p = p_1\left(\frac{x_1}{x_3 + x_4}\right) p_2\left(\frac{x_2}{x_3 + x_4}\right) p_3(x_7) p_4(x_8) p_5(v_2) \quad . \tag{12}$$

The general shapes of the functions appearing in Eq. (12) are indicated in Fig. 4. The function $p_1(\cdot)$ in Fig. 4 shows the relationship to the degree of satiation expressed as the ratio of the energy stored in the stomach to the sum of the energies stored in the body, the latter being taken as a measure of the size of the fish.

The significance of the function $p_1(\cdot)$ is that it contrib-

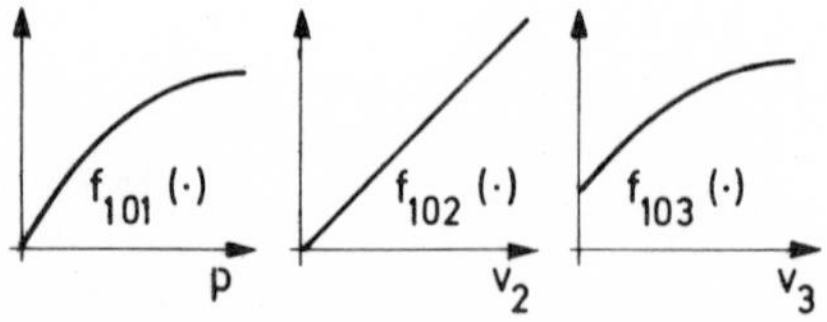


*FIG. 3. Functional relationship determining $f_{10}(\cdot)$.*

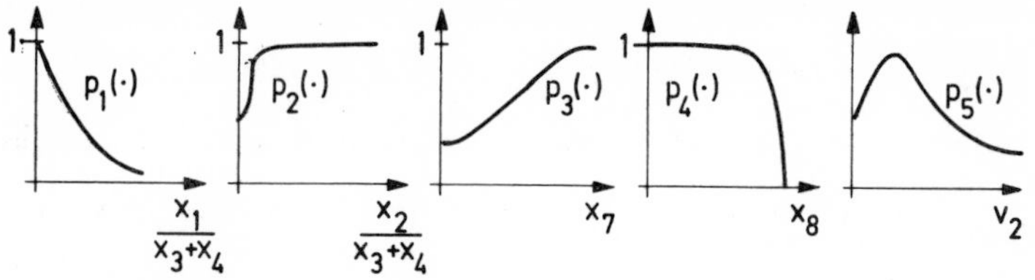


*FIG. 4. Functional relationships determining the intensity of foraging* p.

utes to the intensity of foraging with a "gain" that is maximal when the fish is hungry and zero when the fish is satiated.

The function $p_2(\cdot)$ shows that the "physical vigor" described by the relative energy content of the blood contributes with a gain that drops to a low value when the energy in the blood is low. Furthermore the two hormonal states $x_7$ and $x_8$ have influences as given by $p_3(\cdot)$ and $p_4(\cdot)$. The hormonal state $x_7$ is related to "aggressiveness" and $x_8$ is the sexual hormone. When the level of the latter hormone rises above a certain value, the interest in foraging drops to zero. The dependence on the concentration of food ($v_2$) given by the function $p_5(\cdot)$ has a maximal value at some level of food concentration because the fish discovers that there is little yield from the search at lower concentrations, and at higher concentrations the need for searching is low. Using the intensity of foraging described by Eq. (12), in Eq. (11), it is seen that the intake of food $f_{10}(\cdot)$ depends on the concentration of food ($v_2$) in such a way that it approaches saturation when the concentration of food increases. The amount of energy used to acquire the food decreases, however.

It should be emphasized that one can suggest a number of analytical functions that can produce curves with shapes as indicated in Fig. 4, however, the parameters used to obtain any given curve must be estimated on the basis of experimental observations. Therefore, it is advantageous to use as few such parameters as possible.

The proposed forms for the remaining functions appearing in Eqs. (2) - (6) are based on elementary physiological principles:

$$f_{01}(\cdot) = f_{011}(x_1)f_{012}(x_3 + x_4)f_{013}(v_1), \quad (v_1 = \text{temperature}) \tag{13}$$

$$f_{21}(\cdot) = f_{211}(x_1)f_{212}\left(\frac{x_2}{x_3 + x_4}\right)f_{213}(v_1) \tag{14}$$

$$f_{32}(\cdot) = f_{321}(x_3)f_{322}\left(\frac{x_2}{x_3 + x_4}\right)f_{323}(x_7) \times f_{324}(x_8)f_{325}(v_1) \tag{15}$$

$$f_{42}(\cdot) = f_{421}(x_4)f_{422}\left(\frac{x_2}{x_3 + x_4}\right)f_{423}(v_1) \tag{16}$$

$$f_{24}(\cdot) = f_{241}(x_4)f_{242}\left(\frac{x_2}{x_3 + x_4}\right)f_{243}(x_6) \times f_{244}(x_7)f_{245}(x_8)f_{246}(v_1) \tag{17}$$

$$f_{52}(\cdot) = f_{521}(x_3 + x_4)f_{522}\left(\frac{x_2}{x_3 + x_4}\right) \times f_{523}\left(\frac{x_5}{x_3 + x_4}\right)f_{524}(x_8)f_{525}(v_1) \tag{18}$$

$$f_{25}(\cdot) = f_{251}\left(\frac{x_5}{x_3 + x_4}\right)f_{252}\left(\frac{x_2}{x_3 + x_4}\right)f_{253}(v_1) \tag{19}$$

These functional relationships are illustrated in Fig. 5.

The only functions shown in Fig. 2 which remain to be described are those representing the energy loss. The use of well-established physiological principles leads to

$$f_{021}(\cdot) = f_{0211}(x_3 + x_4 + x_5)f_{0212}(v_1) = \text{standard metabolism} \tag{20}$$

$$f_{022}(\cdot) = f_{021}(\cdot)f_{0221}(p, v_1) = \text{foraging losses} \tag{21}$$

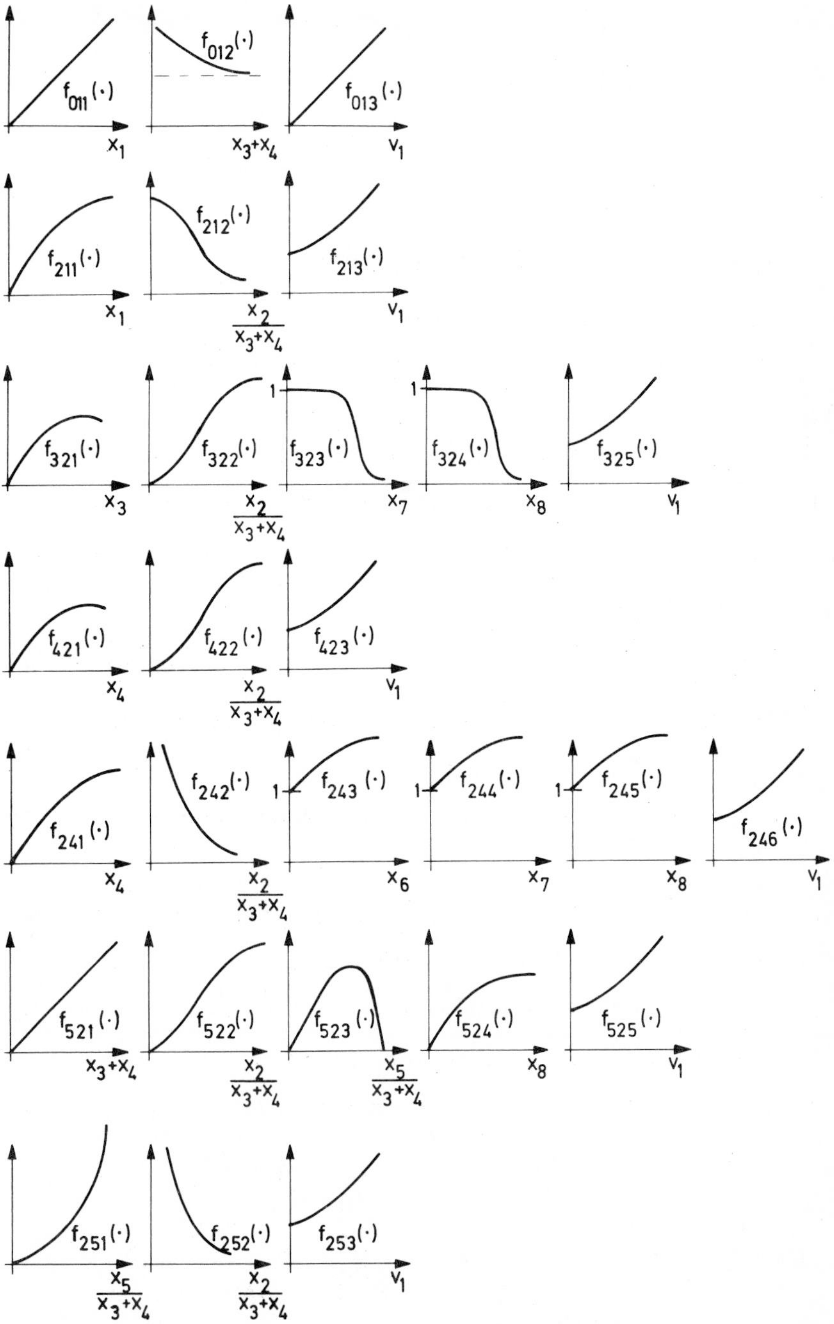


*FIG. 5. Functional relationships determining* $f_{21}(\cdot)$, $f_{32}(\cdot)$, $f_{42}(\cdot)$, $f_{24}(\cdot)$, $f_{52}(\cdot)$, *and* $f_{25}(\cdot)$.

$$f_{023}(\cdot) = \alpha_{021}(\cdot) f_{0231}(|\dot{\underset{\sim}{r}}|, v_1) = \text{migration losses}$$

$$|\dot{\underset{\sim}{r}}| = \text{speed of migration} \tag{22}$$

$$f_{024}(\cdot) = \alpha_{024} f_{21}(\cdot) = \text{deaminization losses}$$

$$(\alpha_{024} \approx 0{,}27) \tag{23}$$

The total respiration is the sum of these four functions, the elements of which are shown in Fig. 6.

Details of the functions appearing in Eqs. (13) - (23) are given in [37].

The mathematical model, consisting of Eqs. (2) - (9), containing a total of eight state variables that describe the biological state of the fish can now be expressed as a nonlinear, vector differential equation

$$\dot{\underset{\sim}{x}} = \underset{\sim}{f}(\underset{\sim}{x}, \underset{\sim}{v}, |\dot{\underset{\sim}{r}}|, t) \quad . \tag{24}$$

The state variables of Eq. (24) are all significant in determining the behavior of the fish.

To illustrate how the different variables of Eq. (24) are related, a simulation of a somewhat simplified version of the model is shown in Fig. 7 [39]. The external excitations are assumed to be limited to the temperature ($v_1$) and the concentration of food ($v_2$) that follow known time patterns over a period of two years. Integrating the model, with arbitrary initial conditions, gives the time functions of $x_1$, $x_2$, $x_3$,

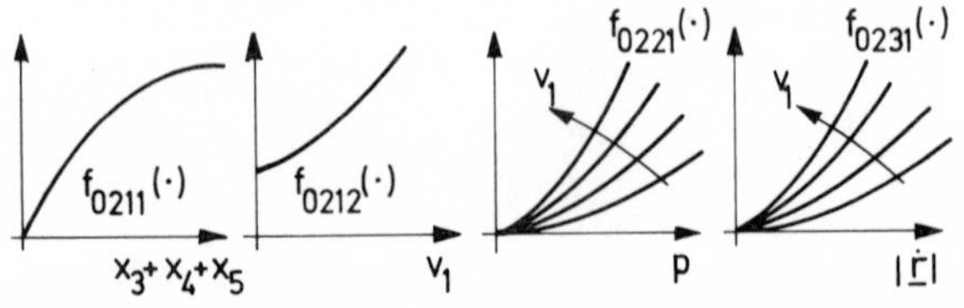


*FIG. 6. Functional relationships determining* $f_{02}(\cdot)$.

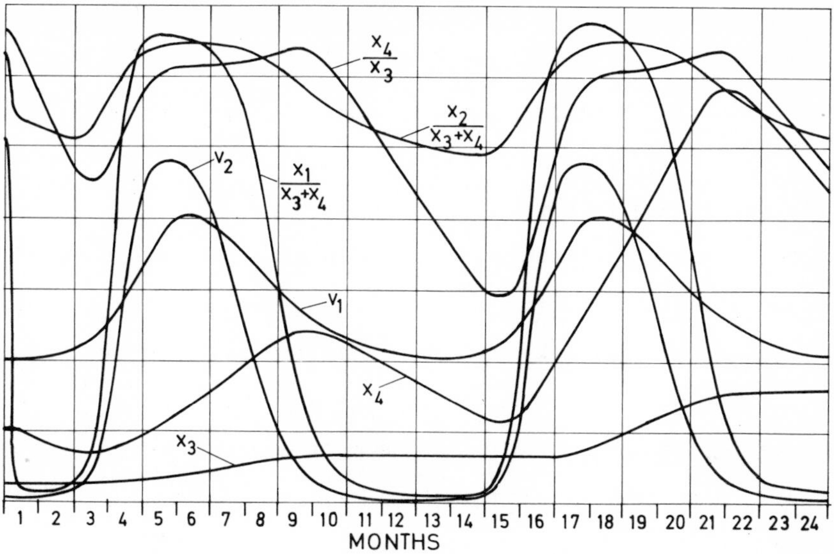


*FIG. 7. Computer simulation of the biological state of a fish over a period of two years.*

and $x_4$. The ratio $x_4/x_3$ is a "condition factor" that gives a measure of the shape of the fish and shows that the fish is chubby in September and slender in March.

### *B. THE ENVIRONMENTAL STATE*

As described in the introduction and in Fig. 1 the environment contains all the physical, chemical, and biological phenomena that are external to the fish and that influence its biological state and behavior. In the system illustrated in Fig. 1, most of these quantities will be calculated by the mathematical models. However, even with this system of models there will be external driving forces such as the conditions of the atmosphere (sun, wind, etc.), the influence of other animals not included in the models (fish, birds, benthos, etc.), and the influence of human activities (fishing, pollution, etc.). From the point of view of the modeling undertaken here, all of these quantities belong to the environment of the fish and are described by the vector $\underset{\sim}{v}(t,\underset{\sim}{z})$. In the previous section three of these environmental variables have already been introduced: $v_1$, temperature; $v_2$, concentration of food; and $v_3$, light intensity.

Consideration of the concentration and aggressiveness of predators is also important in modeling fish behavior. These will be accounted for by $v_4$, the intensity of predators.

Furthermore, most species of marine fish are sensitive to changes in the chemical properties of the water such as salinity and concentration of dissolved oxygen: $v_5$, salinity; $v_6$, concentration of dissolved oxygen.

Chemical agents in the water that produce olfactory stimuli have a strong influence upon fish behavior during certain phases of the life cycle. Such agents may have either physical or biological origins and usually move together with the water masses away from their sources. This is a possible explanation why many species of fish are known to migrate counter-current. Let $v_7$ be the concentration of any single olfactory agent.

All of these environmental variables ($v_1$ through $v_7$) are scalar quantities: that is, they carry no directional information. Other physical phenomena such as water velocity, electrical current, low frequency sound waves, and light rays must be represented by vectors in three-dimensional space, and thus have three components. One such vectoral quantity is described by $v_8$, the east component of the vector quantity; $v_9$, the north component of the vector quantity; and $v_{10}$, the down component of the vector quantity.

A fish moving freely in the water will experience changes in the environment for three reasons; first, because of its own changes in location; second, because of the time variations in the environmental state; and third, because changes in its own biological state will produce apparent stimuli even though the environment does not change.

### *C. MODELING FISH MIGRATION*

Historically there has been a tendency to substitute superstition and mystique for knowledge in the explanation of the mechanisms governing fish migration. Even in modern biological literature there are examples of studies that attempt to discover supernatural abilities in fish. However, because of the principles presented in this chapter the author agrees with those who do not believe that it is necessary to assume supernatural abilities in order to explain observed fish behavior. Questions as "what navigational principles does a fish use in order to find its way across the great oceans to its spawning grounds at a proper time?" are "loaded."

The opinion expressed in this contribution is that fish movement at any time is the outcome of a rather simple-minded local process of maximizing "comfort" which is a result of the interaction between the individual fish and its total environment. Since the fish experiences a dynamic change in its biological state governed by the internal biochemical processes and influenced by the environment, it will sense the environmental state differently at different times. It is believed that this leads to changing preferences and movement in the preferred directions. A species will have a higher probability of survival if it has a sequence of biochemical events that leads it into a self-repeating "limit cycle" such that it succeeds in reproducing. If it does not have such a sequence, it will not continue to exist.

The concept of maximization of comfort (or minimization of pain) is introduced as a generalized term because it does not necessarily refer to comfort (or pain) in the meaning used by man, but rather some kind of nervous sensation that the fish more or less unconsciously either maximizes or minimizes. It is well known that many subsystems in animals are not controlled consciously by the brain, but governed by endocrine phenomena in the body. Since a number of competing comforts (or pains) may be present at the same time, it is suggested that the fish actually finds an optimum compromise response to the various influences. This compromise is governed by the biological sensitivity that is associated with each of the different kinds of comfort. For instance, it is known that fish will dislike water that is above a certain temperature as well as water that is cold. The two kinds of pain that result from these extremes may have entirely different physiological causes, but the fish finds a way of comparing and evaluating them. Also sensations resulting from clearly unrelated influences must be evaluated by the fish so as to yield some optimal compromise as, for instance, between a high concentration of food and the presence of predators.

*1. Sensing Mechanisms Active in Fish Migration*

Most species of fishes are well equipped with sophisticated sensing organs that enable them to receive information from the environment [41].

*Vision* is one of the most important senses of the fish [42]. It plays a dominant role in close navigation, in schooling, and in feeding. Experiments show that a fish that has a tendency to imitate the actions of others in the same group is able to learn of something desirable outside its own field of vision by observing other fishes. The general intensity of light plays an important role because it determines to what extent the fish can locate its prey and also because many species seem to control their vertical migration according to the light intensity [43]. A third phenomenon in vision is the "Sun compass," which can be given a theoretical interpretation in terms of comfort. A number of biochemical processes in the fish are synchronized with the daily oscillations of light caused by the earth's rotation. Since many nervous sensations are related to endocrine phenomena, it is reasonable to believe that the sensitivity to certain aspects of light in the morning could be different from that at night. This could lead to directional migration as discussed later.

*Olfaction* (sense of smell) is very well developed in most fishes and plays an important role in fish navigation. Numerous investigations of olfactory sensing have been made [44] - [50]. Since olfaction is connected with the concentration of chemical substances in the water, the stimuli are derived from a scalar field of information. Searching in order to derive the gradient of this scalar field is assumed to be the principal method of navigation even though it is conceivable that the fish may be capable of deriving information from the microstructure of the flow patterns in the water carrying the stimuli. This, however, would only reduce the need for the directional perturbations performed by the fish.

*Hearing* acoustical signals in the water is important to most species of fish. The mechanisms involved in hearing have now been quite well identified [51] - [53]. The otholites and the lateral line are both active in hearing and it has been shown that the sense of hearing has some directionality. In studying the role of hearing in the directional migration of fish, one could distinguish between sound that carries specific information (e.g., predators, food source, etc.) and sound that is just general noise being either annoying or attractive to the fish. A specific sound will most probably give a strong and rapid response, whereas general noise will tend to produce a slow movement toward or away from the source. Specific sound sources are difficult to model in pelagic migration except in cases where the locations of the sound source are known. General noise sources, however, are quite common in the ocean and lend themselves well to modeling in terms of comfort. Such is the case when water masses flow at different speeds and form interfaces producing shear forces, and thereby generate sound patterns that either have attractive or repulsive influences upon fish.

*Temperature* is sensed by all species of fish both directly by means of specialized organs (ampullae of Lorenzini and the lateral line) and indirectly through its influence on metabolic processes. Both of these methods of sensing are conveniently modeled in terms of comfort. There is strong evidence from observations of major species of pelagic fishes that a changing temperature preference due to metabolic or seasonal changes may cause the fish to change depth. Since ocean currents often have directions near the surface that are different from those near the bottom, this may produce a movement of the fish along with the water masses. This creates the same net effect as migration, but it is a passive movement. In fact it is conceivable that long distance migration in the ocean can be accomplished with only one or two changes of depth.

Sensing of *water velocity* definitely occurs in most species of fish, but it is not well established which organs are most active. The detectors involved are probably the lateral line, the otholites, and pressure-sensing organs near the body surface. Vision may also play a role as a fish detects the relative movement of the bottom or other objects. It is also conceivable that electrical potentials generated by water movement are active [54].

*Electrosensing* has long been known to be an important means of navigation in some species, such as sharks. There is, however, great disagreement among researchers as to the significance of electrosensing in the migration of conventional fishes.

### 2. *Fish Behavior in Terms of Optimization*

In the modern literature on biological systems, there are a number of contributions dealing with optimization as a mechanism in behavior [55] - [58]. It is helpful to distinguish between at least three levels of optimization:

(1) Momentary optimization by the fish with no concern for the future.

(2) Dynamic optimization derived from maximizing a measure of success integrated over a period of time (day, month, year).

(3) Genetic optimization of behavioral parameters through natural selection of the species most fitted for survival.

Of these three levels of optimization (1) and (3) are the most usual in nature. Momentary optimization may well be regarded as dynamic optimization over a short-time interval (seconds, minutes). For example, the time interval of optimization may be related to the time it takes for the animal to move across its range of visual perception. This means that most animals would not have much ability to plan their behavior (e.g., storing food in the summer for consumption in the winter). Level (3) allows for long-range optimization since a natural selection process would lead to a biochemically controlled tendency to do something in one time interval which could be beneficial in another. In this case the animal is not optimizing consciously, but the mere fact that it exists is the outcome of an optimization performed through natural selection. Level (3) is a parametric optimization that extends over a large number of generations of the species. In the following we will assume that only momentary optimization [level (1)] is performed by the fish itself.

The "comfort" is expressed mathematically as a scalar field $C(\underset{\sim}{x},\underset{\sim}{v})$ which is a function of the biological state of the fish and of the environmental state [40]. For practical purposes it may be advantageous to expand $C(\cdot)$ in some way in order to isolate the influences of the different environmental factors. One form of expansion is

$$C(\underset{\sim}{x},\underset{\sim}{v}) = C_1(\underset{\sim}{x},\underset{\sim}{v}) + C_2(\underset{\sim}{x},\underset{\sim}{v}) + \ldots + C_m(\underset{\sim}{x},\underset{\sim}{v}) \quad , \tag{25}$$

in which each of the "subcomforts" are primarily dependent upon one or few of the environmental variables.

$C_1$ is taken to be the dependence of the comfort upon temperature (Fig. 8). In addition to the dependence upon temperature, Fig. 8 also suggests that $C_1(\cdot)$ depends on one of the biological state variables ($x_7$). The effect of $x_7$ is that the preference for a certain narrow temperature range becomes less

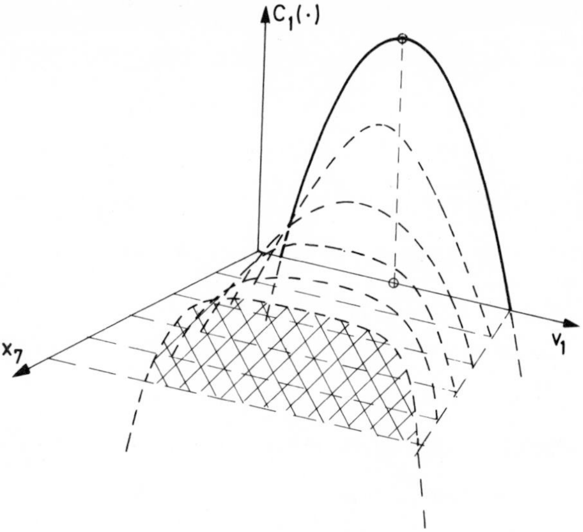


*FIG. 8. The comfort associated with temperature.*

pronounced and that the contribution to the total comfort becomes less. At both ends of the temperature scale, $C_1(\cdot)$ assumes large negative values depicting the existence of "pains."

The contribution of the concentration of food ($v_2$), is given by $C_2(\cdot)$, and is shown in Fig. 9. It is seen that $C_2(\cdot)$

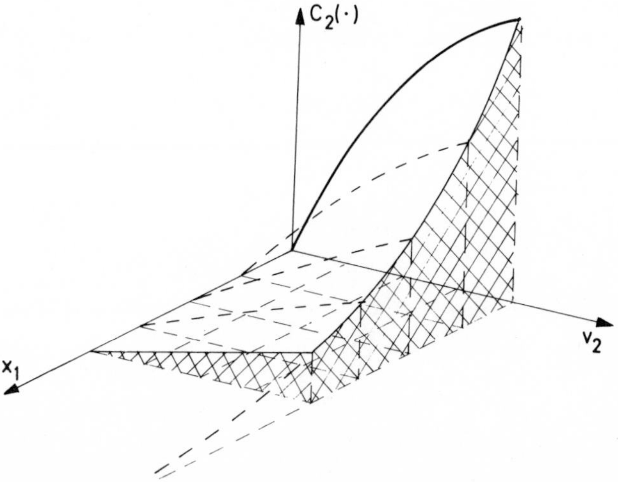


*FIG. 9. The comfort associated with the concentration of food.*

is also a function of the degree of satiation, the biological state variable $x_1$, in such a way that the motivation for food is gradually reduced to nearly zero when satiation is reached. It is also obvious that $C_2(\cdot)$ cannot assume negative values.

The contribution to the comfort associated with the general light intensity ($v_3$) is shown in Fig. 10 as $C_3(\cdot)$. Here it is indicated that one of the biological states ($x_8$) is contributing in the way that it shifts the preferred light intensity and modifies the sensitivity to changes in light. The relationships depicted in Fig. 10 are interpretations of information available in the literature, but it is not known by the author if there has been experimentation to isolate such relationships.

Figure 11 suggests that the contribution $C_4(\cdot)$ relates to the intensity of predators ($v_4$) in such a way that the comfort drops very sharply when the intensity of predators increases above a certain value, high values of $v_4$ give highly negative comfort. No other variable is considered here to be of

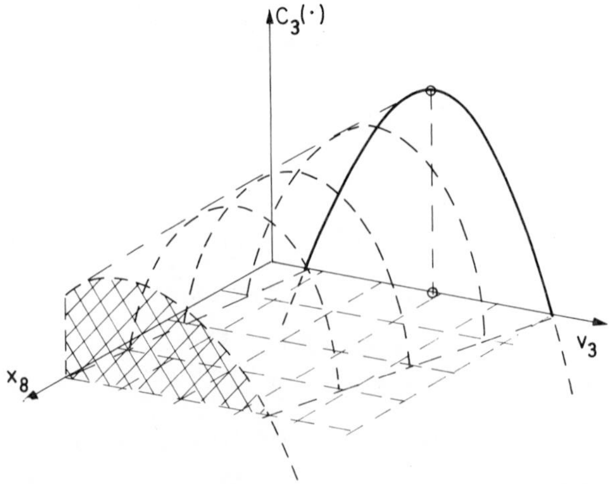


*FIG. 10. The comfort associated with the general light intensity.*

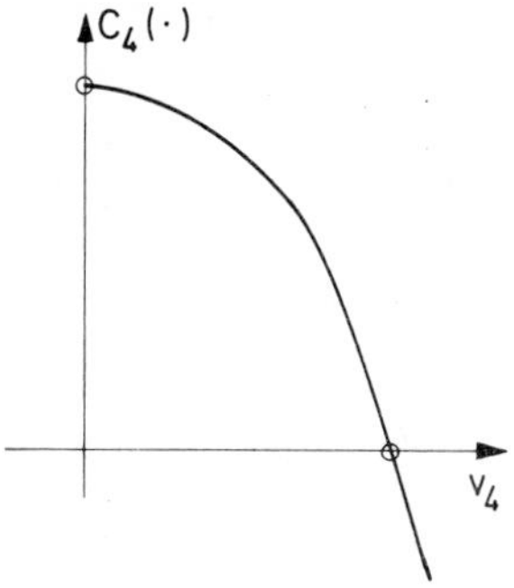


*FIG. 11. The comfort associated with the intensity of predators.*

importance to $C_4(\cdot)$, even though it is conceivable that an increase in $x_7$ may reduce the discomfort somewhat. The intensity of predators ($v_4$) is acknowledged to be a function of the concentration of predators, the aggressiveness of predators, and the location of predators relative to the fish. Some care should be exercised in separately modeling these particular phenomena.

A discomfort (negative comfort) associated with the deviation from any particular direction of migration governed by a directional (vectoral) source of stimuli such as electrical potential, light rays (sun clock), and low frequency sound is represented by $C_5(\cdot)$. The sun clock phenomenon is typical of the interaction between a geophysical oscillator (the sun) and a biological oscillator (synchronized with light intensity) and will be used as an example of the function $C_5(\cdot)$. To illustrate this, suppose that the fish prefers to swim toward the sun, but when the sun is not up, the fish rests. If the sun rises in the east and sets in the west, this will give a pattern of migration as shown by curve a in Fig. 12. With constant swimming speed the pattern of migration will be semicircular, always terminating at the north-south line at night,

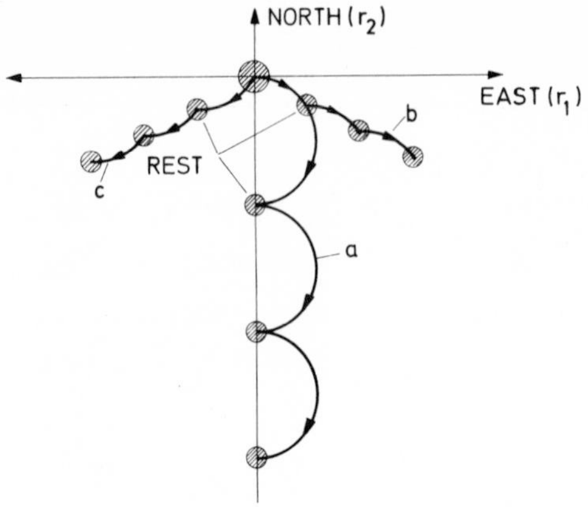


*FIG. 12. Examples of "sun-clock" navigation.*

thus giving a net migration toward the south. Now, suppose that the fish swims toward the sun during a relatively short period in the day and rests during the remaining time. Then the direction of migration will depend upon the period of the day during which it swims. If it swims toward the sun only in the morning, the result will be a migration primarily to the east as indicated by curve b in Fig. 12. If it tends to swim toward the sun only for a short period at the end of the day, the migration will be in a westerly direction as indicated by curve c in Fig. 12. In each of these cases the distance traveled is shorter than for the case of curve a because it is assumed that the fish rests most of the day. It is important to note that the tendency to go toward the sun, say, in the morning, is not conscious, but is governed by elementary biochemical processes. Thus, the fish does not calculate when it must go toward the sun in order to migrate in a certain direction; rather, the fish migrates in a certain direction because its elementary chemical processes make it to toward the sun during a certain interval of its daily biochemical cycle.

This phenomenon can be interpreted mathematically in terms of maximization of comfort by employing the scalar function:

$$C_5(\underset{\sim}{x},\underset{\sim}{v}) = s_5(\underset{\sim}{x})\underset{\sim}{\alpha}^T\underset{\sim}{r} \quad , \tag{26}$$

where

$$\underset{\sim}{\alpha}^T = [\alpha_0 \cos\theta, -\alpha_0 \sin\theta, 0] \tag{27}$$

and $\theta = \omega t$ = angle between east and the direction of the sun; $\omega$ the earth's rotational frequency; $\alpha_0$ is constant; and $s_5(\underset{\sim}{x})$ is the periodic sensitivity function governed by the biological state (biological clock) as sketched in Fig. 13.

If we assume the fish searches in the field of comfort in such a way as to move toward the optimum with a speed that is proportional to the gradient of the comfort with respect to location, we can derive a differential equation of movement which is as follows:

$$\dot{\underset{\sim}{r}} = \frac{\partial C(\underset{\sim}{x},\underset{\sim}{v})}{\partial \underset{\sim}{r}} \quad , \tag{28}$$

where $\underset{\sim}{r}$ is the coordinate vector relative to the earth of the fish in the water masses.

$$\underset{\sim}{r} = \begin{bmatrix} r_1 \\ r_2 \\ r_3 \end{bmatrix} = \begin{matrix} \text{east} \\ \text{north} \\ \text{down} \end{matrix} \quad .$$

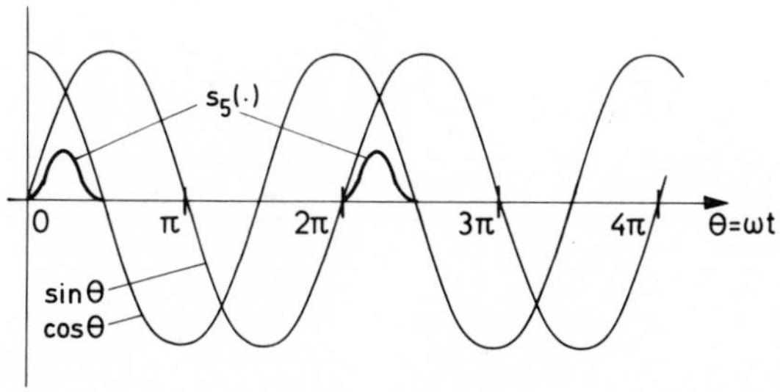


*FIG. 13. The periodic sensitivity function governed by the biological clock.*

The quantity $\dot{\underset{\sim}{r}}$ is the average velocity, not the velocity of the searching perturbations necessary to locate and remain on the gradient. If the water masses are also moving, the total velocity of the fish movement will be the sum of these two velocities.

The velocity of migration due to the sun clock only will become,

$$\dot{\underset{\sim}{r}} = \frac{\partial C_5(\cdot)}{\partial \underset{\sim}{r}} = s_5(\underset{\sim}{x})\underset{\sim}{\alpha} \quad . \tag{29}$$

Equation (29) states that the velocity of migration in the horizontal plane is proportional to the magnitude of the sensitivity function $s_5(\cdot)$ and points in the direction of the sun as illustrated in Fig. 14.

The examples of elementary comfort functions $C_1(\cdot)$ through $C_5(\cdot)$ offered above are simplified for the sake of clarity. In an actual case one may find it advantageous not to make an expansion as shown in Eq. (25), but rather to describe the total comfort as a function of all the actual variables.

The state of the fish can now be defined as an augmented vector containing the biological state $\underset{\sim}{x}$ derived from Eq. (24) and the location $\underset{\sim}{r}$ derived from Eq. (28). Thus, the total model of the fish behavior will be

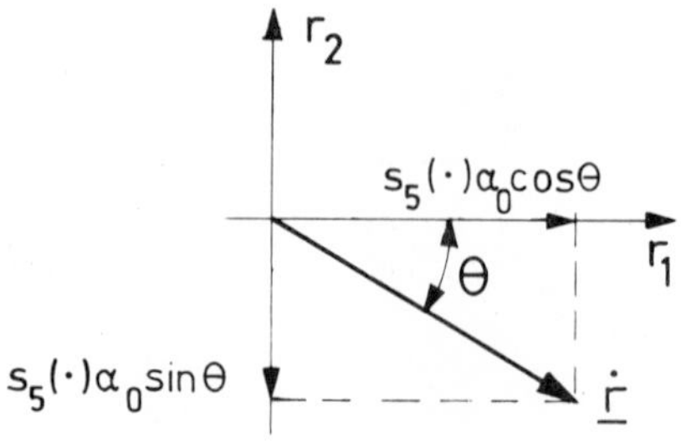


*FIG. 14. The direction of migration.*

$$\begin{bmatrix} \dot{\underset{\sim}{x}} \\ \dot{\underset{\sim}{r}} \end{bmatrix} = \begin{bmatrix} \underset{\sim}{f}(\underset{\sim}{x}, \underset{\sim}{v}, |\dot{\underset{\sim}{r}}|, t) \\ \dfrac{\partial C(\underset{\sim}{x}, \underset{\sim}{v})}{\partial \underset{\sim}{r}} \end{bmatrix} . \tag{30}$$

The system represented by Eq. (30) is illustrated in Fig. 15. By introducing a total system state vector:

$$\underset{\sim}{x}^t = \begin{bmatrix} \underset{\sim}{x} \\ \underset{\sim}{r} \end{bmatrix} , \tag{31}$$

Eq. (30) can be written in a compact form

$$\dot{\underset{\sim}{x}}^t = \underset{\sim}{F}(\underset{\sim}{x}^t, \underset{\sim}{v}, t) . \tag{32}$$

For computational purposes, the differential equation of Eqs. (30) - (32) must be discretized:

$$\begin{aligned} \underset{\sim}{x}^t(k+1) &= \underset{\sim}{x}^t(k) + \Delta t \underset{\sim}{F}(\underset{\sim}{x}^t(k), \underset{\sim}{v}(k), k) \\ &= \underset{\sim}{\mathcal{G}}(\underset{\sim}{x}^t(k), \underset{\sim}{v}(k), k) , \end{aligned} \tag{33}$$

where $\Delta t$ is the time increment; $k$ the increment number, and

$$\underset{\sim}{v}(k) = \underset{\sim}{v}(t, \underset{\sim}{z}) \Big|_{\substack{t = k\Delta t \\ \underset{\sim}{z} = \underset{\sim}{r}}} .$$

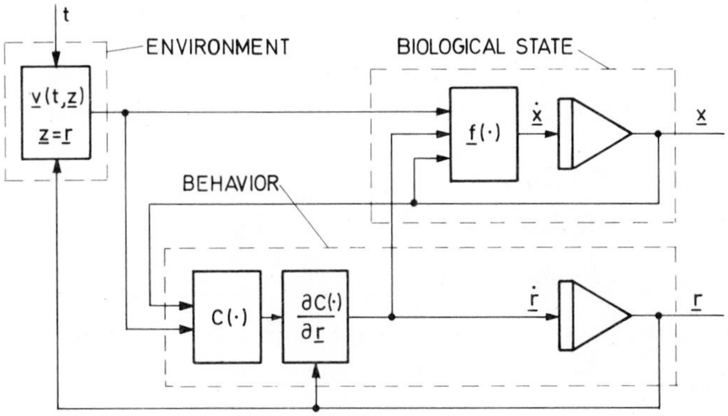


*FIG. 15. Block diagram of the total system model.*

## III. ESTIMATION AND PREDICTION OF FISH BEHAVIOR

The models derived in the previous paragraphs can be used in a scheme for estimating the present and future behavior of the fish [40]. Such an estimation scheme would in a real case also include models of the other subsystems of the total ecosystem (physical and chemical oceanography, zooplankton, fish population, etc.).

An estimation system would utilize the model of Eq. (33). When solving Eq. (33) or integrating Eq. (32) with given initial conditions, the accuracy of the result depends on many factors such as the validity of the hypotheses built into the model, the accuracy of the parameters used and the variance of the estimate of the initial conditions. The estimated future values of the system state will be associated with an uncertainty that can also be estimated to some extent. Under all circumstances it is necessary to acquire measurements in the real system to update the initial conditions and to discover errors in the model parameters and perhaps in the model structure itself. Without measurements of the real system, the state variables of the model will soon diverge from those of the real system. In many cases of migrating fish, research vessels are able to take samples of individuals in a school and to locate the schools by means of echo-sounding equipment. Thus, many of the variables constituting $\underset{\sim}{x}^t$ may be measured at certain times, but an uncertainty must be associated with each measurement to account for sampling errors. The measurements are modeled by

$$\underset{\sim}{y}(k) = \underset{\sim}{G}(\underset{\sim}{x}^t(k)) + \underset{\sim}{w}(k) \quad , \tag{34}$$

where $\underset{\sim}{G}(\cdot)$ may well be a simple linear, diagonal transformation.

Standard Kalman-filtering techniques using Eqs. (33) and (34) can produce an estimate of the present state as shown in Fig. 16, where $\bar{\underset{\sim}{x}}^t(k)$ is the *a priori* estimate of total system state; $\hat{\underset{\sim}{x}}^t(k)$, *a posteriori* estimate of total system state; $\bar{\underset{\sim}{y}}^t(k)$, estimated measurement; and $K(k)$, Kalman filter gain matrix.

The well-established recursive algorithms for determining $K(k)$ are reviewed:

$$K(k) = \hat{X}(k)D^{\mathrm{T}}(k)W(k)^{-1} \quad ; \tag{35}$$

$$\hat{X}(k) = (\bar{X}(k)^{-1} + D^{\mathrm{T}}(k)W(k)^{-1}D(k)^{-1} \quad ; \tag{36}$$

$$\bar{X}(k+1) = \phi(k)\ \hat{X}(k)\phi^{\mathrm{T}}(k) + \Omega(k)V(k)\Omega^{\mathrm{T}}(k) \quad ; \tag{37}$$

where

$$\phi(k) = \frac{\partial \underset{\sim}{\mathcal{F}}(\cdot)}{\partial \hat{\underset{\sim}{x}}^t(k)} , \quad D(k) = \frac{\partial \underset{\sim}{G}(\cdot)}{\partial \bar{\underset{\sim}{x}}^t(k)} , \quad \Omega(k) = \frac{\partial \underset{\sim}{\mathcal{F}}(\cdot)}{\partial \underset{\sim}{v}(k)} \tag{38}$$

and

$$\hat{X}(k) = \mathrm{cov}(\underset{\sim}{x}^t(k) - \hat{\underset{\sim}{x}}^t(k)), \quad \bar{X}(k) = \mathrm{cov}(\underset{\sim}{x}^t(k) - \bar{\underset{\sim}{x}}^t(k)) ,$$

$$W(k) = \mathrm{cov}(\underset{\sim}{w}(k)), \quad V(k) = \mathrm{cov}(\underset{\sim}{v}(k)) \quad . \tag{39}$$

In the above expressions it has been assumed that the measurements are made at the time instants $k$. This is not

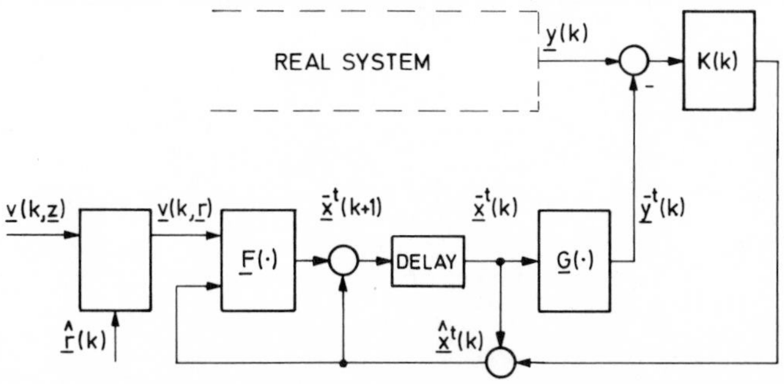


*FIG. 16. Block diagram of total system state estimator.*

realistic in a practical case, but if it is assumed that the sampling interval is a multiple of $k$, it is trivial to compute modified values of the Kalman filter gain matrix.

The scheme shown in Fig. 16 produces an estimate of the present state of the system.

When a *prediction* is desired, the model is used to continue the calculations starting with the estimated present state. Note that future values of the environmental state will drive the model of the state of the fish since it has been assumed that the environment is estimated using a scheme similar to that used for the other subsystems. If future values of the environment cannot be predicted, then even a mathematical prediction of the fish behavior would be uncertain.

Figure 17 shows a possible outcome of the estimation and prediction of the migration of two aggregates of pelagic fish

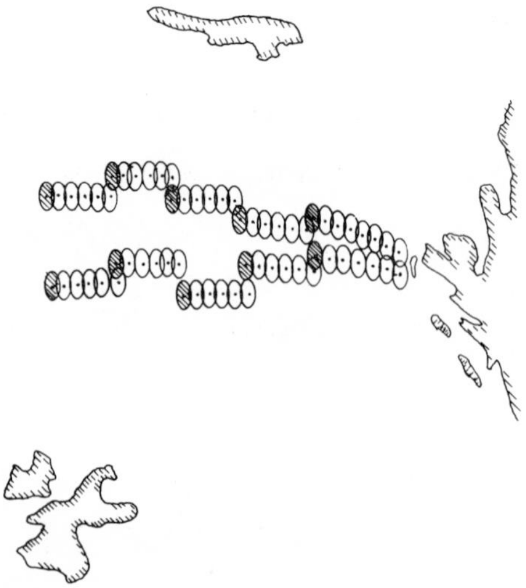

*FIG. 17. Estimation and prediction of the migration of two aggregates of pelagic fish in an ocean.*

in an ocean. The small circles indicate that the aggregates have a certain horizontal distribution.

## IV. CONTROL OF FISH BEHAVIOR

Control of fish behavior can be exercised if it is possible to generate artificial stimuli which, when combined with natural vectoral stimuli will produce a new vectoral stimulus that has the desired direction and magnitude [59]. Figure 18a indicates how two individual scalar fields of comfort $C_1(\cdot)$ and $C_2(\cdot)$ combine to produce the total stimulus vector leading to fish movement. In Fig. 18b, a similar argument is used with a control stimulus vector $\underset{\sim}{u}$ applied to the fish together with

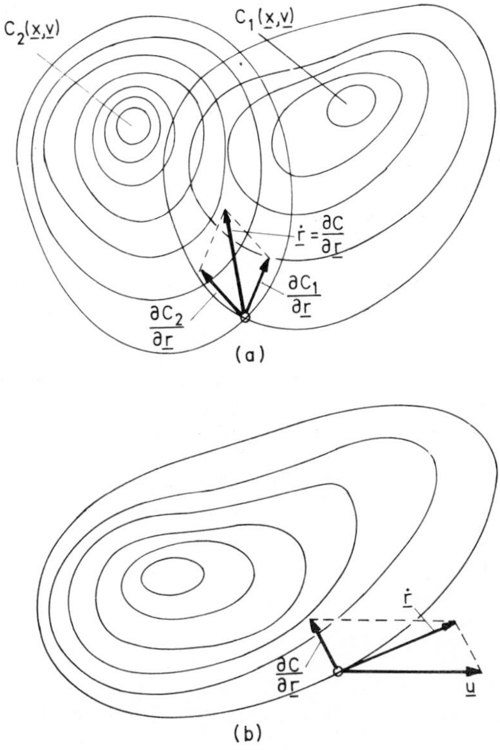


*FIG. 18. Combining the gradients of natural fields of stimuli with a control stimulus.*

the gradient vector of the natural field of comfort. Adding these two vectors together as shown produces the net control vector that is proportional to the velocity of movement $\dot{\underset{\sim}{r}}$. The magnitude and direction of $\underset{\sim}{u}$, therefore, are dependent upon the magnitudes and directions of the vectors of the natural stimuli. In order to generate a proper control vector, an estimate of the natural stimuli $\partial C(\cdot)/\partial \underset{\sim}{r}$ must be obtained.

Active control of fish behavior can, in principle, be achieved in a number of different ways:

Direct control by generating and transmitting artificial stimuli that either attract or repel the fish.

Indirect control by modifying the biological state (e.g., hormonal factors) so that the natural stimuli will appear to be different.

Direct control by modifying the environment, thereby producing modified stimuli.

Although only the first category will be further discussed, significant applications of the second can certainly be suggested [59].

Practical systems for the control of fish behavior require that the methods of generating control stimuli satisfy certain basic specifications. The system must have a reasonable cost and must be easy to install and maintain. The influence of the stimuli must extend over the required distance from the source and be propagated with sufficient speed. The stimuli must be such that they maintain strong attractive or repulsive actions independent of time and the biological state of the fish.

No known single source of stimuli satisfies all these

requirements simultaneously. However, by combining different kinds of stimuli and utilizing the ability of the fish to learn, one can arrive at multiple sources of stimuli that satisfy all of the requirements.

One important aspect of behavioral control is the utilization of the strong tendency among most species to imitate the behavior of other individuals. Imitation is one of the basic phenomena among schooling fishes. Observers of fish behavior know that if one or a few fishes rushing in some direction are seen by other fishes, the others will immediately join them. These in turn cause others still farther away to do likewise. Consequently, a "chain reaction" is established and the transmission of information may far exceed the limited range of direct visual contact [1].

### *A. CONTROL STIMULI*

Only a brief summary of different control stimuli can be presented here. A more thorough discussion is presented by Balchen [59].

Natural and artificial stimuli transmitted to the fish and detected by the fish can be categorized in a number of different ways. For example, one can distinguish between attractive and repulsive stimuli according to the direction of the response of the fish relative to the source of the stimuli. Another distinction is between directional (vectoral) and nondirectional (scalar) stimuli. Also, stimuli may be categorized according to their physical properties such as acoustical, olfactory, visual, electrical, etc.

### 1. *Acoustical Stimuli*

Sound waves in the water are easy to generate; they propagate fast and far in the water masses and they are easily detected by most species of fish. Therefore, acoustical stimuli are the most important for the control of fish.

Acoustical stimuli may be used in at least two different ways: For direct attraction or repulsion, depending on the information contained in the signal; as a message signal in a conditioning system using another strong stimulus as the actual attraction or repulsion. Examples of these two types of acoustical stimuli will be discussed later.

### 2. *Visual Stimuli*

Visual stimuli may be generated by objects in the water or in the air or by light sources. They may be either attractive or repulsive depending on the circumstances and the species. In a manner similar to that for acoustical stimuli, visual stimuli may be used in two ways; that is, either directly as a control stimulus or as a message signal in a conditioning system.

### 3. *Electrical Stimuli*

The first British patent for a method of electrical fishing was issued in 1863. Since then many systems have been suggested and tested, but the commercial importance of electrical fishing does not yet seem to be very great [60].

Ordinary species of fish without specialized electrical receptor organs may sense electrical potentials of rather low value (10 mV/cm), especially if the potential is pulsating in the frequency range 0,5-10 pulses/sec. The fact that a fish senses the electrical potential does not mean, however, that it will react by swimming. When the electrical field is increased beyond a certain limit, the nervous sensation is one of discomfort. Thus, it will be frightened away from the source if it has previously learned the direction in which to swim to escape from the source. A further increase in the electrical field gives rise to electrotaxis. In a constant or pulsating direct current field, the fish will tend to swim toward the positive electrode (anode) while in this physiological state. If the electrical field is increased still further, the fish will experience electronarcosis which is a state of immobility resulting from muscular slackening.

Electrical stimuli may be used as a direct control of fish movement by electrotaxis or as a source of discomfort to be combined with some other signal (visual or acoustical) in a conditioning system.

## B. *SYSTEMS FOR CONTROL OF FISH BEHAVIOR*

Two examples of systems for control of fish behavior will be discussed in the following paragraphs.

### 1. *Control of Schooling Species (e.g., herring) by Acoustical Repulsion*

Many schooling species are repelled by low frequency sound and a system for pelagic control of a school can be designed as shown in the block diagram of Fig. 19. Experiments with such a system have been reported in [61], [62]. In this system a low-frequency acoustic generator is installed on board a vessel, which can be maneuvered as desired with respect to

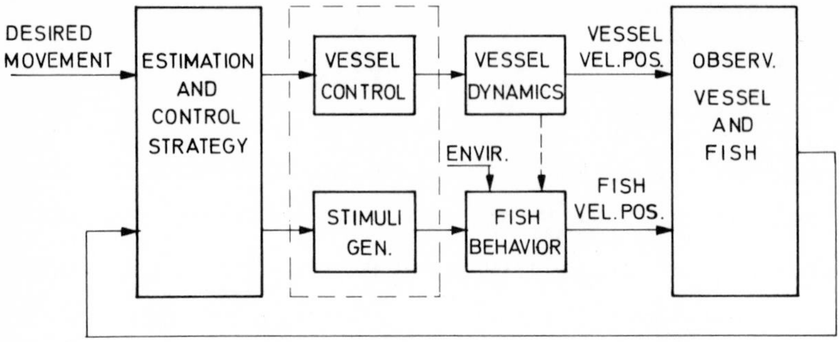


*FIG. 19. Block diagram of a system for control of schooling species by acoustical repulsion.*

velocity and position relative to the fish. It is assumed that the vessel is equipped with means for measuring the vessel velocity and position as well as the position of the school of fish (horizontal and depth coordinates) relative to the vessel. The school of fish in its uncontrolled mode of behavior will be subject to a vector of natural stimuli. The acoustic generator will produce a vectoral control stimulus with a direction away from the source. Using the concept of vector summation as shown in Fig. 18, the net direction and speed of movement can be determined. Certainly, geometric summation of the two vectors is an oversimplification because the occurrence of the artificial stimulus may modify the significance of the natural stimuli to some extent.

Since the source of the stimuli is located on board the vessel, it must be moved to the appropriate position. This presents a problem when the magnitude of the natural stimulus vector and the distance from the vessel to the school are great because the vessel must make large movements to change the direction of the control stimulus.

The estimation and control strategies shown in Fig. 19 must include components for estimation of the field of natural stimuli based upon observation of the school movement. Furthermore, it must include means for predicting the response of the school to control stimuli as well as the stimuli produced by the noise of the vessel maneuvering. It must also contain a model of the vessel dynamics to make it possible to predict vessel behavior resulting from control actions and disturbances from the environment (wind, current). In order to produce the proper control stimulus, it is necessary to have a model of the stimulus generation that describes the stimulus received by the fish as a function of distance, water conditions, etc. Finally, a description (model) of the desired behavior (movement) is necessary, which incorporates information about the expected natural stimuli at future locations. The components mentioned above must be present in some form in any system having similar objectives, whether the control and estimation functions are performed by control devices, a human, or an animal.

*2. The Use of Conditioning (Learning) for Harvesting of Pollack [Gadus (Pollachius) Virens] Held in Long-Term Storage in a Bay*

Many fisheries show pronounced peaks in the catch at certain times of the year. Since the capacity of the fish processing industry (freezing and canning) is limited, there is a need for a large-scale buffer storage system that can keep fish alive over a long period (0 - 6 months). Experiments using natural fjords and bays to store fish, particularly pollack [*Gadus (Pollachius) Virens*] have been in progress in Norway for some time [63]. So far the experiments have been very promising with respect to both the degree of survival and the quality of the fish at the end of the storage period. Two major problems have been encountered in long term storage: fencing off the inlet to the bay; and harvesting the fish.

Only a system for harvesting fish will be discussed here, but similar principles can be applied to the fencing problem. When trying to devise a system for attracting fish, one is faced with the problem of finding a signal that is both far-reaching and has a strong and persistent attraction. Conditioning fish with small portions of food combined with an acoustical signal in a convenient frequency range has proven to be a very powerful technique. Since fish do not receive any food during long-term storage, the motivation for food is strong. On the basis of these features, the following system, shown in Fig. 20, has been developed and tested.

A number of pneumatically operated and electronically controlled "conditioning units" are located submerged in the bay at distances of about 50 - 100 meters. Each unit is equipped with a loudspeaker and an apportioning valve from which about 50 - 500 $cm^3$ of food is ejected following a sequence of sound pulses in the frequency range 200 - 400 Hz. These events propagate from unit to unit with a time spacing of 15 - 120 min. After a very short time (about 1 h), the fish learns to combine the acoustical signal with the appearance of food and rushes toward the source of the sound to feed on the very small amount of food. Consequently, a schooling is artificially initiated from unit to unit attracting fish from appreciable distances. Over a period of some days or weeks, most fish in the bay will become conditioned and be attracted whenever the attraction of the conditioning system is greater than that of other sources. It has been observed, and it follows logically

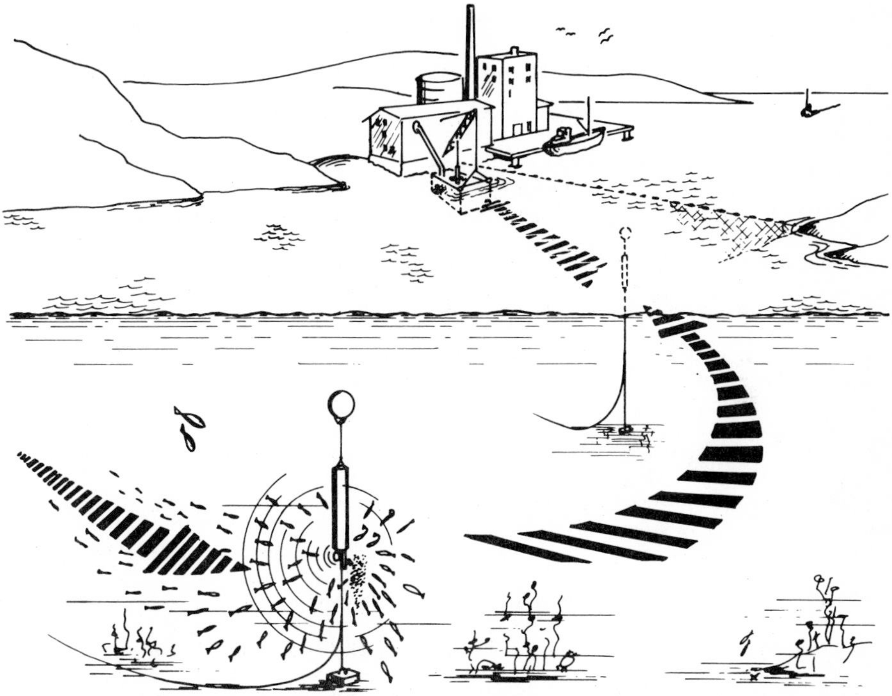

*FIG. 20. A system for harvesting of fish in long-term storage.*

from the theory presented previously, that the density of fish around a conditioning unit will rise to a maximal value determined by the balance between the attraction of the food and the repulsion due to the competition among the fish in the school. It has also been observed that adding a strong light source to the conditioning unit during the dark winter season improves the attraction of the fish. This is probably caused by the improved ability to detect the food particles by vision. Also some direct attraction to the light (caused by curiosity) must be expected.

In a system for harvesting the fish, one of the conditioning units is located inside a trap that closes automatically after the fish have entered it. The trap is then either emptied into the transportation system of the factory or into a well boat for transportation to the factory.

A variety of applications of this principle can be envisaged such as farming of fish in coastal waters (fjords, bays) under "guidance" and harvesting of natural fish population having a density that is too low for efficient use of traditional fishing gear. A major advantage of such a system is that it does not kill the fish.

## V. CONCLUSIONS

The principles of mathematical modeling and state estimation have been applied to the description of certain aspects of fish biology and behavior, and some examples of control of fish behavior have been outlined. This type of application of modern system theory is in its infancy, and it is believed that such techniques can be used in studying many important problems connected with the proper utilization of marine biological resources.

## ACKNOWLEDGMENT

The research program, which is the basis for this contribution, was sponsored jointly by the Royal Norwegian Council for Scientific and Industrial Research and the Norwegian Fisheries Research Council. The author acknowledges this support and also the support of those colleagues who have critically reviewed the material.

REFERENCES

1. J. G. BALCHEN, "Feedback Control of Schooling Fish," *Proc. IFAC 5th World Congr.* Pt. 1. Instr. Soc. Am., Pittsburgh, Pennsylvania, 1972.
2. L. VON BERTALANFFY, "A Quantitative Theory of Organic Growth,"*Hum. Biol. 10*, 181-213 (1938).
3. L. VON BERTALANFFY, "Quantitative Loss in Metabolism and Growth," *Quant. Rev. Biol. 32*, 217-31 (1957).
4. S. P. BASU, "Active Respiration of Fish in Relation to Ambient Concentration of Oxygen and Carbondioxide," *J. Fish. Res. Board Can. 16*, 175-212 (1959).
5. F. W. H. BEAMISH, "Respiration of Fishes with Special Emphasis on Standard Oxygen Consumption. II. Influence of Weight and Temperature on Respiration of Several Species," *Can. J. Zool. 42*, 177-188 (1964).
6. F. W. H. BEAMISH, "Seasonal Changes in the Standard Rate of Oxygen Consumption of Fishes," *Can. J. Zool. 42*, 189-94 (1964).
7. F. W. H. BEAMISH, "Endurance of Some North West Atlantic Fishes," *J. Fish. Res. Board Can. 23*, 341-7 (1966).
8. E. C. BLACK, "Energy Stores and Metabolism in Relation to Muscular Activity in Fishes," *H. R. MacMillan Lecture in Fisheries*, pp. 51-67, Univ. of British Columbia, 1958.
9. A. B. BOWERS and F. G. T. HOLIDAY, "Histological Changes in the Gonade Associated with the Reproductive Cycle of the Herring *(Clupea Harengus L.)*," *Mar. Res. 5*, 16 pp. (1961).
10. E. A. BRAY and L. A. CAMPFIELD, "Metabolic Factors in the Control of Energy Stores," *Metabolism 24*, No. 1, 99-117 (1975).
11. J. R. BRETT, "Respiratory Metabolism and Swimming Performance of Young Sockeye Salmon," *J. Fish. Board Can. 21*, 1183-1226 (1964).
12. J. R. BRETT, "Swimming Energies of Salmon," *Sci. Am. 213*, 80-85 (1965).
13. J. R. BRETT, "Fish - The Energy Cost of Living," *in* "Marine Agriculture: (McNeil, ec.), pp. 37-52, Oregon State University Press, Corvallis, Oregon, 1970.
14. J. R. BRETT, "Energetic Responses of Salmon to Temperature (A Study of some Thermal Relations in the Physiology and Fresh Water Ecology of Sockeye Salmon *(Oncorhynchus nerka)*," *Am. Zool. 11*, 99-113 (1971).
15. J. R. BRETT, "Satiation Time, Appetite, and Maximum Food Intake of Sockeye Salmon *(Oncorhynchus Nerka)*," *J. Fish. Board Can. 26*, 409-415 (1971).
16. J. R. BRETT, J. E. SHELBOURN, and C. T. SHOOP, "Growth Rate and Body Composition of Fingerling Sockeye Salmon *(Oncorhynchus nerka)* in Relation to Temperature and Ration Size," *J. Fish. Board Can. 26*, 2363-2394 (1969).
17. J. R. BRETT and E. A. HIGGS, "Effect of Temperature on the Rate of Gastric Digestion in Fingerling Sockeye Salmon *(Oncorhynchus nerka)*," *J. Fish. Board Can. 27*, 1767-79 (1970).
18. H. BROCKERHOFF, "Digestion of Fat by Cod," *J. Fish. Board Can. 23*, 1835-39 (1966).
19. D. R. IDLER and W. A. CLEMENS, "The Energy Expenditures of Frazer River Sockeye Salmon During the Spawning

Migration to Chilko and Stuart Lakes," *Int. Pacific Salmon Fisheries Report*, Progress Report No. 6, 88 pp, 1956.
20. R. JONES, "The Rate of Elimination of Food from Stomachs of Haddock (M. ae), Cod (G. m.), and Whiting (M. m.)," *J. Cons. Int. Explor. Mer. 35*(3), 225-243 (1974).
21. R. JONES and J. R. G. HISLOP, "Investigations into the Growth (M. ae.) and Whiting (M. m.) in Aquaria," *J. Cons. Int. Explor. Mer. 34*(2), 144-189 (1972).
22. M. KLEIBER, "The Fire of Life," *Wiley, New York*, 454 pp, 1961.
23. S. J. LOCKWOOD, "The Use of the von Bertalanffy Growth Equation to Describe the Seasonal Growth of Fish," *J. Cons. Int. Explor. Mer. 35*(2), 175-179 (1974).
24. J. E. PALOHEIMO and L. M. DICKIE, "Food and Growth of Fishes. I. A Growth Curve Derived from Experimental Data," *J. Fish. Board Can. 22*, 521-542 (1965).
25. J. E. PALOHEIMO and L. M. DICKIE, "Food and Growth of Fishes. II. Effect of Food and Temperature," *J. Fish. Board Can. 23*, 869-908 (1966).
26. J. E. PALOHEIMO and L. M. DICKIE, "Food and Growth of Fishes. III. Relation Among Food, Bodysize, Growth Efficiency," *J. Fish. Board Can. 23*, 1209-48 (1966).
27. A. M. PHILLIPS, Jr., "Nutrition, Digestion, and Energy Utilization," *in* "Fish Physiology I" (Hoar and Randall, eds.), Ch. 7, *Academic Press, New York*, 1969.
28. G. E. SHULMAN, "Life Cycles of Fish. Physiology and Biochemistry" (translated from Russian), *Halsted Press (John Wiley)*, 1974.
29. S. SOSKIN, "Carbohydrate Metabolism; Correlation of Physiological, Biochemical, and Clinical Aspects," *Univ. Chicago Press, Chicago, Illinois* (1952).
30. C. C. TAYLOR, "Growth Equations with Metabolic Parameters," *J. Cons. Int. Explor. Mer. 27*(3), 270-286 (1962).
31. A. WEATHERLEY, "Growth and Ecology of Fish Populations," *Academic Press, New York*, 293 pp, 1972.
32. G. G. WINBERG, "Rate of Metabolism and Food Requirements of Fishes," *Fisheries Res. Board Can. Transl. Ser. 194*, 1-202 (1960).
33. D. M. EGGERS, "A Synthesis of the Feeding Behavior and Growth of Juvenile Sockeye Salmon in the Limnetic Environment," Ph.D. Thesis, Univ. of Washington, 217 pp, 1975.
34. S. E. KERR, "A Simulation Model of Lake Trout Growth," *J. Fish. Board Can. 28*, 815-819 (1971).
35. J. G. BALCHEN, "Mathematical Modelling of Fish Behavior. Principles and Applications," *Proc. IFAC 6th World Congr., Boston/Cambridge*, Instr. Soc. Am., Pittsburgh, Pennsylvania 1975.
36. D. SLAGSTAD, T. WESTGÅRD, K. OLSEN, S. SÆLID, and J. G. BALCHEN, "Mathematical Modelling of Population, Quality, Migration, and Distribution of Important Species of Fish in an Ocean" (in Norwegian), Report STF48 A75050, Found. of Scient. Ind. Research (SINTEF), Norway, 1975.
37. J. G. BALCHEN, "Modelling of the Biological State of Fishes," Report STF48 A76023. Found of Scient. Ind. Research (SINTEF), Trondheim, Norway, 1976.
38. R. AASLID, J. DISTEFANO, and J. G. BALCHEN, "Modeling of the Hormonal State of Fishes," Report STF48 A76081, Found. of Scient. Ind. Research (SINTEF), Norway, 1976.
39. H. BERNTSEN, "Simulation of a Model of the Biological State

of Fishes," Report STF48 A76053 (in Norwegian), Found. of Scient. Ind. Research (SINTEF), Trondheim, Norway, 1976.

40. J. G. BALCHEN, "Principles of Migration in Fishes," Report STF48 A76045, Found. of Scient. Ind. Research (SINTEF), Trondheim, Norway, 1976.

41. M. FONTAINE, "Physiological Mechanisms in the Migration of Marine and Amphihaline Fish," *Adv, Mar, Biol. 13*, 241-355 (1975).

42. V. R. PROTASOV, "Vision and Near Orientation of Fish," Israel Program for Scientific Translations, Jerusalem, 1970.

43. D. W. NARVER, "Diel Vertical Movements and Feeding of Under-yearling Sockeye Salmon and Limnetic Zooplankton in Babine Lake, British Columbia," *J. Fish. Res. Board Can. 27*, 281-316 (1970).

44. P. M. J. WOODHEAD, "The Behavior of Fish in Relation to Light in the Sea," *Oceangr. Mar. Biol. An. Rev. 4*, 337-403 (1966).

45. T. J. HARA, "An ElectroPhysiological Basis for Olfactory Discrimination in Homing Salmon: A Review," *J. Fish. Res. Board Can. 27*, 565-586 (1970).

46. A. D. HASLER, "Odour Perception and Orientation in Fishes," *J. Fish. Res. Board Can. 11*, 107-29 (1954).

47. K. OSHIMA, W. E. HAHN, and A. GORBMAN, "Olfactory Discrimination of Natural Waters by Salmon," *J. Fish. Res. Board Can. 26*, 2111-2123 (1969).

48. K. OSHIMA, W. E. HAHN, and A. GORBMAN, "Electroencephalographic Olfactory Response in Adult Salmon to Waters Traversed in Homing Migration," *J. Fish. Res. Board Can. 26*, 2124-2133 (1969).

49. A. M. SUTTERLIN, "Chemical Attraction of Some Marine Fish in Their Natural Habitat," *J. Fish. Res. Board Can. 32*, 729-738 (1975).

50. H. TEICHMANN, "Die Chemorezeption der Fische," *Ergebn. Biol. 25*, 177-205 (1962).

51. W. N. TAVOLGA and J. WODINSKY, "Auditory Capacities in Fishes. Pure Tone Thresholds in Nine Species of Teleosts," *Bull. Am. Mus. Nat. Hist. 126*, 179-239 (1963).

52. O. SAND and A. D. HAWKINS, "Acoustic Properties of the Cod Swimbladder," *J. Exp. Biol. 58*, 797-820 (1973).

53. S. DIJKGRAAF, "The Functioning and Significance of the Lateral-line Organs," *Biol. Rev. 38*, 51-105 (1962).

54. F. R. HARDEN JONES, "Fish Migration," *Edward Arnold Ltd., London*, 1968.

55. R. BECKETT and K. CHANG, "An Evaluation of Kinematics of Gait by Minimum Energy," *J. Biomech. 1*, 147-159 (1968).

56. D. J. RAPPORT, "An Optimization Model of Food Selection," *The American Naturalist 150*, No. 946, 575-88 (1971).

57. R. R. ROSEN, "Dynamical Systems Theory in Biology," Vol. 1, *Wiley, New York*, 1971.

58. D. M. WARE, "Growth, Metabolism, and Optimal Swimming Speed of Pelagic Fish," *J. Fish. Res. Board Can. 32*, 33-41 (1975).

59. J. G. BALCHEN, "Control of Fish Behavior," Report STF48 A76082, Found. of Scient. Ind. Research (SINTEF), Trondheim, Norway, 1976.

60. V. G. STERNIN, I. V. NIKONOROV, and Yu. K. BUMEISTER, "Electrical Fishing, Theory and Practice" (translated

from Russian), Israel Program for Scientific Translations, Jerusalem, 1976.

61. J. DALEN and H. O. THORSEN, "Controlling Schools of Herring. Experiments 1972," Report 73-19-T, Div. Eng. Cybernetics, The Norwegian Inst. Technol., 1973 (in Norwegian).

62. J. DALEN, "Controlling Schools of Herring. Experiments 1973," Report 73-173-T, Div. Eng. Cybernetics, The Norwegian Inst. Technol., 1973 (in Norwegian).

63. P. BRATLAND, S. KRISHNAN, and G. SUNDNES, "Studies on the Long-Time Storage of Living Saithe, *Pollachius Virens*," *Fisk. Dir. Skr. Ser. Hav. Unders, Norway 16*, 279-300 (1976).

# Modeling for Process Control

MORTON M. DENN

*Department of Chemical Engineering*
*University of Delaware*
*Newark, Delaware*






ISBN 0-12-012715-6

## I. INTRODUCTION

Dynamic models of process units represent an essential compromise between the simplicity that is desirable for application and the complexity that is representative of the real physical phenomena. Units like chemical reactors and distillation columns often contain complex flows of streams undergoing phase change and large numbers of chemical reactions. Detailed descriptions of the phenomena will rarely be available. The essence of good modeling is the skillful retention of the dominant physicochemical phenomena while ignoring the secondary effects, and the degree of detail required will vary depending on the application.

In this chapter we shall outline the procedures that are used to obtain process models. We shall emphasize models that are based on fundamental physical and chemical principles, rather than empirical models that are based entirely on input-output analysis of existing units. While ultimate application requires a combination of both approaches, only the fundamental models can be extrapolated beyond the particular conditions under which data were gathered.

## II. PRINCIPLES OF MODELING

The approach to process modeling that we shall take follows that of Russell and Denn [1]. Process models are constructed by application of the three fundamental conservation principles: conservation of mass, momentum, and energy.

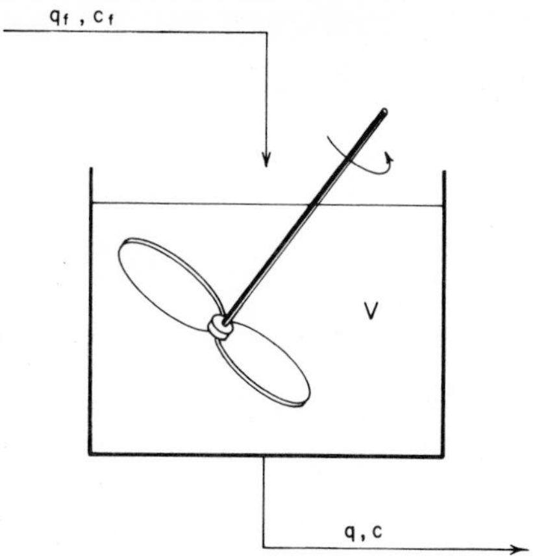


*FIG. 1. Schematic of a continuous-flow well-stirred tank.*

The conservation principles alone are never sufficient to provide an adequate model, however. Consider the elementary process shown schematically in Fig. 1. Here, a stream containing a dissolved solute flows continuously into a tank containing a solution of the same material, possibly at a different concentration. An effluent stream is withdrawn continuously. The feed and effluent flow rates, $q_f$ and $q$, respectively, and the concentration of the feed stream $c_f$ are presumed known as functions of time. It is desired to compute the effluent concentration and the liquid volume holdup in the tank as functions of time.

The first decision to be made is the selection of the fundamental conservation principles to be applied. Conservation of mass is clearly relevant. If the system is presumed to be isothermal, then for a liquid phase system at moderate pressures, the conservation of energy is of secondary importance. The principle of conservation of momentum is quite important, but as we shall demonstrate shortly, explicit consideration of the momentum equation can often be circumvented by making appropriate assumptions about the flow patterns.

The principle of conservation of mass is simply a bookkeeping statement: *The rate of change of mass within a specified region of space equals the rate at which mass enters minus the rate at which mass leaves.* It is evident that two more decisions must be made at this point. We must identify those measurable variables that characterize the mass; these are the *state variables*, which for the system under discussion are concentration, density, and volume. We must also identify the region of space to which the principle of conservation of mass is to be applied. Such a region is usually called a *control volume*.

The selection of the control volume is intimately related to our perception of the momentum transport within the system. The simplest assumption that can be made is that agitation is sufficient to make the system completely homogeneous. In that case there is a unique value of each of the state variables at any time, and the entire tank can be taken as the control volume. This *perfect mixing assumption* removes the need to consider spatial variation and leads to a lumped parameter model. The perfect mixing assumption also eliminates any need to consider the principle of conservation of momentum in modeling the process. If the perfect mixing assumption cannot be made, then a control volume must be selected in which each of the state variables has a unique value; in some cases, as shown subsequently, this volume must be a differential segment of the process unit.

The total mass in the tank at any time is $\rho V$, where the liquid density $\rho$ is the same at all points because of the mixing assumption. Mass enters at a rate $\rho_f q_f$, where $\rho_f$ is the density of the feed stream, and mass leaves at a rate $\rho q$. Note that, because of the perfect mixing assumption, the density of the effluent stream at any time is the same as the density of the liquid at all points in the vessel. The principle of mass conservation is, therefore,

$$d(\rho V)/dt = \rho_f q_f - \rho q \quad . \tag{1}$$

Conservation of mass also applies to each component species. The total amount of solute in the tank is $cV$, where $c$ is the concentration. The rates at which solute enters and leaves the control volume are, respectively, $c_f q_f$ and $cq$. Hence,

$$d(cV)/dt = c_f q_f - cq \quad . \tag{2}$$

We now reach the final step in the modeling process. These two equations exhaust the independent information available in the conservation principles. Yet, assuming that the properties of the feed stream are known at all times, there are three unknowns: $V$, $\rho$, and $c$. The resolution in this case is, of course, transparent: there is a relationship between the concentration and the density. This relationship is not derivable from the conservation principles at any macroscopic level and is unique to the particular solute-solvent system being processed. Such a relation between state variables that is unique to the particular system is called a *constitutive equation* (or, in a thermodynamics context, an equation of state). Selection of appropriate constitutive equations is one of the most important aspects of process modeling, and it is here that judicious experimentation is often required.

It is convenient to express the empirical density-concentration constitutive equation in the form

$$\rho(c) = \rho_0 + \rho_1 c + \phi(c) \quad , \tag{3}$$

where $\rho_0$ and $\rho_1$ are constants and $\phi(c)$ contains all of the nonlinearities. Equations (1) to (3) form the complete dynamic model. A further important simplification is possible, however, by eliminating the density from Eq. (1). Substitution of Eq. (1) into Eq. (3), followed by subtraction of Eq. (2) multiplied by $\rho_0$, leads to

$$\frac{dV}{dt} = \frac{\rho_0 + \phi(c) - c_f\phi'(c)}{\rho_0 + \phi(c) - c\phi'(c)} q_f - q \quad . \tag{4}$$

The prime denotes differentiation with respect to $c$. If the density data are linear over the interval $(c_f, c)$, then Eq. (4) reduces to

$$dV/dt = q_f - q \quad . \tag{5}$$

This is the equation that would be obtained by simply assuming that all densities in Eq. (1) are the same. In formulating process models, it is often assumed that all liquid streams have equal densities and that Eq. (5) is applicable, despite the fact that densities of common liquids may vary by 40% or more. This gross approximation usually suffices because it would be unusual for the coefficient of $q_f$ in Eq. (4) ever to differ significantly from unity.

## III. CHEMICALLY REACTING SYSTEMS

### A. *CONSERVATION OF MASS*

We now consider the case in which the feed stream in Fig. 1 contains a reactive solute, A, and the tank is a continuous flow reactor in which the chemical reaction

$$A \rightarrow \nu B \tag{6}$$

takes place; that is, one molecule of species A reacts to form $\nu$ molecules of species B. This simplified reaction scheme is sufficient to illustrate the general approach to representing the dynamics of vessels designed for more complex chemical reactions. We again assume perfect mixing.

The principle of conservation of mass as applied to the total mass is unchanged by the presence of the chemical reaction. Equation (1) thus follows immediately, and Eq. (5) is obtained if the density is linear in each concentration and a second condition is also satisfied: the rate of change of density with respect to each concentration must be proportional in the same way to the molecular weight of the species [1]. The equations for species A and B must account for the fact that A disappears from the control volume and B appears because of chemical reaction as well as flow. If concentration is measured in molar units and we denote the rate of disappearance of A per unit volume because of reaction as $r$, then it follows that the rate of appearance of B because of reaction is $\nu r$. The equations of conservation of mass for each species are then

$$d(c_A V)/dt = q_f c_{Af} - q c_A - Vr \quad , \tag{7a}$$

$$d(c_B V)/dt = q_f c_{Bf} - q c_B + \nu V r \quad . \tag{7b}$$

Here, $c_A$ and $c_B$ denote molar concentrations of A and B, respectively, and subscript f denotes properties of the feed stream.

It is necessary to establish the constitutive equation defining the dependence of the intrinsic reaction rate $r$ on the concentration. Establishing such rate expressions is a major interface activity between chemistry and chemical engineering. The relevant experiments are usually done in laboratory batch reactors, for which $q_f = q = 0$. In such a reactor, it then follows that $dc_A/dt = -r$; the rate expression so evaluated, however, is a functional relationship that is generally valid for all reactor configurations. There is considerable confusion over this elementary point, particularly in the sanitary engineering literature, where the rate expression $r$ is sometimes written as a time derivative in flow systems that are operating at a steady state.

The rate will depend on both $c_A$ and $c_B$ for a reversible chemical reaction. If the reaction is irreversible (i.e., the reaction $\nu B \rightarrow A$ does not occur to a measurable extent), then $r$ will be a function only of $c_A$ and Eqs. (7a) and (7b) are uncoupled. The simplest form for a reaction rate is linear, or *first order*,

$$r = kc_A \quad . \tag{8}$$

The rate constant $k$ will be temperature dependent. Few systems are truly first order, but a linear rate expression often suffices. The dynamical properties that we wish to illustrate do not depend critically on the functional form of $r(c_A)$, so we will use Eq. (8) in order to exploit its algebraic simplicity, in which case Eq. (7a) becomes

$$\frac{d(c_A V)}{dt} = q_f \, c_{Af} - qc_A - kVc_A \quad . \tag{9}$$

It is essential that the best available kinetic rate expressions be used in any simulation of a real reactor, because the details of the system performance may be sensitive to the rate function; the fluid catalytic cracker discussed below is a good example of such sensitivity.

*B. CONSERVATION OF ENERGY*

The rates of chemical reactions are highly temperature dependent, so temperature control is an important consideration in reactor analysis. Temperature dynamics are taken into account through the principle of conservation of energy, which is also a bookkeeping statement: *The rate of change of energy within a specified region of space equals the rate at which energy enters minus the rate at which energy leaves.* Energy enters or leaves by flow and by heating or cooling and by the performance of work. The principle of conservation of energy is nearly always applied incorrectly to flowing reacting systems in texts on process dynamics; this is true even in such otherwise excellent books as those by Gould [2] and Douglas [3]. Typically, a term is introduced which purports to account for "generation because of chemical reaction;" through a series of compensating errors, the correct equation for a homogeneous liquid phase system is then obtained. This approach works with certainty only when the correct equation is known, and the literature abounds with reactor models containing incorrect temperature equations. The method usually outlined in process dynamics texts can lead to particularly serious errors when a change of phase takes place within the reactor.

Proper application of the principle of energy conservation requires the use of some fundamental thermodynamic relations. A complete treatment is beyond the scope of this chapter and is covered in textbooks [1,4], but the subject is of sufficient importance in process modeling that we shall go into some detail here. In the absence of important kinetic and potential energy contributions the total energy is approximated by the internal energy $U$, so the energy equation is

$$dU/dt = \rho_f q_f(\underline{U}_f + p_{f/\rho_f}) - \rho q(\underline{U}_e + p_{e/\rho}) + Q + W \quad . \qquad (10)$$

The underbar denotes per unit mass. $p$ is the pressure, $Q$ the rate at which heat is added to the system, and $W$ the rate at which work is done on the system. The subscript "e" denotes the effluent, and it is necessary to distinguish between some of the thermodynamic properties of the material in the reactor and the effluent. The terms $qp$ represent that portion of the work required to move the liquid streams into and out of the reactor. The remaining work rate $W$ is often called shaft work.

It is usually convenient to work in terms of the *enthalpy*, defined

$$H = U + pV \quad . \qquad (11)$$

For a liquid, $dH/dt$ and $dU/dt$ are nearly the same; this is equivalent to the statement that heat capacities at constant pressure and at constant volume are numerically equal for liquids. Equation (10) can therefore be written approximately as

$$dH/dt = \rho_f \; q_f \; \underline{H}_f(T_f) - \rho q\underline{H}(T) + Q + W \quad . \qquad (12)$$

This equation is not valid for vapors and gases, though it is often applied. The temperature of each stream is explicitly noted.

Enthalpy is a function of temperature, pressure,and composition, but the pressure dependence can usually be neglected for liquids under normal processing conditions. Hence, it is not necessary to distinguish between the enthalpy of the effluent stream and the liquid in the reactor. Two partial derivatives of the enthalpy have processing significance and can be measured experimentally. The *heat capacity at constant pressure per unit mass* $\underline{c}_p$, is defined

$$\underline{c}_p = \frac{1}{\rho V} \left.\frac{\partial H}{\partial T}\right)_{p,\ \text{all}\ n_i\ =\ \text{constant}} \tag{13}$$

$n_i$ is the number of moles of species $i$, equal to $c_i V$. The *partial molar enthalpy* $\tilde{H}_i$ is defined

$$\tilde{H}_i = \left.\frac{\partial H}{\partial n_i}\right)_{T,p,n_j\ =\ \text{constant},\ i \neq j} \tag{14}$$

Finally, the *Gibbs-Duhem equation* states that

$$H = \sum n_i \tilde{H}_i \tag{15a}$$

or, equivalently,

$$\underline{H} = \frac{1}{\rho} \sum c_i \tilde{H}_i \quad . \tag{15b}$$

The first step in developing an equation for the system temperature from Eq. (12) is to express all enthalpies in terms of the same reference temperature; the reactor temperature is the most convenient reference. It follows from Eq. (13) that

$$\underline{H}_f(T) = \underline{H}_f(T_f) + \int_{T_f}^{T} \underline{c}_{pf}(T)\ dT \quad . \tag{16a}$$

For simplicity we shall assume that $\underline{c}_{pf}$ is independent of temperature and write

$$\underline{H}_f(T) = \underline{H}_f(T_f) + \underline{c}_{pf}[T - T_f] \quad . \tag{16b}$$

Equation (12) can then be written

$$dH/dt = \rho_f q_f \underline{c}_{pf}[T_f - T] + \rho_f q_f \underline{H}_f(T) - \rho q \underline{H}(T) + Q + W \ . \tag{17}$$

Next, the derivative $dH/dt$ is expanded by means of the chain rule,

$$\frac{dH}{dt} = \frac{\partial H}{\partial T}\frac{dT}{dt} + \sum_i \frac{\partial H}{\partial n_i}\frac{dn_i}{dt} \tag{18a}$$

or

$$\frac{dH}{dt} = \rho V \underline{c}_p \frac{dT}{dt} + \sum \tilde{H}_i \frac{d(c_i V)}{dt} \quad . \tag{18b}$$

The summation in Eq. (18) may include nonreactive species as well as A and B. Combination of Eqs. (17) and (18) with the mass balance Eqs. (7) then leads to a temperature equation,

$$\rho V \underline{c}_p \frac{dT}{dt} = \rho_f q_f \underline{c}_{pf} [T_f - T] - [\nu \tilde{H}_B - \tilde{H}_A] Vr + Q + W \quad . \tag{19}$$

There is an assumption here that the partial molar enthalpy of a given species is the same in both the feed mixture and the mixture in the reactor. The grouping of partial molar enthalpies $\nu\tilde{H}_B - \tilde{H}_A$ is known as the *enthalpy of reaction*, or the heat of reaction, and it can be measured in a calorimeter or computed from tabulated heat of formation or heat of combustion data. Thus, there is no need to measure individual partial molar enthalpies. The final working form of Eq. (19) is usually written

$$\rho V \underline{c}_p \frac{dT}{dt} = \rho_f q_f \underline{c}_{pf} [T_f - T] + (-\Delta H_R) Vr + Q + W \quad . \tag{20}$$

The enthalpy of reaction $\Delta H_R$ is negative for exothermic, or heat releasing reactions.

Equation (20) is sometimes written

$$\underline{c}_p \frac{d}{dt} \rho V T = \rho_f q_f \underline{c}_{pf} T_f - \rho q \underline{c}_p T + (-\Delta H_R) Vr + Q + W \quad .$$

it is readily established that this form neglects a term $\rho_f q_f (\underline{c}_p - \underline{c}_{pf}) T$ and is therefore equivalent to assuming that the heat capacities of feed and effluent stream are the same. The derivative is sometimes written $d(\rho V c_p T)/dt$, which is completely wrong unless $\underline{c}_p$ is a constant. At least one textbook, which we shall not cite here, contains a worked example showing the "importance" of the appearance of $\underline{c}_p$ within the differentiation for a highly temperature-dependent heat capacity.

The temperature dependence of parameters in a reaction rate expression usually follows the Arrhenius relationship of $\exp(-E/RT)$, where $T$ is absolute temperature, $R$ the gas constant, and $E$ an "activation energy." Thus, for the first-order reaction defined by Eq. (8), we would have

$$r = kc_A = k_o \exp(-E/RT)\, c_A \quad , \tag{21}$$

which leads to the coupling between the mass and energy equations.

The heat transfer to the reactor may come from a jacket, which can also be treated as a well-mixed vessel. Using the subscript "j" to denote "jacket" the jacket equation can be written

$$\rho_j V_j \underline{c}_{pj} \frac{dT_j}{dt} = \rho_j q_j \underline{c}_{pj} [T_{jf} - T_j] - Q \quad . \tag{22}$$

Equation (22) assumes constant heat capacity and no accumulation of coolant in the jacket. The heat transfer term $-Q$ is equal to the negative of the reactor heat transfer, indicating no losses to the surroundings.

Finally, the rate of heat transfer is usually expressed in terms of a heat transfer coefficient $h$ and the available heat transfer area $a$ as

$$Q = -\,ha[T - T_j] \quad . \tag{23}$$

The heat transfer coefficient represents a composite of resistances to heat transfer in the reactor wall and between the wall and both bulk fluid phases. Equation (23) *defines* the heat transfer coefficient; a further constitutive expression is needed relating $h$ to system properties and flow conditions.

## C. REACTOR DYNAMICS

For the case of an irreversible, first-order chemical reaction, Eqs. (9) and (20) through (23) define the complete dynamic model. It is instructive to consider first the steady state, and for convenience we will take $\rho_j = \rho$, $\underline{c}_{pf} = \underline{c}_p = \underline{c}_{pj}$. The steady state is then defined by three system parameters:

$$\theta = V/q \quad ,$$

$$J = -\Delta H_R / \rho \underline{c}_p \quad ,$$

$$u = \frac{q_j}{q} / [1 + \frac{\rho \underline{c}_p q_j}{ha}] \quad .$$

The steady-state system equations can be manipulated into the equation

$$T - T_f + u(T - T_{cf}) = k_o \theta [J c_{Af} + (T_f - T) + u(T_{cf} - T)] \exp(-E/RT) \quad . \tag{24}$$

The left-hand side of this equation can be interpreted as the rate at which reaction heat is removed at steady state by flow and heat transfer, while the right-hand side is the rate at which reaction heat is produced. The two sides are plotted versus $T$ in Fig. 2; the intersections represent the possible steady states. For some parameter values there may be three intersections, indicating multiple steady states and potential instabilities. The relative slopes at the intermediate steady state indicate that small perturbations causing insufficient removal of reaction heat will lead to an increased rate of heat production, while small perturbations causing excess heat removal will lead to a decreased rate of production. This is

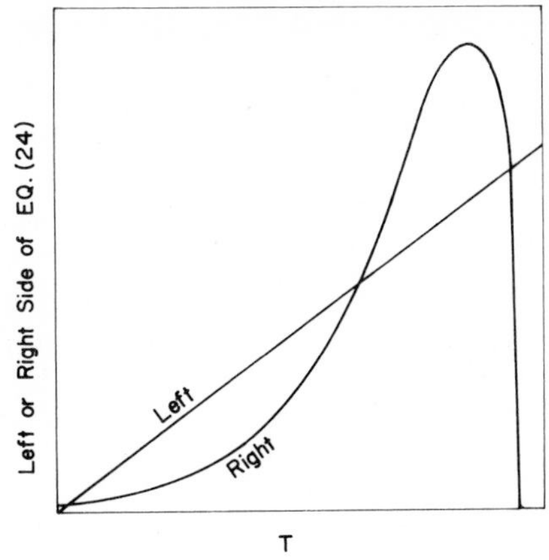


*FIG. 2. Plot of left-hand side (heat removal rate) and right-hand side (heat production rate) of Eq. (24) versus temperature. Steady states of the system are defined by intersection of the two curves.*

an unstable configuration, and the steady-state equations are sufficient to demonstrate that the intermediate steady state cannot be maintained in the absence of control.

The reactor dynamics in the neighborhood of a steady state are obtained by linearization of the nonlinear system differential equations. For illustrative purposes, it is convenient to consider cases in which the response time of the jacket is rapid compared to the reactor response time. In that case the jacket dynamics may be neglected and the system may be taken to be approximately second order. The dominant time constants $\tau_1$ and $\tau_2$ are then readily shown to be solutions to the equations

$$\frac{\tau_1 + \tau_2}{\theta} = \frac{2 + u + k_o\theta \exp(-E/RT)[1 - (Jc_A E/RT^2)]}{1 + u + k_o\theta \exp(-E/RT)[1 + u - (Jc_A E/RT^2]} \tag{25a}$$

$$\frac{\tau_1\tau_2}{\theta} = \{1 + u + k_o\theta \exp(-E/RT)[1 + u - \frac{Jc_A E}{RT^2}]\}^{-1} \quad . \tag{25b}$$

The time constants are strong functions of the steady-state operating conditions and the reaction kinetic parameters.

The necessary and sufficient condition for stability of a steady state to small disturbances is positivity of both the numerator and denominator in Eq. (25a).

## IV. EFFECT OF NONPERFECT MIXING

The calculated dynamical response of a stirred vessel depends critically on the validity of the perfect mixing assumption. The effect of imperfect mixing is most readily seen by considering the flow system illustrated in Fig. 3. This might be a schematic diagram of a baffled tank, in which $\beta V$ represents the portion of the tank directly influenced by the mixer and $(1-\beta)V$ the semi-stagnant region near the baffles, while $\lambda q$ represents the rate of interchange between the two regions. If we assume that each section is perfectly mixed, then the dynamical equations for a solute in the absence of chemical reaction are

$$\beta V \frac{dc_A}{ct} = qc_{Af} - qc_A + \lambda q \xi_A - \lambda q c_A \quad , \tag{26a}$$

$$(1 - \beta) V \frac{d\xi_A}{dt} = \lambda q c_A - \lambda q \xi_A \quad . \tag{26b}$$

Here, $\xi_A$ denotes the concentration of solute A in the stagnant zone.

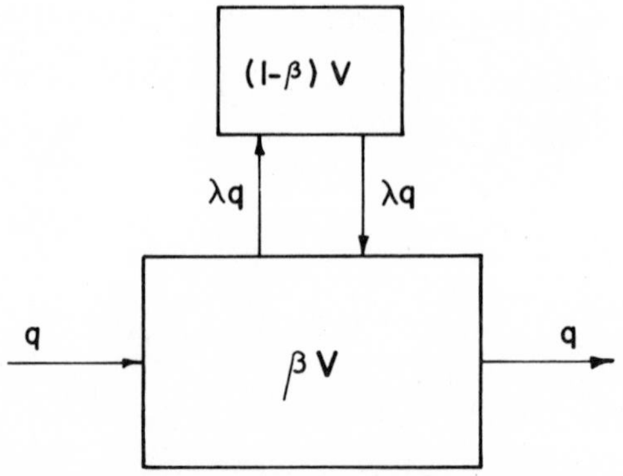


*FIG. 3. Schematic of an imperfectly mixed system.*

The time constants for this linear system are readily computed to be

$$\frac{\tau_1}{\theta}, \frac{\tau_2}{\theta} = \frac{2\beta(1-\beta)}{\lambda + 1 - \beta} \{1 \pm [1 - \frac{4\beta(1-\beta)\lambda}{(\lambda + 1 - \beta)^2}]^{1/2}\}^{-1} \quad . \quad (27)$$

The essential structure is revealed by two limiting cases. In the first, interchange is rapid between the two sections and $\lambda >> 1 - \beta$; in that limit,

$$\lambda >> 1 - \beta: \quad \frac{\tau_1}{\theta} \to 1, \quad \frac{\tau_2}{\theta} \to \frac{\beta(1-\beta)}{\lambda} << 1 \quad .$$

Here, the stagnant region equilibrates rapidly with the mixed core and the overall dynamics are unaffected by the segregation. The other limit is one of slow interchange, with $\lambda << 1 - \beta$; in that limit,

$$\lambda << 1 - \beta: \quad \frac{\tau_1}{\theta} \to \beta, \quad \frac{\tau_2}{\theta} \to \frac{1-\beta}{\lambda} >> 1 \quad .$$

In this latter case the dynamical response is dominated by the rate of interchange with the segregated region and the response time is considerably lengthened.

Models of continuous flow reactors must account for possible zones within the system. Complete mixing will be difficult to achieve in high viscosity systems such as polymerization reactors, and perfect overall mixing cannot be achieved at any viscosity level if the length to diameter ratio of the reaction vessel is greater than two or three. Incomplete mixing is usually accounted for by dividing the total volume into smaller, completely mixed regions, with interconnections between the regions.

## V. TUBULAR REACTORS AND HEAT EXCHANGERS

Many chemical reactions take place in tubular devices, in

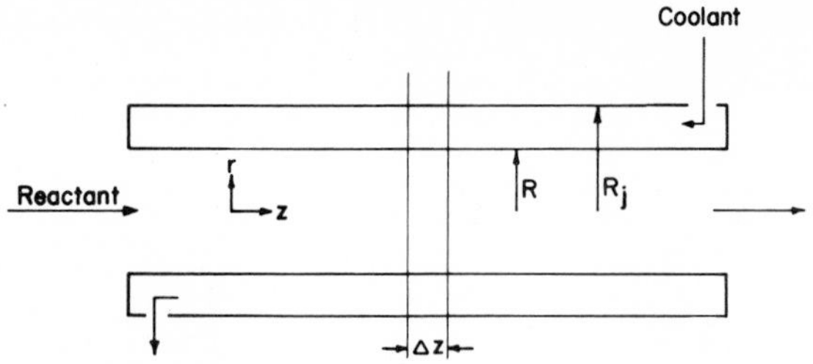


*FIG. 4. Schematic of a tubular reactor with a counterflow cooling jacket.*

which the reactor length is many times the diameter. The reactor may contain only the reacting fluid, or it may be packed with an inert or reactive solid material. Heat exchange frequently takes place through the wall, with the heat exchange fluid moving countercurrent to the reacting stream. Such a system is shown schematically in Fig. 4. A shell-and-tube heat exchanger has the same configuration except that the inner stream is not reactive.

For simplicity we first consider the case of an unpacked reactor with the single reaction A → νB, Eq. (6). If the fluid motion is turbulent, requiring a Reynolds number in excess of 2100, then we may often neglect radial variations and consider only variations in time and axial ($z$) position. Because of the continuous variation in the $z$-direction, the control volume must be taken as a differential segment of length $\Delta z$ in order to ensure a unique value of the state variables in the control volume. The equation of conservation of mass applied to the control volume with area $\pi R^2$ and length $\Delta z$ is then

$$\frac{\partial}{\partial t}\,\rho\pi R^2\Delta z = \rho\pi R^2 v\Big|_z - \rho\pi R^2 v\Big|_{z+\Delta z} \quad .$$

$v$ is the mean axial velocity, so the volumetric flow rate at any point is $\pi R^2 v$ and the mass flow rate is $\rho\pi R^2 v$. Dividing by

$\pi\Delta z$ and taking the limit as $\Delta z$ approaches zero then gives

$$\frac{\partial}{\partial t}(\rho R^2) = -\frac{\partial}{\partial z}(\rho R^2 v) \quad . \tag{28a}$$

For a tube of constant cross section, the $R^2$ term can be removed, giving

$$\frac{\partial \rho}{\partial t} = -\frac{\partial \rho v}{\partial z} \quad . \tag{28b}$$

For an incompressible fluid the density is constant, leading to the result that velocity at any time is independent of position.

When we consider the component species, there are three ways in which they can enter or leave the control volume. We have already considered convective flow and chemical reaction in modeling the well-mixed system; here we must also consider dispersion relative to the mean motion. Dispersion may result from molecular diffusion or it may result from a diffusive-like random turbulent eddying. in either case the dispersion is modeled by Fick's law of diffusion, which is a constitutive equation relating the diffusive flux to the concentration gradient:

$$\text{diffusive flux} = -\mathcal{D}(\partial c/\partial z) \tag{29}$$

The diffusivity $\mathcal{D}$ is a complex function of composition for multicomponent molecular diffusion [5], but it is usually approximated as a constant. Similarly, $\mathcal{D}$ is often approximated as a constant for turbulent eddy diffusion. The equation for conservation of mass for species A is then

$$\frac{\partial}{\partial t}\pi R^2 c_A = \pi R^2 v c_A\Big|_z - \pi R^2 v c_A\Big|_{z+\Delta z} + \left(-\pi R^2 \mathcal{D}_A \frac{\partial c_A}{\partial z}\right)\Big|_z - \left(-\pi R^2 \mathcal{D}_A \frac{\partial c_A}{\partial z}\right)\Big|_{z+\Delta z} - \pi R^2 \Delta z r \quad .$$

Dividing by $\pi R^2 \Delta z$ and taking the limit $\Delta z \to 0$, with $R$ and $\mathcal{D}_A$ constant for simplicity, we obtain

$$\frac{\partial c_A}{\partial t} = - \frac{\partial}{\partial z} v c_A + \mathcal{D}_A \frac{\partial^2 c_A}{\partial z^2} - r \quad . \tag{30}$$

For a first-order irreversible reaction, $r$ is given by Eqs. (8) and (21). It is important to note that the reaction rate depends only on the chemistry and not on the reactor configuration.

In many applications it can be shown that the dispersion term is negligible compared to the convective flux. Furthermore, incompressibility can generally be assumed. In that case Eq. (31) simplifies to

$$\frac{\partial c_A}{\partial t} + v \frac{\partial c_A}{\partial z} = - r \quad . \tag{31}$$

For laminar flow, where there is a substantial velocity gradient in the radial direction, the different residence times along different fluid streamlines give rise to an *apparent* dispersion that can be modeled by Eq. (30) with $\mathcal{D}_A$ replaced by a *Taylor dispersion* coefficient. This important effect is discussed by Gould [2] and in this volume by Aris, and we shall not consider dispersion effects further here.

For the energy equation the dispersion is a consequence of conduction, turbulence, Taylor dispersion, and perhaps radiation in a high temperature system. The derivation of the energy equation parallels that for the well-mixed reactor; assuming incompressibility and negligible dispersion the equation for the temperature is

$$\frac{\partial T}{\partial t} + v \frac{\partial T}{\partial z} = Jr + \frac{2h}{\rho c_p R} (T_j - T) \quad . \tag{32}$$

The equation for the cooling jacket is

$$\frac{\partial T_j}{\partial t} + v_j \frac{\partial T_j}{\partial z} = - \frac{2hR}{\rho_j c_{pj}(R_j^2 - R^2)} (T_j - T) \quad . \tag{33}$$

For countercurrent flow $v_j$ is negative. The equations for a shell-and-tube heat exchanger are obtained by setting $r$ to zero in Eq. (32). It is to be noted that Eqs. (31) - (33) are wave equations. The countercurrent nature of the system means that waves propagating with the flow will reflect and propagate against the flow as well, resulting in interactions and correspondingly complex dynamical response. The dynamics of such systems are discussed in standard texts [2,3,6]; Friedly [6] has shown that the transfer function of the effluent in a single pass heat exchanger may be approximated by the form

$$G(s) = \frac{k}{1 + \tau s} [1 - \exp\{-(a + bs)\}] \quad . \tag{34}$$

The parameters $k$, $\tau$, $a$, and $b$ are specified functions of the parameters in Eqs. (32) and (33). Friedly has also calculated approximate transfer functions for multipass heat exchangers.

When the reactor has a solid packing, then it is necessary to modify Eqs. (31) and (32) so that only the void (fluid) volume is included. Terms must also be added, which account for mass and heat transfer between the solid packing and the bulk fluid, and mass and energy balances must be written for the solid particles. It is often assumed that the temperature and concentration fields within the particle are uniform and that all resistance to mass and heat transfer is lumped at the surface; if the packing is a catalyst, then nonuniformities can be lumped into an *effectiveness factor* [4], which is computable from geometic and physicochemical parameters.

Dynamic models for packed reactors have been developed by Liu and Amundson [7] and Crider and Foss [8]. The former considered a catalytic reaction, in which all reactions take place in the solid phase, while the latter considered an inert solid phase with reaction in the fluid. Silverstein and Shinnar [9] have shown that the dynamical response of both systems is similar. The effect of the solid phase can be understood qualitatively by reference to Fig. 3, in which the portion $\beta V$ may be looked on as the fluid phase and $(1 - \beta)V$ as the solid. If a concentration or thermal pulse passes into the fluid phase, then it is transferred in part into the solid, which acts as a sink; because of the finite rate of interchange between solid and fluid, the solid subsequently acts as a source for the fluid phase. Hence, thermal and mass components that have different transfer rates between fluid and solid phase will pass through the system at different rates. When solutes are nonreactive, but are adsorbed on the solid phase at different rates, the result is a chromatographic separation. In a packed reactor the concentration and thermal waves travel at different velocities and interact in a complex manner. One consequence of this interacting wave behavior is an initial decrease in effluent temperature following a step increase in feed temperature, with a subsequent approach to a higher effluent temperature steady state. Sinai and Foss [10] have studied the dynamics of a packed reactor experimentally and show good agreement with the transfer functions developed by Crider and Foss. Friedly [11] has given approximate transfer functions for the limiting case of rapid heat exchange between solid and fluid in a packed adiabatic reactor; the transfer function between effluent and inlet temperature is

$$G(s) = \frac{K_1[1 - \exp(-as)]}{(1 + \tau_1 s)(1 + \tau_2 s)} + K_2 \exp(-as) \quad . \tag{35}$$

The parameters are all expressible in terms of system parameters appearing in Eqs. (31) and (32).

Lumped parameter approximations are sometimes used to describe the distributed tubular devices. Reactors with dispersion can be treated as a series of interconnected stirred tanks, where the number of tanks is equal to twice the Peclet number [12]. The use of orthogonal collocation to obtain an efficient low-order lumped model to the partial differential equations has been demonstrated by Michelsen *et al.* [13]. Jutan *et al.* [14] have applied collocation to obtain a seventh-order lumped model of an experimental reactor in which both axial flow and radial dispersion must be considered.

## VI. APPLICATION TO COMMERCIAL REACTORS

### *A. STABILITY CONSIDERATIONS*

The possibility of the existence of multiple steady states in a flow reactor seems to have been first described in 1918 by Liljenroth [15] with regard to the ammonia reactor, and the phenomenon is well known in combustion [16]. The manner of intersection of the two curves in Fig. 2 suggests that a small change in operating conditions could cause one or the other of the stable intersections to vanish, resulting in a large decrease (blowout) or increase (runaway) of the reactor temperature. This is a potential problem with the ammonia reactor; Baddour *et al.* [17] have shown, for example, that the optimal operating point for a TVA ammonia reactor is close to transition to unstable operation.

Similar potential problems exist in other commercial systems. Westerterp [18] has shown that the manufacture of phthalic anhydride may correspond to a reactor with five steady states, three of which are stable, with optimal operation at the intermediate stable state. Because of the relative slopes of curves like those in Fig. 2, the stability region separating the optimal state from the neighboring unstable states is small. Silverstein and Shinnar [9] have made a similar observation regarding the hydrocracking process. Van Heerden [19] has identified possible regions of multiple steady states in the ammonia and phthalic anyhdride processes as well as carbon monoxide conversion, oxidation of sulfur dioxide, and in an ethylene flame, and Degnan [20] has experimentally and theoretically demonstrated the existence of multiple steady states for carbon monoxide conversion in a monolithic reactor-heat exchanger. Multiple steady states may play a role in the operation of the fluid catalytic cracker, which is discussed in the next section.

### *B. DYNAMICS OF A FLUID CATALYTIC CRACKER (FCC)*

The fluid catalytic cracker has been the subject of many reported modeling studies, undoubtedly because of the central role that the FCC plays in petroleum refinery operation. FCC models are known to be prevalent throughout the oil industry for use in determining detailed changes in product characteristics resulting from changes in feedstock, catalyst, and operating levels, but proprietary considerations have generally kept the details of such models from appearing in the open literature. The dynamic models that have been published focus on overall unit responses, and they do adequately reflect the

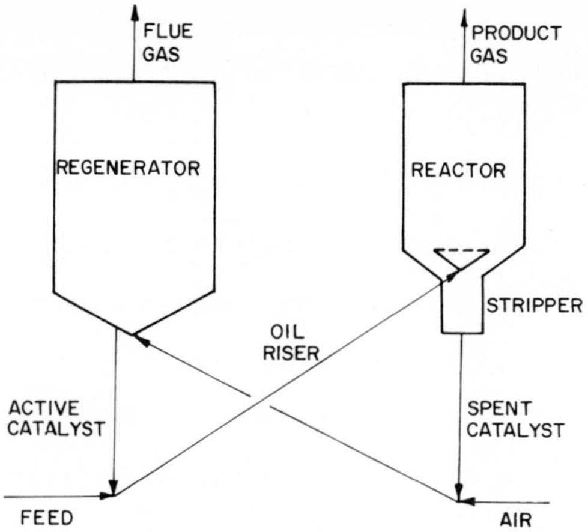


*FIG. 5. Schematic of the fluid catalytic cracking process.*

qualitative behavior of the FCC process.

The FCC process is shown schematically in Fig. 5. High boiling, high molecular weight gas oil is fed into the bottom of the riser, where it flows cocurrently with hot catalyst into the reactor. The cracking reaction, which mostly takes place in the riser, converts the heavy hydrocarbons to low molecular weight hydrocarbons such as gasoline. During the course of the reaction, carbon is deposited on the catalyst, causing the catalyst to lose its activity. Spent catalyst is separated in the reactor by gravity and cyclones and falls through a stripping section, where entrained oil is removed with steam, and the catalyst then passes into the air riser. The spent catalyst is blown with air into the fluidized bed regenerator, where the coke is burned from the catalyst. One of the important considerations in FCC operation is maintaining a satisfactory carbon monoxide-carbon dioxide-oxygen level in the flue gas from the regenerator, since too high a carbon monoxide level could lead to afterburning and dangerously high temperature excursions. The overall thermal balance on the system is important, because heat generated by the exothermic coke burning reaction is required to sustain the endothermic cracking reaction.

Kurihara's model [21] is typical of those that have been published. Each reactor is taken to be a perfectly mixed system, with the riser included as part of the cracking reactor, and solids and gases are assumed to be at the same temperature. A component mass balance is written only on the carbon, leading to the following mass and energy balances. Catalyst holdup in each reactor is assumed to be constant, and the original nomenclature is retained to facilitate reading the published literature.

*Cracking reactor:*

Carbon: $$H_{ra} \frac{dC_{sc}}{dt} = 50\ R_{cf} + 60\ R_{rc}\,(C_{re} - C_{sc}) \quad . \tag{36a}$$

Catalytic carbon: $$H_{ra} \frac{dC_{cat}}{dt} = 50\ K_{cc}P_{ra}H_{ra} - 60\ R_{rc}C_{cat} \quad . \tag{36b}$$

Energy: $$S_c H_{ra} \frac{dT_{ra}}{dt} = 60\ S_c R_{rc}(T_{rg} - T_{ra}) - 0.875\ S_f D_{tf} R_{tf}(T_{ra} - T_{fp}) - 0.875\ (\Delta H_{fv}) D_{tf} R_{tf} - 0.5\ (\Delta H_{cr}) R_{oc} \quad . \tag{36c}$$

*Regenerator:*

Carbon: $$H_{rg} \frac{dC_{rc}}{dt} = 60\ R_{rc}(C_{sc} - C_{rc}) - 50\ R_{cb} \quad . \tag{36d}$$

Energy: $$S_c H_{rg} \frac{dT_{rg}}{dt} = -\ 60\ S_c R_{rc}(T_{rg} - T_{ra}) - 0.5\ S_a R_{ai}(T_{rg} - T_{ai}) + 0.5\ (\Delta H_{rg}) R_{cb} \quad . \tag{36e}$$

Nomenclature is given in Table I together with typical values. Numerical factors in the model equation reflect the choice of units. Oxygen in the flue gas is computed from an algebraic equation. The constitutive equations for the coke burning rate $R_{cb}$, the carbon forming rate $R_{cf}$, and the oil cracking rate $R_{oc}$ are not repeated here. The specific dependence of these rates on the other process variables represents the difficult part of the model formulation, and differences between models will largely depend on the accuracy with which the rate equations reflect the true phenomena occurring within the reactor.

The uniqueness and stability of steady states of the FCC have been studied analytically by Iscol [22] and Lee and Kugelman [23]. Iscol concluded that the normal operating state of the FCC could be unstable to small perturbations, but with a very slow transient in the neighborhood of the steady state. Lee and Kugelman concluded that the normal operating region has a unique, stable steady state. The essential difference between the two models is in the constitutive expression for the coke formation rate $R_{cf}$ and demonstrates the sensitivity of the model to the constitutive equation. The cat cracker does have a steady state beyond the normal operating range, corresponding to high regenerator temperature and complete carbon monoxide conversion. The Amoco Oil UltraCat process [24] operates at this state and produces a higher gasoline yield than the conventional operating mode. Amoco has apparently developed proprietary technology that enables them to approach this stable state in a safe manner.

The dynamic model has been used in several multivariable control studies in order to determine and test algorithms for

TABLE I

Nomenclature and Units for Fluid Catalytic Cracker Model

| | | |
|---|---|---|
| $C_{cat}$ | Catalytic carbon on spent catalyst (0.875)[a] | wt % |
| $C_{rc}$ | Carbon on regenerated catalyst (0.64) | wt % |
| $C_{sc}$ | Total carbon on spent catalyst (1.475) | wt % |
| $D_{tf}$ | Density of total feed (7.0) | lb/gal |
| $H_{ra}$ | Reactor catalyst holdup (60) | ton |
| $H_{rg}$ | Regenerator catalyst holdup (200) | ton |
| $K_{cc}$ | Velocity constant for catalytic carbon formation | |
| $O_{fg}$ | Oxygen in flue gas (0.2) | mol % |
| $P_{ra}$ | Reactor pressure (40) | psia |
| $P_{rg}$ | Regenerator pressure (25) | psia |
| $R_{ai}$ | Air rate (400) | M lb/hr |
| $R_{cb}$ | Coke burning rate | M lb/hr |
| $R_{cf}$ | Total carbon forming rate | M lb/hr |
| $R_{oc}$ | Gas-oil cracking rate | M lb/hr |
| $R_{rc}$ | Catalyst circulation rate (40) | ton/min |
| $R_{tf}$ | Total feed rate (100) | M bbl/day |
| $S_a$ | Specific heat of air (0.3) | Btu/lb °F |
| $S_c$ | Specific heat of catalyst (0.3) | Btu/lb °F |
| $S_f$ | Specific heat of feed (0.7) | Btu/lb °F |
| $T_{ai}$ | Air inlet temperature (175) | °F |
| $T_{fp}$ | Feed preheater temperature (744) | °F |
| $T_{ra}$ | Reactor temperature (930) | °F |
| $T_{rg}$ | Regenerator temperature (1155) | °F |
| $t$ | Time | hr |
| $\Delta H_{cr}$ | Heat of cracking (160) | Btu/lb |
| $\Delta H_{fv}$ | Heat of feed vaporization (60) | Btu/lb |
| $\Delta H_{rg}$ | Heat of regeneration (10700) | Btu/lb |

[a]Typical values are in parentheses.

manipulation of the two primary control variables, the flow rates of air and regenerated catalyst [21,25-28]. Kurihara's work [21], using optimal control theory, is of particular interest here, since it leads to a control structure that focuses entirely on the regenerator, whereas conventional control schemes take measurements from both reactors. The Kurihara scheme and a patented Mobil modification are discussed critically by Lee and Weekman [28] and Shinnar [29]; the former include simulation results that contain information that is not part of the model described here and hence were presumably obtained using an unpublished Mobil model.

The fact that the regenerator dynamics dominate the process response follows immediately from the model and the parameters in Table I. At the simplest level of analysis it can be observed that regenerator holdup is large compared to reactor holdup. Kurihara therefore made a pseudosteady state assumption in the reactor and set the three reactor time derivatives to zero, reducing the system to second order in time. Figure 6 shows a comparison by Isaacs [25] between the open loop step responses of the full and reduced order models, and it is evident that neglecting the reactor dynamics has little influence on the response. A similar conclusion follows from consideration of the linearized model equations. Nakano [29] has computed the relative open loop eigenvalues in the Kurihara model to be (1,25,264,327,477), while Schuldt and Smith [26] have reported relative open loop eigenvalues in a similar model to be (1,10.7,179,234,351). Clearly one, or at most two temporal modes dominate the open loop transient, and it can be shown [30] that these modes are themselves dominated by the regenerator variables.

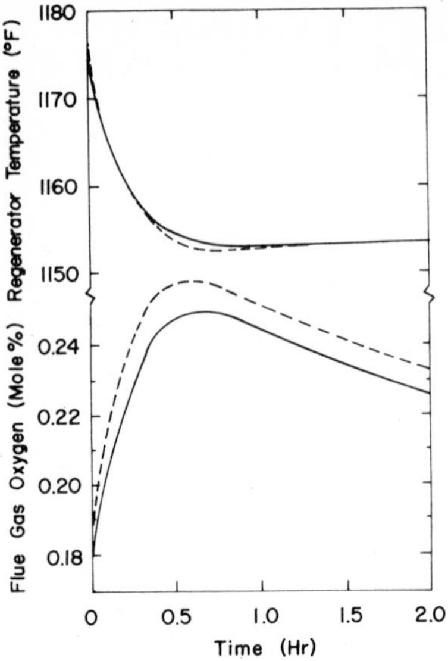


*FIG. 6. Open loop step response of the FCC [27]. The solid line is the full fifth-order model and the dashed line is the second-order model that includes only regenerator dynamics.*

Detailed modeling of the FCC requires accounting for the many chemical reactions that occur. It is impossible to keep track of (or even identify) all relevant species, but there has been considerable success in lumping classes of compounds with similar properties into a single pseudocompound. Component mass balances and rate expressions need only then be written for the lumped variables. A successful lumping scheme for the FCC is described by Jacob *et al.* [31]. Lumping of components is also discussed by Aris in this volume.

## VII. ACTIVATED SLUDGE PROCESS

The activated sludge process for wastewater treatment has been studied a great deal in recent years, and it provides a nice example of the influence of process structure on the dynamics. The essential parts of the process are shown in Fig. 7. The aerator is a biochemical reactor in which dissolved organics

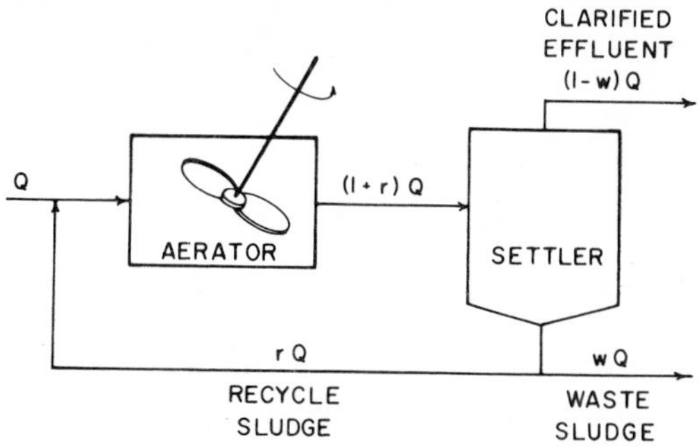


*FIG. 7. Schematic of activated sludge process for waste water treatment.*

are decomposed by suspended organisms. The organisms are separated in the settler and recycled to the reactor, with a small portion being wasted to maintain a steady state. Clarified effluent is removed overhead in the settler. Olsson [32] prepared a detailed review of activated sludge dynamics and control.

The aerator can be modeled in the first approximation as a perfectly mixed reactor with two components, substrate and organisms; the former is sometimes described as biological oxygen demand, or BOD, and represents a lumping of all dissolved organics and other oxygen-consuming materials into a single pseudocompound. Dissolved oxygen is clearly a relevant variable as well, but biological growth (reaction) rates are essentially independent of oxygen concentration above a critical value, which is nearly always maintained. The aerator mass balance then consists of two equations:

$$V \frac{dC_{\mathrm{x}}}{dt} = rQC_{\mathrm{xr}} - (1 + r)QC_{\mathrm{x}} + Vr \quad , \tag{37a}$$

$$V \frac{dC_{\mathrm{s}}}{dt} = QC_{\mathrm{sf}} + rQC_{\mathrm{sr}} - (1 + r)QC_{\mathrm{s}} - \frac{1}{Y} VR \quad . \tag{37b}$$

Subscripts x and s refer to organism and substrate in the reactor, respectively, and f and r to the feed and settler underflow. $Y$ is the "yield factor," relating conversion to organism growth, and the rate can usually be represented by the Monod equation

$$R = \frac{k_1 C_s C_x}{1 + k_2 C_s} \quad . \tag{38}$$

At steady state, an organism balance on the separator, assuming no organisms in the clarified effluent, gives

$$\text{steady state:} \quad 0 = (1 + r)QC_x - (r + w)QC_{xr} \quad . \tag{39}$$

Note that complete "washout" of organisms, $C_x = C_{xr} = 0$, $C_s = C_{sf}$ is a possible steady-state solution to the system equations.

Dynamic modeling of a continuous settler is not yet a well-developed art. There are theoretical reasons to expect the settler to consist of discrete layers, with constant concentration of suspended solids in each layer, under steady conditions, and dynamic models usually assume that the same behavior will occur during transients. Referring to Fig. 8, the thickness of the $n$th layer is $h_n$ and the velocity with

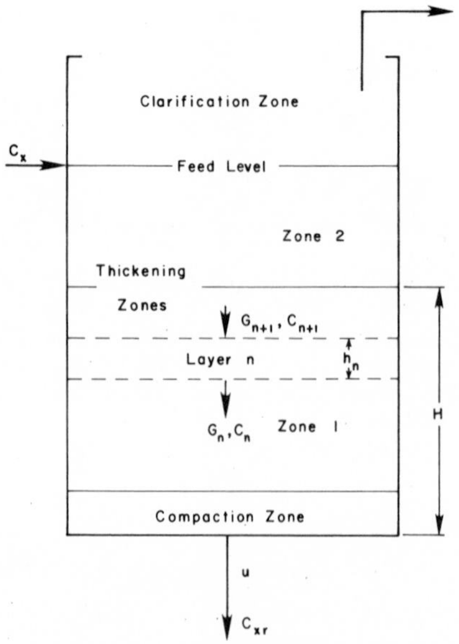


*FIG. 8. Schematic of the continuous settler.*

which its lower interface moves is $\delta_n$. $C_n$ is the solids concentration in the layer and the flux of solids leaving the layer is $G_n$. The solids flux consists of two terms. The first is a uniform convective motion equal to the downward fluid velocity $u$ times the concentration; $u$ equals $(r + w)Q/A$, where $A$ is the settler area. The second contribution to the flux results from solids settling *relative* to the mean motion. This settling velocity $v$ is a function of concentration $C_n$ and can be measured in a batch experiment. Thus,

$$G_n = C_n[u + v(C_n)] \quad . \tag{40}$$

A typical flux curve is shown in Fig. 9. The location of the minimum, known as the limiting flux and limiting concentration, defines the concentration in a steady experiment. Details of settler design are discussed in [33].

The mass balance equations for the $n$th layer are

$$dh_n/dt = \delta_{n-1} - \delta_n \quad , \tag{41}$$

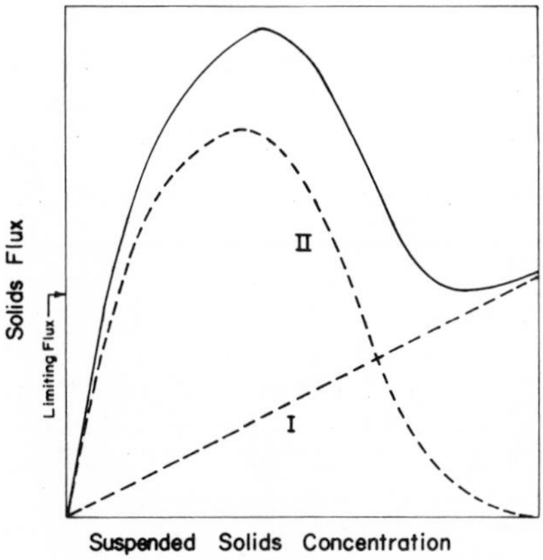


*FIG. 9. Typical flux curve. The dashed line I is the flux from convective flow and the dashed line II is the flux from settling relative to the convective flow. The solid line is the total flux, which is the sum of the two.*

$$dC_n h_n/dt = (G_{n+1} - C_{n+1}\,\delta_n) - (G_n - C_n\,\delta_{n-1}) \tag{42}$$

with

$$\delta_n = \frac{G_{n+1} - G_n}{C_{n+1} - C_n} \quad . \tag{43}$$

Dynamic models that account for the appearance and disappearance of layers have been developed by Tracy and Keinath [34] and Chi and Howell [35], with the latter extended by Attir, Denn, and Petty [36,37]. A new layer may be formed when the underflow rate or feed flow or concentration is changed. When the underflow changes, a new layer is started at an appropriate location within the thickening section at a concentration equal to the limiting concentration corresponding to that velocity. When the feed changes, a new layer may form at the top of the thickening section. A layer is removed if $h_n$ reaches zero. Under some conditions discrete concentration layers are not permitted by the second law of thermodynamics [37], and some modifications must be made. Details are contained in the original papers. Momentum transport is ignored except for a case in which an adverse concentration gradient develops; i.e., $C_n > C_m$, $n < m$. This is known to be an unstable configuration in which vertical circulations are induced, and in such a case the layers are assumed to mix.

Under normal settler operation, the clarified effluent contains only a small amount of suspended solids (the model predicts an overhead concentration of zero) and the top of the sludge blanket separating the thickening region from the clarification zone is at an intermediate position. Under these conditions the settler acts as a buffer and underflow concentration is relatively insensitive to changes in the reactor. If the

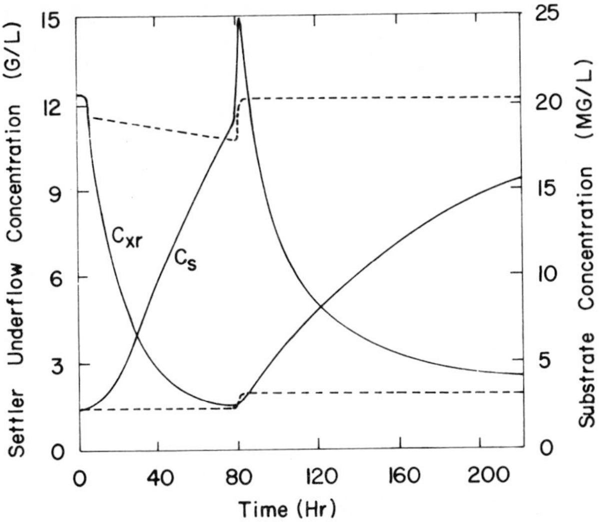


*Fig. 10. Dynamical response of an activated sludge process to 25% step reduction in flow rate at t = 2 h and a step increase to the original flow rate at 80 h [36]. Solid lines: recycle stream under ratio control. Broken lines: recycle flow rate uncharged.*

sludge blanket drops to the small compression zone at the bottom of the settler, then the system is said to be *underloaded*, and the buffering capacity is lost. Figure 10 shows some simulation results [36] for a reactor-settler system that is operating initially at steady state, but with a low sludge blanket. The feed flow rate $Q$ is reduced by 25% at $t = 2$ h, then increased to the original value at $t = 80$ h, with the underflow velocity and recycle ratio $r$ kept constant (i.e., the reactor is under ratio control). The settler goes rapidly to underloaded conditions, and the transient response time is seen to be of the order of several days, though reactor and settler holdups are of the order of a few hours.

This sluggish transient behavior is seen only under severely underloaded conditions. The broken lines in Fig. 10 show the response for the same flow changes, but with the absolute amount of recycle maintained constant. For this case the settler is just slightly underloaded and the sludge blanket recovers an intermediate height rather quickly following the stepup in flow rate. The radically different behavior results from a difference of only about 6% in the residence time of the reactor under recycle.

The influence of recycle on overall system transient response has received little attention except for a paper by Gilliland *et al.* [38]. As these authors have shown, the response times can be greatly increased. The calculation shown here is a particularly dramatic illustration.

## VIII. MELT DRAWING

In all of the cases considered thus far, we have made *a priori* assumptions about the momentum transport in order to avoid the necessity of solving the equations for conservation of momentum. Polymer and glass melt drawing operations represent processes in which the system dynamics are determined in large measure by the momentum equation; such operations are widespread in fiber and sheet manufacturing processes.

We will focus on the process of continuous drawing of a circular filament, which is shown schematically in Fig. 11. An extruded melt filament is solidified and subsequently taken up at a downstream position. The takeup velocity is faster than the velocity of extrusion, so the filament is drawn down in the melt zone; the area reduction ratio or *draw ratio* $D_R$ is equal to the ratio of take-up to extrusion velocity.

The diameter of the drawn filament may be affected by fluctuations in throughput, melt temperature, and conditions of cooling. For simplicity we will consider only an idealized situation in which physical properties are independent of

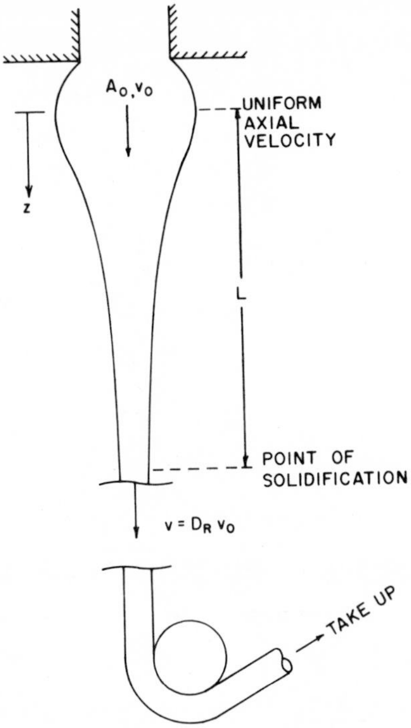


*FIG. 11. Schematic of the melt spinning process.*

temperature and in which solidification occurs at a fixed point that is externally controlled; these are not conditions that are characteristic of most commercial processing but they do simulate some laboratory experiments.

The equation of conservation of mass for this distributed system is identical to Eq. (28a) for constant density but changing area:

$$\frac{\partial A}{\partial t} + \frac{\partial}{\partial z}(Av) = 0 \quad . \qquad (44)$$

$A$ is the cross-sectional area and $v$ is the mean velocity in the axial ($z$) direction. We will write the momentum equation for conditions characteristic of polymer filament drawing, where inertia, surface tension, and air drag effects may often be neglected. For those portions of the filament where the rate of change of curvature is not substantial, the force $F$ is the same at all positions on the filament, and the momentum equation is simply

$$A[S_{zz} - \frac{1}{2}(S_{xx} + S_{yy})] = F(t) \quad . \tag{45}$$

$S_{xx}$, $S_{yy}$, and $S_{zz}$ are the normal stresses generated within the fluid by the motion, and they are assumed to be independent of radius at any axial position. The detailed derivation of Eq. (45) and a discussion of the assumptions is given by Matovich and Pearson [39] and Kase and Matsuo [40].

The relation between the stresses and the deformation field is determined by the rheological constitutive equation of the particular fluid. Most polymer melts are viscoelastic and demonstrate a complex, time and history dependent stress-deformation rate behavior. The spinline dynamics are greatly influenced by the relative magnitudes of the characteristic response times of the polymer melt and the residence time of a fluid element in the melt zone on the spinline. For simplicity, because many of the essential dynamical features are still retained, we will assure that the fluid is an incompressible Newtonian liquid, in which the nonisotropic part of the stress has a constant proportionality to the deformation rate:

$$\begin{aligned} S_{zz} &= -p + 2\mu \cdot \frac{\partial v_z}{\partial z} \; ; \\ S_{xx} &= -p + 2\mu \frac{\partial v_x}{\partial x} \; ; \\ S_{yy} &= -p + 2\mu \frac{\partial v_y}{\partial y} \; ; \end{aligned} \tag{46}$$

where $p$ is the isotropic pressure and $\mu$ is the viscosity. It is true for any incompressible fluid that

$$\frac{\partial v_x}{\partial x} + \frac{\partial v_y}{\partial y} + \frac{\partial v_z}{\partial z} = 0 \quad , \tag{47}$$

in which case Eq. (45) can be written as

$$3\mu\, A\, \frac{\partial v}{\partial z} = F(t) \quad . \tag{48}$$

The boundary conditions are

$$\begin{aligned} A &= A_0, \quad v = v_0 \quad \text{at} \quad z = 0 \quad , \\ v &= D_R\, v_0 \qquad\quad\ \text{at} \quad z = L \quad . \end{aligned} \tag{49}$$

The steady-state solution to these partial differential equations is

$$v = v_0 \exp(z \ln D_R/L) \quad , \tag{50a}$$

$$A = A_0 \exp(-z \ln D_R/L) \quad , \tag{50b}$$

$$F = 3\mu \ln D_R\, A_0\, v_0/L \quad . \tag{50c}$$

The linearized transient behavior is then determined by the perturbation equations

$$\frac{\partial a}{\partial t} + \frac{\partial}{\partial z}\left[v_0 \exp\left(\frac{z \ln D_R}{L}\right) a + A_0 \exp\left(\frac{-z \ln D_R}{L}\right) u\right] = 0 \tag{51a}$$

$$\frac{\partial}{\partial z}\left[\frac{v_0 \ln D_R}{L} \exp\left(\frac{z \ln D_R}{L}\right) a + A_0 \exp\left(\frac{-z \ln D_R}{L}\right) \frac{\partial u}{\partial z}\right] = 0 \quad , \tag{51b}$$

where $a$ and $u$ are perturbations in area and velocity, respectively. The take-up speed is presumed to be constant, so

$$u = 0 \quad \text{at} \quad z = L \quad . \tag{52}$$

Fluctuations in $u$ and $a$ are possible at $z = 0$.

Equation (51) can be reduced to quadratures and solved. The frequency response of the uptake area to forcing in the initial area and initial velocity is given in part by Pearson and Matovich [41]; Fig. 12 shows a calculation at $D_R = 15$ by George [42]. The amplitude ratio shows a series of peaks, with

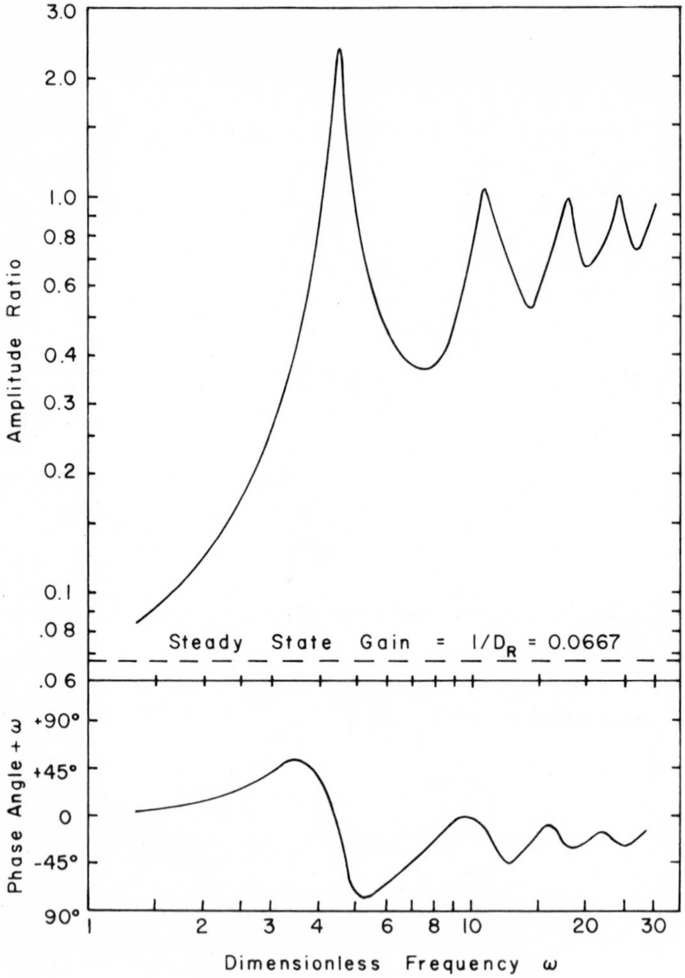


*FIG. 12. Frequency response of takeup area to changes in initial area,* $D_R = 15$ *[42].*

the following limiting behavior for area forcing:

$$\lim_{\omega \to 0} |G(i\omega)| = D_R^{-1} \quad ;$$

$$\lim_{\omega \to \infty} |G(i\omega)| = 1 \quad . \tag{53}$$

The amplitude ratio of the first resonant peak becomes infinite at a draw ratio of 20.21, and the open loop system in unstable at higher draw ratios. The high and low frequency asymptotic response and first peak are approximated by a transfer function of the form

$$G(s) = \frac{\exp(-\tau_R s)(\tau_R^2 s^2 + 2\lambda\omega_n \tau_R s/\sqrt{D_R} + \omega_n^2/D_R)}{\tau_R^2 s^2 + 2\zeta\omega_n \tau_R s + \omega_n^2} . \tag{54}$$

$\tau_R$ is the residence time in the melt zone,

$$\tau_R = \int_0^L \frac{dz}{v} = \frac{L(D_R - 1)}{v_0 \ln D_R} . \tag{55}$$

$\zeta$ is a function of $D_R$ which is positive for $D_R < 20.21$ and negative for $D_R > 20.21$. The dimensionless natural frequency $\omega_n$ is approximately 4.5.

The system exhibits sustained oscillations for draw ratios greater than 20.21. This limit cycle behavior is known as "draw resonance" and can be estimated by an approximate solution to Eqs. (44) and (48) which considers the nonlinear response of the first spatial mode [43]. Draw resonance is observed experimentally at a draw ratio of about 20 for silicone oils and polyethylene terephthalate melts, both of which exhibit approximately Newtonian fluid behavior. The critical draw ratio for the onset of sustained resonance is usually less for non-Newtonian fluids, and there is stabilization at high draw ratios as a result of both fluid viscoelasticity and cooling. The theoretical and experimental status of polymer melt spinning stability and sensitivity has been reviewed recently by Petrie and Denn [44] and Kase and Denn [45].

## IX. DISTILLATION COLUMN DYNAMICS

The detailed modeling of distillation column dynamics has received a great deal of attention in the literature and is treated in standard texts [2,46,47]. We shall briefly describe

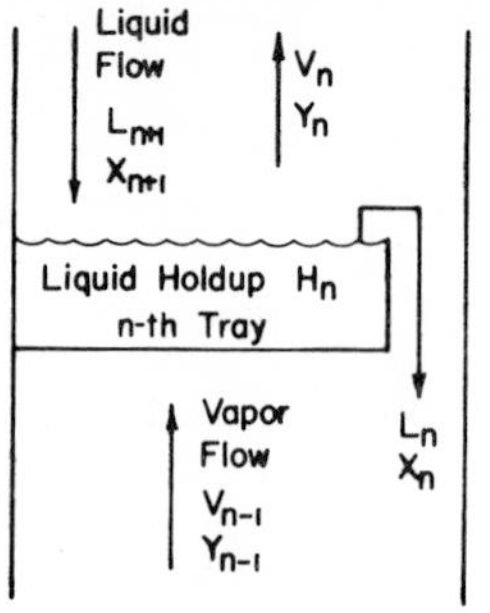


*FIG. 13. Schematic of a plate distillation column.*

here the results of the earliest analyses of column dynamics, because these are adequate to illustrate the general approach and they reveal some essential dynamical features.

A schematic of a plate distillation column is shown in Fig. 13. Holdups and flow rates are expressed in molar units, and $X_n$ and $Y_n$ represent the mole fraction of a volatile component in the liquid and vapor phase, respectively. It is assumed that holdup in the vapor phase is negligible. In that case, the overall and component mass balances are, respectively,

$$\frac{dH_n}{dt} = V_{n-1} + L_{n+1} - V_n - L_n \quad , \tag{56}$$

$$\frac{dX_nH_n}{dt} = V_{n-1}Y_{n-1} + L_{n+1}X_{n+1} - V_nY_n - L_nX_n \quad , \tag{57}$$

There is a temperature and pressure gradient across the column, but use of the energy equation may be avoided in the first approximation by assuming that the column is adiabatic and that the enthalpy of vaporization of the volatile component is constant. Together with the assumption of no vapor holdup, these assumptions imply that $V_{n-1} = V_n = V$, a constant. We will further neglect column hydraulics and assume that liquid flow rates are constant and that liquid holdups are constant and equal. These assumptions are not at all necessary, but they are convenient for our purposes here. The mass balance then simplifies to

$$(H/L)(dX_n/dt) = (V/L)(Y_{n-1} - Y_n) + X_{n+1} - X_n \quad . \tag{58}$$

At this point we require a constitutive equation relating the vapor composition to that in the liquid. The simplest assumption is that vapor and liquid at a tray are in equilibrium,

$$Y_n = \phi(X_n) \quad . \tag{59}$$

This is a constitutive relation that can be obtained by batch experimentation. More generally, we would assume that the deviation from equilibrium is expressible in terms of a *plate efficiency*, which is measured experimentally at steady state and assumed to apply equally well in the transient. Equation (58) is then, finally,

$$\frac{H}{L} \frac{dX_n}{dt} = X_{n+1} - [\frac{V}{L} \phi(X_n) + X_n] + \frac{V}{L} \phi(X_{n-1}) \quad . \tag{59}$$

This is a second-order difference-first-order differential equation. Two boundary conditions are needed in $n$, but it is not necessary for us to go into these.

The linearized perturbation equations about steady-state operation are

$$\frac{H}{L} \frac{dx_n}{dt} = x_{n+1} - [\frac{V}{L} \phi'(X_n) + 1]x_n + \frac{V}{L} \phi'(X_{n-1})x_{n-1} \quad . \tag{60}$$

For the case in which $\phi(X)$ is nearly linear, this is a system of constant coefficient linear difference-differential equations, and an analytical solution is possible [48]. An alternative approach which does not require the assumption of linearity, following Jackson and Pigford [49] and Pigford [50], is to write Eq. (60) as

$$\frac{H}{L}\ \frac{dx_n}{dt} = x_{n+1} - 2x_n + x_{n-1} + [1 - \frac{V}{L}\ \phi'(X_n)]x_n$$
$$- [1 - \frac{V}{L}\ \phi'(X_{n-1})]x_{n-1} \quad . \tag{61}$$

If we consider a column with a large number of trays and think of $n$ as a node point of a continuous position variable, then the term $x_{n+1} - 2x_n + x_{n-1}$ divided by the square of the plate spacing represents a finite difference approximation to a second partial derivative in space, and the remaining terms divided by the spacing represent a difference approximation to a first partial derivative. Equation (61) is therefore the difference approximation to the parabolic partial differential equation

$$\frac{N^2 H}{L}\ \frac{\partial x}{\partial t} = \frac{\partial^2 x}{\partial z^2} + N\ \frac{\partial}{\partial z}\ \{[1 - \frac{V}{L}\ \phi'(X)]x\} \quad , \tag{62}$$

where $N$ is the total number of trays, and $z$ is normalized to range from zero at the bottom of the column to unity at the top.

Approximate analytical solutions to Eq. (62) can be obtained for appropriate boundary conditions by use of the Rayleigh-Ritz method [50]. The structure of the solution is revealed by considering a special case that is often approximately true, $\phi' V/L \approx 1$. In that case Eq. (62) simplifies to the diffusion equation,

$$\frac{N^2 H}{L}\ \frac{\partial x}{\partial t} = \frac{\partial^2 x}{\partial z^2} \quad . \tag{63}$$

The eigenvelues of this equation are real, negative, and widely spaced. The first eigenvalue is proportional to the inverse square of the number of trays. Thus, we obtain the result that the response of the plate column with a large number of trays will be approximately first order, with a time constant proportional to the square of the number of trays; for appropriate boundary conditions, the time constant is approximately $HN^2/6L$. Solutions including the "convective" term $1 - \phi'(X)V/L$ show the same $N^2$ dependence of the dominant time constant, but the coefficient depends on the relative volatilities of the components to be separated [50].

## X. CONCLUDING REMARKS

Although we have focused on dynamic process models, in keeping with the theme of this volume, it is certainly true that most industrial models in use today that are based on fundamental principles are steady-state models. A steady-state model of a complex process requires far less expenditure of time and money to construct and test than a dynamic model, and the dollar return from improved design and operating conditions may be easier to justify than the potential for improved control. The sensitivity information that can be derived from a steady-state model may sometimes be adequate for control purposes in processes with large time constants. We have already seen that some stability information can be derived directly from the steady-state stirred reactor equation, for example.

The examples used to illustrate process modeling techniques were chosen for their simplicity and diversity. The essential point that must be emphasized is that the utility of a process model is determined in large degree by the assumptions about momentum transport and the constitutive equations for reaction, transport, and thermodynamic phenomena. This is illustrated by the differing views on the stability of a fluid catalytic cracker that stem from different carbon formation rates and the different dynamical response of a melt spinning process with Newtonian and viscoelastic polymer melts. In practical terms, this means that it is unlikely that a meaningful fundamental model will be constructed without substantial input from engineers and scientists with specific training and experience in the area being studied.

Finally this chapter relates directly to the question of the present status of the implementation of advanced methods in process control. Additional and related material can be found in the papers and discussion recorded in the proceedings of a recent research conference on the subject [51], and the case study of evaporator dynamics and control by Fisher and Seborg [52] is particularly recommended.

ACKNOWLEDGMENT

Some of the preparation of this chapter was done during a period in residence at the University of Sydney, Sydney, New South Wales, Australia, and Monash University, Clayton, Victoria, Australia. I am grateful to Profs. R. I. Tanner and R. W. Prince of the University of Sydney and to Prof. O. W. Potter and Dr. D. V. Boger of Monash University for their hospitality.

REFERENCES

1. T. W. F. RUSSELL and M. M. DENN, "Introduction to Chemical Engineering Analysis," John Wiley and Sons, New York, 1972.
2. L. A. GOULD, "Chemical Process Control," Addison-Wesley, Reading, Massachusetts, 1969.
3. J. M. DOUGLAS, "Process Dynamics and Control," in two

volumes, Prentice-Hall, Englewood Cliffs, New Jersey, 1972.
4. R. ARIS, "Introduction to the Analysis of Chemical Reactors," Prentice-Hall, Englewood Cliffs, New Jersey, 1965; "Elementary Chemical Reactor Analysis," Prentice-Hall, Englewood Cliffs, New Jersey, 1969.
5. R. B. BIRD, W. E. STEWART, and E. N. LIGHTFOOT, "Transport Phenomena," John Wiley and Sons, New York, 1960.
6. J. C. FRIEDLY, "Dynamic Behavior of Processes," Prentice-Hall, Englewood Cliffs, New Jersey, 1972.
7. S. L. LIU and N. R. AMUNDSON, *Ind. Eng. Chem. Fundamentals 1*, 200 (1962).
8. J. E. CRIDER and A. S. FOSS, *Am. Inst. Chem. Eng. J. 12*, 514 (1966).
9. J. SILVERSTEIN and R. SHINNAR, paper presented at 70th Annual Meeting, Am. Inst. Chem. Eng., New York, Nov., 1977.
10. J. SINAI and A. S. FOSS, *Am. Inst. Chem. Eng. J. 16*, 658 (1970).
11. J. FRIEDLY, *Preprints 1967, Joint Automatic Control Conference*, Philadelphia, 1967, p. 216.
12. J. COSTE, D. RUDD, and N. R. AMUNDSON, *Can. J. Chem. Eng. 39*, 149 (1961).
13. M. L. MICHELSEN, H. B. VAKIL, and A. S. FOSS, *Am. Inst. Chem. Eng. J. 12*, 323 (1973).
14. A. JUTAN, J. P. TREMBLAY, J. F. MacGREGOR, and J. D. WRIGHT, *Am. Inst. Chem. Eng. J. 23*, 732 (1977).
15. F. G. LILJENROTH, *Chem. Met. Eng. 19*, 287 (1918).
16. D. A. FRANK-KAMENETSKII, "Diffusion and Heat Transfer in Chemical Kinetics," 2nd ed., Plenum, New York, 1969.
17. R. F. BADDOUR, P. L. T. BRIAN, B. A. LOGEAIS, and J. P. EYMERY, *Chem. Eng. Sci. 20*, 281, 297 (1965).
18. K. R. WESTERTERP, *Chem. Eng. Sci. 17*, 423 (1962)
19. C. VAN HEERDEN, *Chem. Eng. Sci. 8*, 133 (1958).
20. T. F. DEGNAN,Jr., "The Monolithic Reactor - Heat Exchanger," Ph.D. Dissertation, Univ. of Delaware, Newark, 1977.
21. H. KURIHARA, "Optimal Control of Fluid Catalytic Cracking Processes," Sc.D. dissertation, M.I.T., Cambridge, Massachusetts, 1967. Summarized in L. A. GOULD, L. B. EVANS, and H. KURIHARA, *Automatica 6*, 695 (1970).
22. L. ISCOL, *Preprints 1970 Joint Automatic Control Conference*, Atlanta, 1970, p. 602.
23. W. LEE and A. M. KUGELMAN, *Ind. Eng. Chem. Proc. Des. Dev. 12*, 197 (1973).
24. W. D. FORD, R. C. REINEMAN, I. A. VASALOS, and R. J. FAHRIG, "Modeling Catalytic Cracking Reactors," presented at 69th Annual Meeting Am. Inst. Chem. Eng., New York, December, 1976.
25. N. NAKANO, "Modal Control of a Catalytic Cracker," M.S. Thesis, M.I.T., Cambridge, Massachusetts, 1971.
26. S. B. SCHULDT and F. B. SMITH, Jr., *Preprints 1971 Joint Automatic Control Conference*, St. Louis, 1971, p. 270.
27. B. ISAACS, "Linear-Quadratic Optimal Control of Fluidized Catalytic Cracking Processes," B.Ch.E. Thesis, Univ. of Delaware, Newark, 1974.
28. W. LEE and V. W. WEEKMAN, Jr., *Am. Inst. Chem. Eng. J. 22*, 27 (1976).
29. R. SHINNAR, in A. S. FOSS and M. M. DENN, eds., "Chemical Process Control," *Am. Inst. Chem. Eng. 72*, No. 159, 167 (1977).

30. C. GEORGAKIS, personal communication, 1976.
31. S. M. JACOB, B. GROSS, S. E. VOLTZ, and V. W. WEEKMAN, Jr., *Am. Inst. Chem. Eng. J. 22*, 701 (1976).
32. G. OLSSON, in A. S. FOSS and M. M. DENN (eds.), "Chemical Process Control," *Am. Inst. Chem. Eng. Symp. Ser. 72*, No. 159, 52 (1977).
33. R. I. DICK, *Journal Sanitary Eng. Div., Proc. Am. Soc. Civil Eng. 96*, 423 (1970).
34. K. D. TRACY and T. M. KEINATH, *Am. Inst. Chem. Eng. Symp. Ser. 70*, No. 136, 291 (1974).
35. J. CHI, "Optimal Control Policies for the Activated Sludge Process," Ph.D. Dissertation, State Univ. of New York at Buffalo, 1974.
36. U. ATTIR and M. M. DENN, *Am. Inst. Chem. Eng. J. 24*, 693 (1978).
37. U. ATTIR, M. M. DENN, and C. A. PETTY, in G. F. BENNETT (ed.), "Water 1976," *Am. Inst. Chem. Eng. Symp. Ser. 73*, No. 167, 49 (1977).
38. L. R. GILLILAND, L. A. GOULD, and T. J. BOYLE, *Preprints 1970 Joint Automatic Control Conference*, Stanford, 1970, p. 140.
39. M. A. MATOVICH and J. R. A. PEARSON, *Ind. Eng. Chem. Fundamentals 8*, 512 (1969).
40. S. KASE and T. MATSUO, *J. Polymer Sci., Part A 3*, 2541 (1965).
41. J. R. A. PEARSON and M. MATOVICH, *Ind. Eng. Chem. Fundamentals 8*, 605 (1969).
42. H. H. GEORGE, personal communication, 1976.
43. R. J. FISHER and M. M. DENN, *Chem. Eng. Sci. 30*, 1129 (1975).
44. C. J. S. PETRIE and M. M. DENN, *Am. Inst. Chem. Eng. 22*, 209 (1976).
45. S. KASE and M. M. DENN, *Preprints 1978 Joint Automatic Control Conference*, Philadelphia, 1978, p. II-71.
46. W. L. LUYBEN, "Process Modeling, Simulation, and Control for Chemical Engineers," McGraw-Hill, New York, 1973.
47. O. RADEMAKER, J. E. RIJNSDORP, and A. MAARLEVELD, "Dynamics and Control of Continuous Distillation Units," Elsevier, New York, 1975.
48. D. E. LAMB, R. L. PIGFORD, and D. W. T. RIPPIN, *Chem. Eng. Prog. Symp. Ser. 57*, No. 36, 132 (1961).
49. R. F. JACKSON and R. L. PIGFORD, *Ind. Eng. Chem. 48*, 1020 (1956).
50. R. L. PIGFORD, *Dechema Monogr. 53*, 217 (1965).
51. A. S. FOSS and M. M. DENN, eds., "Chemical Process Control," *Am. Inst. Chem. Eng. Symp. Ser. 72*, No. 159 (1977).
52. D. G. FISHER and D. E. SEBORG, "Multivariable Computer Control: A Case Study," North Holland, 1976.

# Water Resource Systems Models

WILLIAM W-G. YEH AND LEONARD BECKER

*Engineering Systems Department*
*University of California*
*Los Angeles, California*






ISBN 0-12-012715-6

## I. INTRODUCTION: TECHNIQUES OF SYSTEMS MODELING AND DYNAMIC SYSTEMS IN WATER RESOURCES

During the last decade, one of the most important advances made in the field of water resources is the adoption of optimization and optimal control techniques for planning, design, and management of complex water resources systems. The analysis of a complex water resources system involves thousands of decision variables and constraints. Once the planning, management, or design objective has been determined, the problem lends itself to solution techniques developed in the fields of optimal control and operations research. These two fields are closely related in that both deal with optimization. Mathematical programming techniques such as linear and nonlinear programming developed in operations research are primarily for solving algebraic types of equations, while techniques such as maximum principle and calculus of variations are for optimizing functionals subject to differential equations. Applications of all techniques to water resources systems require a thorough understanding and considered judgment. More important is that in actual application to a specific problem, it requires expertise in the problem itself as well as a thorough knowledge about the limitations of the technique.

In this chapter applications of systems modeling to water resources will be illustrated by three major problems: (1) reservoir control and management, (2) water quality control and management, and (3) inverse problem in groundwater.

Examples are presented to demonstrate specific applications.

## II. APPLICATIONS TO RESERVOIR CONTROL AND MANAGEMENT

### A. *DECISION MODELS FOR REAL-TIME OPERATIONAL PURPOSES*

#### 1. *General Requirements and Characteristics*

The complexities of multipurpose, multifacility systems generally require release decisions to be determined by an optimization model based on some type of mathematical programming. An operational model should operate effectively in "real time," that is, a model should be easily accessible for updating with new information and have relatively low execution time commensurate with the availability of information, the time interval considered, and with low cost. The model should be deterministic to take advantage of present information and to be practically applicable to complex reservoir systems.

In addition to the usually explicitly stated categories of system constraints such as those listed in Table 1 (conservation equations, maximum and minimum storages, maximum and minimum releases, penstock and equipment limitations, and contractual, legal, and institutional obligations arising from the various purposes of the system), there is always the formally unstated but very important constraint of continuing operations. That is, whatever the total time period considered for the optimization, the optimal release policy must result in a system state at the end of the period that is conducive to satisfactory future operations.

A consequence is that both strategic and tactical decisions based on the best possible information are necessary, and an overall decision model can be considered as decomposable into

Table 1

Typical System Constraints

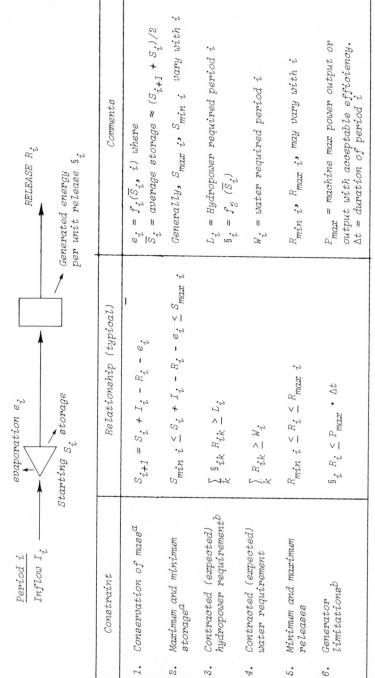

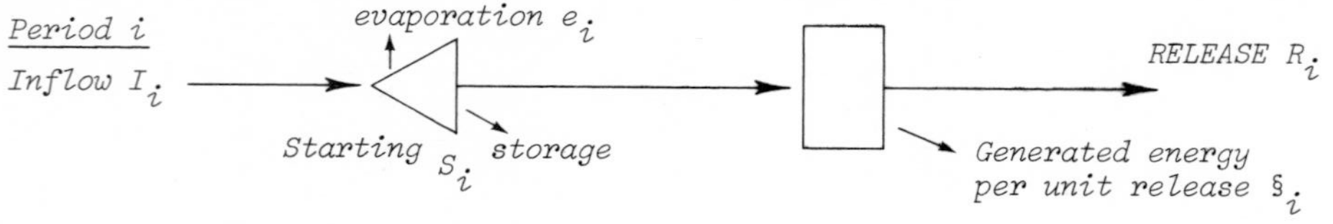


| Constraint | Relationship (typical) | Comments |
|---|---|---|
| 1. Conservation of mass[a] | $S_{i+1} = S_i + I_i - R_i - e_i$ | $e_i = f_i(\overline{S}_i, i)$ where<br>$\overline{S}_i$ = average storage $\approx (S_{i+1} + S_i)/2$ |
| 2. Maximum and minimum storage[a] | $S_{min\ i} \leq S_i + I_i - R_i - e_i \leq S_{max\ i}$ | Generally, $S_{max\ i}$, $S_{min\ i}$ vary with $i$ |
| 3. Contracted (expected) hydropower requirement[b] | $\sum_k §_{ik} R_{ik} \geq L_i$ | $L_i$ = Hydropower required period $i$<br>$§_i = f_2\ (\overline{S}_i)$ |
| 4. Contracted (expected) water requirement | $\sum_k R_{ik} \geq W_i$ | $W_i$ = water required period $i$ |
| 5. Minimum and maximum releases | $R_{min\ i} \leq R_i \leq R_{max\ i}$ | $R_{min\ i}$, $R_{max\ i}$, may vary with $i$ |
| 6. Generator limitations[b] | $§_i\ R_i \leq P_{max} \cdot \Delta t$ | $P_{max}$ = machine max power output or output with acceptable efficiency.<br>$\Delta t$ = duration of period $i$ |

[a] $e_i$ is usually small so that constraints 1 and 2 are nearly linear.

[b] Constraints 3 and 6 are nonlinear unless storage changes are very small.

long- and short-term submodels. These submodels require corresponding reservoir inflow forecasts, and the long-term forecast may be necessarily based on historic patterns but is nevertheless deterministic. Commonly, the long-term model looks ahead over a time span of a year with seasonal, monthly, or weekly time increments. It should be updated after each time increment. The short-term model looks ahead in accordance with the time increments when hydropower is a product, and should also be updated. Figure 1 shows a system of such submodels. Of course, the various submodels go to different levels of detail. The monthly model need not consider regulating reservoirs nor water transit time lags. The daily model must consider these but need not consider individual units in a facility nor the hydroelectric load demand curve for the day. In this case,an hourly submodel (over 24 h) will accept the daily release determinations for each facility without further analysis, but will determine the hour-by-hour releases for each individual unit in each plant.

The objective function to be optimized will depend on the dominant purposes for the particular reservoir system. Usually, however, most purposes and objectives will have been framed within the context of long-term contracts, institutional agreements, and even legislation that are difficult to change in short term; and generally, readily provable benefits for exceeding performance corresponding to contractual obligations are confined to the production of electrical power. An excess of such power will normally bring a return by permitting substitution for more costly modes of generation. Consequently, the objective function will often involve a maximization of hydropower generation while still achieving a satisfactory end-of-

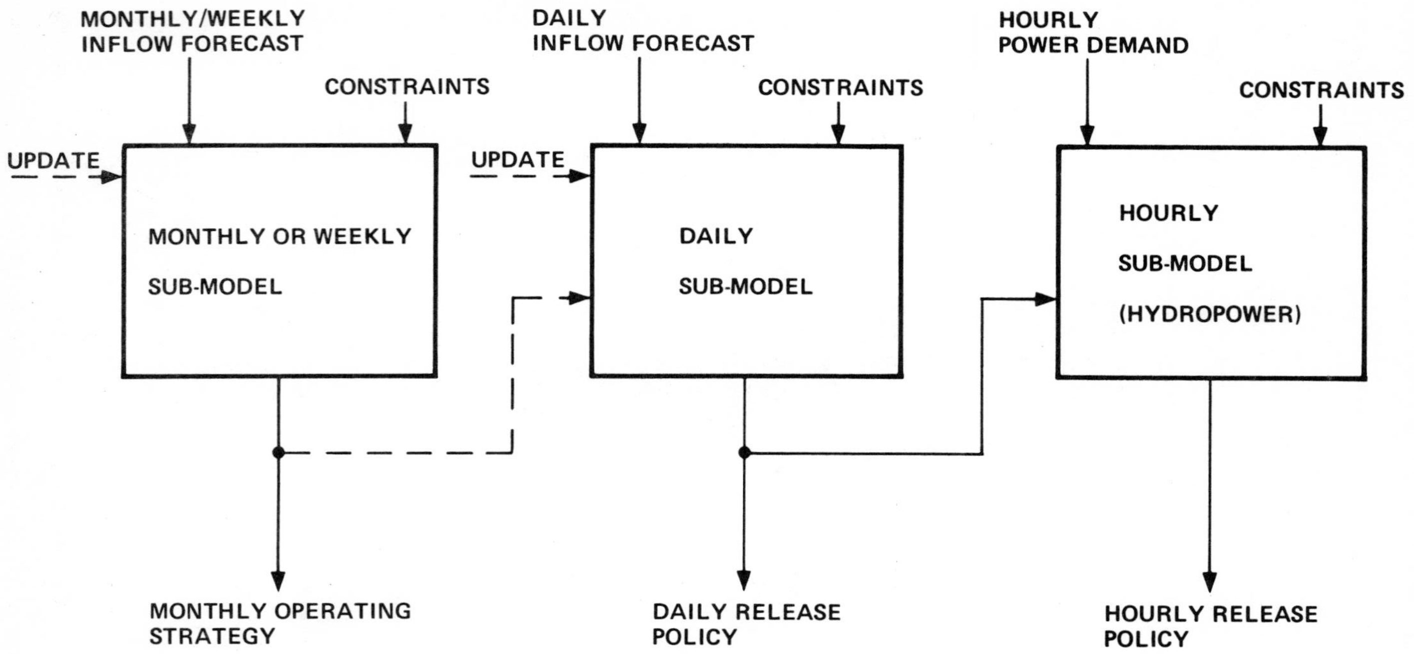


*FIG. 1. Reservoir System Operation Model*

period storage vector. A roughly equivalent form is the satisfaction of all system constraints with a minimum expenditure of the potential energy of the stored water in the reservoirs.

The objective function and power and energy constraints arising from various commitments are nonlinear with respect to reservoir release or reservoir end-of-period storage. However, in many situations, these nonlinearities can be effectively linearized, and linear techniques, which have the capability of handling large numbers of variables and constraints, can be employed.

Whatever the form of the decision model and the technique used, only approximately optimal policies can be obtained, since, almost always, approximations of one sort or another must be made to facilitate practical solutions. All solution techniques being developed for operational use will yield similar answers, if they yield any answer at all. The differences between then are in speed of convergence, computational requirements, the need for an initial feasible policy, and convenience of application. Of course, these differences can be significant. Common problems are related to nonlinearities in constraints and/or objective function and the high dimensionality typical of the reservoir system optimization problem. The latter leads in some applications to a requirement for further problem decomposition.

### 2. *Linear Programming - Dynamic Programming Model*

This model is used for the optimization of operations of the California Central Valley Project (CVP) (Becker and Yeh [1]). System nonlinearities that require linearization with this method are basically a result of the variable head at the turbines consequent to substantial reservoir releases. The average head is the difference between average forebay and tailwater elevations during the time interval under consideration. The former is a function of average reservoir storage, and the latter is a function of the rate of release. Neither of these quantities are known since they involve the variable, reservoir release; however, average head can be effectively obtained by repeated iteration, estimating the average head, optimizing over a relatively small-time period, using the optimal releases to correct the average head, and iterating again over the time period.

A multiple reservoir system will normally give rise to a large number of decision variables (releases) and constraints. Pure linear programming (LP) has been attempted for this type of problem. The major difficulty is that if the linear program is not confined to a small-time period, the above linearization procedure will fail. This difficulty is surmounted by decomposing the optimization problem (for a year, for example) into a series of problems over shorter time periods (months or weeks) and integrating these smaller problems with a form of dynamic programming (DP). That is, the DP is used to select a reservoir storage and release policy path through the specified time periods that is optimal in some sense.

The LP-DP methodology is as follows:

1. Optimal hydropower management within system constraints, including a minimum energy generation requirement can be shown to be approximately equivalent to the minimization of a weighted sum of reservoir releases (Becker and Yeh [1]). The weights may change for each optimization period and each iteration.

2. The minimum energy generation requirement is varied parametrically in discrete steps from whatever contract level may exist to the maximum that can be produced by the system.

3. LP is used to determine for each period the set of optimal solutions corresponding to (a) the parametric values of the minimum energy constraint, and (b) the set of storage vectors at the beginning of each period. The first period will begin, of course, with only one vector; the second period could begin with $n$ storage vectors where $n$ represents the number of discrete steps in the variation of the minimum energy constraint; the third period could begin with $n^2$ storage vectors, and so on.

4. This exponential growth is avoided by using a DP routine with the minimum energy constraint parameter as a single decision variable, the cumulative sum of the energy constraint parameters up to the period $i$ under consideration as a single state variable, and which maximizes a weighted sum of the reservoir end-of-period storages. The DP thus selects a set of optimal policy paths for any period $i$; each path corresponds to a set of discrete values of the cumulative energy constraints.

The recursive equation is

$$f_{i+1}(CE_{i+1}) = \max_{E_{i+1}} \left\{ \sum_k S^k_{i+1}(CE_{i+1}, E_{i+1}) + f_i(CE_i) \right\} , \quad (1)$$

$$i = 0, 1, 2, \ldots, N$$

where $f_{i+1}(CE_{i+1})$ is the storage vector having the maximum component sum for each value of the cumulative energy $CE_{i+1}$ in the period $i + 1$, and $f_0(CE_0)$ is the given beginning storage vector; $E_{i+1}$ is the minimum energy constraint period, $i + 1$; $\sum_k \Delta S^k_{i+1}$ the change in the $k$ storage vector components, period $i + 1$, for the particular values of $CE_{i+1}$ and $E_{i+1}$; and $N$ the number of periods.

5. The above procedures are continued until the final period is reached. If some minimum storage vector is specified for the final period to ensure continuing operations, the number of policy paths is truncated to correspond to that requirement. The operator will choose one of the remaining "optimal" paths that, in his judgment, is best. Each path is optimal in the same sense that for the given value of cumulative generated energy the maximum final storage vector is achieved. Alternatively, for each such storage vector, the maximum energy is generated. The method is illustrated in Fig. 2.

Although a large number of LP solutions are required, the dimensionality of each LP is small and the computer solutions take very little time. In addition, since reservoir constraints

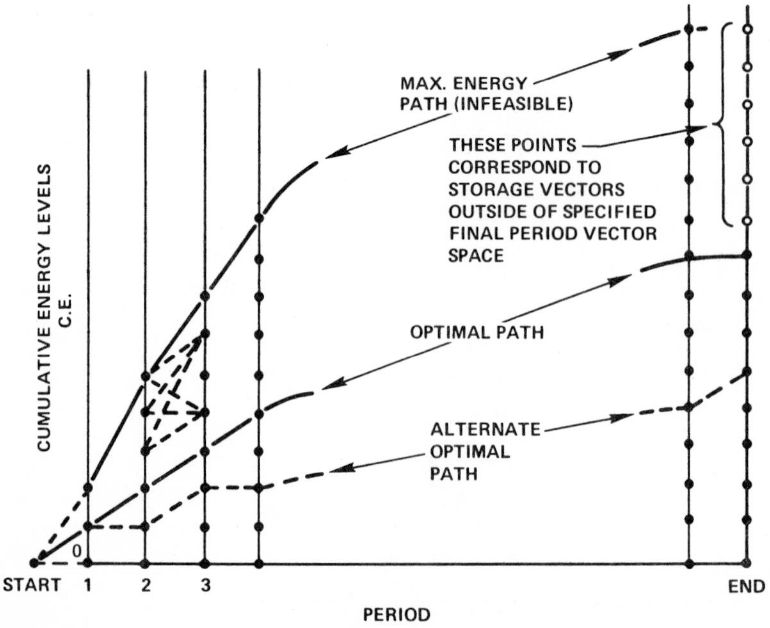


*FIG. 2. Policy path alternatives.*

are usually more numerous than decision variables, solution efficiency is increased by considering dual rather than primal LP. A set of optimal policies was obtained for the CVP for 12 monthly periods in less than 1 min on the IBM 360/91 computer.

### 3. *Penalty Function Model*

This model is used for optimization of operations for the Pacific Northwest hydroelectric system by the Bonneville Power Administration (BPA) (Hicks *et al.* [2]). The system nonlinearities are not linearized; instead most of the constraints (all of the nonlinear constraints) are transferred to the objective function by means of penalty functions, so that the problem becomes almost nonconstrained. It is then solved by the conjugate gradient method of Fletcher and Reeves [3].

The paper by Hicks *et al.* [2] presents sample results for a problem with about 2400 variables and 6200 constraints; however, it is claimed to be practical for problems of the size of 6000 variables and 15,000 constraints that are encountered in the Pacific Northwest system.

The optimization objective is to maximize system energy capability while keeping "acceptable" uniformity in the surplus of power overload for each time interval. This can be stated as

$$\min F(S,R) = \min \{D + w \sum_{j=1}^{J} (D_j - D)^2\} \tag{2}$$

where $S$ and $R$ are the reservoir storage and release vectors, $D_j = L_j - P_j$; $L_j$ the total system load requirement for time interval $j$; $P_j$ the average generated power during time interval $j$; and $w > 0$ is a suitable weight.

$$D = \frac{1}{T} \sum_{j=1}^{J} D_j \, T_j$$

where $T_j$ is the length of time interval $j$ and

$$T = \sum_{j=1}^{J} T_j \quad .$$

Various reservoir and release constraints and power generation limitations (low efficiencies) are considered soft constraints--they may be violated to some extent although penalties will then be imposed. Conservation of mass relationships and maximum and minimum reservoir storages are considered hard constraints, laws of nature that are inviolate. The soft constraints are transferred to the objective function using Courant-type penalty functions. That is, for any bounded variable, $x \leq x_{max}$, for example,

$$V = \{\max[x - x_{max}, 0]\}^2 \tag{3}$$

and the modified objective is

$$\min(F + \sum_i w_i \, V_i) \quad , \tag{4}$$

where the weights $w_i > 0$, and $i$ ranges over all soft constraints.

The conjugate gradient method for solution of the modified objective function requires an initial policy. This is generally obtained from previous experience of actual releases in the last time period. Application of conjugate gradients still requires observance of the hard constraints so that step length in the conjugate directions are limited.

There is no guarantee that a global optimum can be obtained and there may even be some difficulty in obtaining appreciable improvement on the initial policy under some circumstances. The weights $w_i$ are very important to a convergence towards an optimum. Large weights are needed for constraint satisfaction but small weights are more conducive to rapid convergence. Thus, a sequence of iterated solutions in which the weights progress from small to large values are normally required. The rate of progression and relative weightings depend on the experience and skill of the analyst.

BPA experience on a CDC 6400 has been that execution time for a single evaluation is about 400 μsec/variable and that about one evaluation per variable is necessary for optimization. Some results for the sample problem given in Hicks *et al.* [2] are

| | Initial Policy | Final Policy |
|---|---|---|
| Average power surplus | 93.6 MW | 190.7 MW |
| Constraint violations | | |
| $\sum w_i V_i$ | $1 \times 10^9$ | $5.6 \times 10^5$ |
| Computer time = 41 min. | | |

### *B. DECISION MODELS FOR PLANNING PURPOSES*

The design of a system of reservoirs includes the determination of the size of the reservoir to be built at each proposed reservoir site. The design is considered optimal when the net benefits,which are a function of the reservoir sizes and the operating policy, are maximized. The optimization is subject to the requirements of many constraints that must be imposed, including hydrological, economic, social, institutional, political, and legal constraints, as well as the usual physical constraints. In order to calculate the benefits, the system has to be operated and there should be close correspondence between the performance of the system as simulated at the design stage and that attainable after the system is built. Accordingly, one must use at the design stage an operating procedure that is consistent with feasible management of the real system.

Trott and Yeh [4] developed a method that can be used to determine the optimal design of any system of reservoirs with series and parallel connections. For a specific set of reservoir sizes, the return from the system is determined from the optimal operating policy. This policy is determined by decomposing the original multiple-state variable dynamic programming problem by Bellman's method of successive approximation into a series of subproblems of one state variable in such a manner that the sequence of optimizations over the subproblems converge to the solution of the original problem. The costs are assumed to be a function of reservoir size and are computed from storage capacity versus cost curves. A modified gradient technique is then used to determine the set of reservoir sizes that maximize the net benefits, subject to the imposed constraints.

Consider a system composed of $M$ reservoirs with both series and parallel connections. The flow of water between the reservoirs can be described by a set of difference equations (referred to as system equations)

$$S(t + 1) = f[S(t), R(t), Y(t), E(t)] \tag{5}$$

in which $S = [s_1(t),\ldots,s_m(t),\ldots,s_M(t)]$, in which $s_m(t)$ represents the storage level in reservoir $m$ at the beginning of time period $t$; $R = [r_1(t),\ldots,r_m(t),\ldots,r_M(t)]$, in which $r_m(t)$ represents the release policy for reservoir $m$ during time period $t$; $Y = [y_1(t),\ldots,y_m(t),\ldots,y_M(t)]$, in which $y_m(t)$ represents the total inflow into reservoir $m$ during time period $t$, which includes the natural streamflow and release from upstream reservoir; $E = [e_1(t),\ldots,e_m(t),\ldots,e_M(t)]$, in which $e_m(t)$ represents the evaporation loss from reservoir $m$ during time period $t$; $t$ = monthly time period; $m$ = reservoir index; $M$ = total number of reservoirs to be designed in the system; and $f$ is an $M$-dimensional vector of functions.

The net annual benefits are defined as

$$Z = V(S,R) - C(S_{\max}) \tag{6}$$

in which $V(S,R)$ = the annual return by selling commodities produced by the entire system through optimal operation after the system is built; and $C(S_{\max})$ = the equivalent annual cost of constructing the system, in which $S_{\max}$ contains $(s_{\max_1},\ldots, s_{\max_m},\ldots,s_{\max_M})$. Then to design a reservoir system such that the economic efficiency is optimized, it is necessary to

$$\max\{V(S,R) - C(S_{\max})\} \tag{7}$$

subject to

$$\frac{V(S,R)}{C(S_{\max})} \geq 1 \tag{8}$$

and all constraints imposed upon the system.

In order to solve this problem, it is necessary to determine the benefits and the costs accurately for a specific set of reservoir sizes. The return is a function of reservoir size and of the operating policy, which cannot be determined until the system is operated. To do this, the historical monthly streamflows that contain critical periods of droughts are used as the system input.

Assume that each of the $M$ reservoirs has a maximum allowable storage capacity $s_{\max m}$, a minimum pool of $s_{\min m}$, and the time horizon can be divided into $T$ monthly periods. Then, the optimal operating policy associated with this given set of reservoir capacities is the sequence of decisions $R(1)$, $R(2),\ldots,$ $R(T)$, which maximizes the return. The objective function for optimal operation is to

$$\max \sum_{t=1}^{T} V_t[S(t), R(t)] \tag{9}$$

subject to

$$S_{\min} \le S(t) \le S_{\max} \tag{10}$$

$$0 \le R(t) \le S(t) - S_{\min} + Y(t) - E(t) \tag{11}$$

and the system equation is

$$S(t+1) = f[S(t), R(t), Y(t), E(t)] \quad , \tag{12}$$

in which the initial state of the system $S(1)$ is given. These system equations not only describe the movement of water through the system but also describe the transition of the state variable $S$ from stage $t$ to stage $t+1$.

The recursive equation of dynamic programming is

$$\left.\begin{array}{l} f_{t+1}[S(t+1)] = \max\limits_{R(t)}\{V_{t+1}[S(t+1), R(t)] + f_t[S(t)]\} \\ t = 1, 2, \ldots, T \quad \text{and} \quad S(1) \text{ is given.} \end{array}\right\} \tag{13}$$

This is an $M$-dimensional forward dynamic programming problem with $M$ state variables, in which the stage transformation equation is given by the system equation and the decision variable $R$ is a vector of $M$-components. One method to cope with the problem of dimensionality is Bellman's method of successive approximations described by Bellman and Dreyfus [5]. This technique decomposes a problem of high dimensionality into subproblems of lower dimensionality that are solved sequentially.

First, it is necessary to choose an initial feasible sequence for each state variable, $\{s_m^0(t)\}$, $m = 1, 2, \ldots, M$. Then since there are as many decision variables as there are state variables, a subproblem is formed by optimizing over one state variable while requiring the sequence of the other $(M - 1)$-state variables to remain unchanged. This introduces $(M - 1)$-equality constraints on the decision variables, and consequently each subproblem has one state variable and one decision variable. Thus, this $M$-dimensional dynamic programming problem is reduced to the solution of a series of one-dimensional dynamic programming problems. A typical one-dimensional problem is then

$$\max \sum_{t=1} v_m[s_m(t), r_m(t)] \quad , \tag{14}$$

subject to

$$s_{\min_m} \leq s_m(t) \leq s_{\max_m} \tag{15}$$

$$r_m(t) \leq s_m(t) - s_{\min_m} + y_m(t) - e_m(t) \tag{16}$$

and the $(M - 1)$-equality constraints are

$$r_i(t) = s_i(t) - s_i(t + 1) + y_i(t) - e_i(t) \quad ; \quad i = 1, 2, \ldots, M \text{ and for } i \neq m \tag{17}$$

also

$$r_i \geq 0; \quad m - 1, 2, \ldots, M \tag{18}$$

and $s_m(1)$ is given. The recursive equation is

$$f_{t+1}[s_m(t + 1)] = \max_{r_m(t)}\{v_m[s_m(t + 1), r_m(t)] + f_t[s_m(t)]\} \tag{19}$$

$$t = 1, 2, \ldots, T \text{ and } s_m(1) \text{ is given} \quad .$$

Trott and Yeh [4] solved Eq. (19) by incremental dynamic programming. This algorithm further reduces the computer storage requirements needed to solve the standard recursive equation by analyzing only a small subset of all feasible states at each stage. For a given set of maximum reservoir capacities, the optimal operating policies are found by the aforementioned technique. If the cost of the system is considered to be solely a function of reservoir size, then storage capacity versus cost curves can be developed. With these curves the cost of the reservoir can be computed given its size. If the reservoirs are mutually exclusive, then the costs are additive and the cost of the complete system can be obtained by adding the cost of the individual reservoirs. Since the cost and return of the system are now determined, Eq. (6) can be evaluated. The maximum reservoir capacities are now varied using a modified gradient searching procedure to determine the optimal sizes.

Trott and Yeh [4] presented a numerical example using data from the proposed Eel River Ultimate Project in Northern California. They obtained optimal design of s six-reservoir system by maximizing the net benefits. For each set of reservoir sizes considered,the costs were computed from the storage capacity versus costs curves and the benefits were determined by finding the optimal operating policy via incremental dynamic programming and successive approximations. Considerable reduction in computer time and storage requirements were obtained by solving a series of one-dimensional dynamic programming problems instead of a six-dimensional one.

## III. APPLICATIONS TO WATER QUALITY CONTROL AND MANAGEMENT

### *A. SYSTEMS CHARACTERISTICS*

The release of decaying organic material into a flowing stream will tend to change its quality physically, chemically, and biologically. Decomposition of the material will occur, as a consequence of the life processes of various bacterial and other organisms, in various ways and at varying rates, depending on temperature, the compositions, and concentrations of the organic material and decomposed products at the time under consideration, and the concentration of dissolved oxygen in the water. If there is inadequate dissolved oxygen (DO) to satisfy the biochemical oxygen demand (BOD), anaerobic bacteria dominate and putrescent reactions badly pollute the water and most aquatic life is destroyed. Even before this occurs deoxygenation to less than a minimum level will kill fish and many green plants. Important indices of the quality of water at any time are the simultaneous levels of DO and BOD.

Flowing streams tend to natural purification over a period of time as a result of atmospheric reoxygenation and the addition of oxygen generated by green plants during photosynthesis. As with most gases the saturation concentration of oxygen increases with the decrease in temperature; e.g., saturation occurs at 11.3 mg/ℓ at 10°C, and at 9.2 mg/ℓ at 20°C (no chloride ions present and standard atmospheric pressure). The rate of reoxygenation is a function of the DO deficit from saturation, temperature, area of the air-water interface, and turbulence creating mechanisms such as wind. Reoxygenation from green plants depends on the presence of these plants, their access to sunlight, and temperature of the water.

Sedimentation of previous BOD loads may contribute to deoxygenation by creating bottom (or benthal) deposits that may be stirred up and resuspended in the stream in random ways, although rapid deposition of a BOD load may be favorable to the maintenance of water quality in that demands on current DO in the main body of the stream are delayed.

Even if streamflow is considered as steady state and one-dimensional, the water quality states, DO and BOD, are generally complex functions of time and space, and a simplified model of the above phenomena is necessary for a practical analysis of the pollutability of a stream and for adequate control and management of its quality. The most frequently used models are based on the classical work of Streeter and Phelps [6] and its modifications. These models contain parameters that are impractical or impossible to measure *in situ* and difficult to approximate in the laboratory so that they must be estimated from suitable observations of the state variables. The state variables are themselves not easily measured and observations of these variables are generally very noisy.

A basic model of the downstream effects of an upstream pollution load that approximates a number of the water quality processes can be structured as follows:

$$\dot{D} = k_1 L - k_2 D + \eta_1(t) \quad ; \tag{20}$$

$$\dot{L} = k_1' L + \eta_2(t) \quad ; \tag{21}$$

where the dot denotes time differentiation and the stream flows at constant velocity within a reach (a segment of stream with roughly similar properties all along its length), so that distance from the top of a reach can be expressed in time units (days) or $t$.

Also, $D$ is the DO deficit (mg/ℓ) at any point $t$, referred to the saturation concentration, i.e., $D = C_s - C$, where $C_s$ is the DO saturation concentration (mg/ℓ), $C$ the DO concentration at $t$ (mg/ℓ), $L$ the BOD concentration at $t$ (mg/ℓ), $k_1$ the deoxygenation rate (day$^{-1}$), $k_1'$ the BOD decay rate (including sedimentation) (day$^{-1}$), $k_2$ the reaeration rate (day$^{-1}$), and $\eta_1$ and $\eta_2$ include both stochastic disturbances of the DO and BOD dynamics and such irregular phenomena as the stirring up of the benthal deposits.

If $\eta_1$ and $\eta_2$ are ignored, Eqs. (20) and (21) may be combined to eliminate $L$ and yield the relationship

$$D = [k_1 L_0/(k_2 - k_1')](e^{-k_1't} - e^{-k_2 t}) + D_0 e^{-k_2 t} , \tag{22}$$

where $L_0$ is the initial BOD at top of reach (mg/ℓ), and $D_0$ is the initial DO deficit at top of reach (mg/ℓ). Equation (22), the DO "sag" relationship, is listed often in the literature with sedimentation assumed negligible so that $k_1' \approx k_1$. The point of maximum deficit occurs at

$$t_M = \frac{1}{k_1' - k_2} \ln \left\{ \frac{k_1 k_1' L_0}{k_2 k_1 L_0 + k_2(k_1' - k_2) D_0} \right\} \tag{23}$$

and the maximum deficit $D_M$ is

$$D_M = (L_0 k_1/k_2) e^{-k_1' t_M} . \tag{24}$$

Standards of water quality with regard to DO are usually prescribed by water pollution control agencies but requirements can vary widely. Fish requirements are normally most constraining. A typical stream standard is at least 5 mg/ℓ of DO at all times. The greatest difficulty in meeting the standard would be at low water during hot weather.

### B. *WATER QUALITY MANAGEMENT*

Regulatory agencies, in their management of stream and river water quality, may resort to one or more of several alternatives available to them to maintain adequate standards. For example, the agencies may attempt to regulate upstream sources of pollution, if feasible and within their jurisdiction, at economic cost to these sources, or instream aeration augmentation may be considered. The latter alternative is illustrated by Chang and Yeh [7] in their paper on an optimal allocation of artificial aeration along a polluted stream.

It is assumed in the paper that the oxygen sag model of Eq. (22) with $k_1 = k_1'$ is an adequate description of the system without aeration augmentation. Further, the aerators are assumed sufficiently closely spaced along a river reach to preclude the maximum DO deficit point of Eq. (23) existing between them, so that the largest DO deficit between aerators would be at the end of a subreach just before the next aerator. The addition of an aerator at the top of a subreach will modify Eq. (22) for the $i$th subreach to

$$D = \frac{k_1 L_0}{k_2 - k_1} (e^{-k_1 t} - e^{-k_2 t}) + D_0 e^{-k_2 t} - V_i e^{-k_2 (t - t_i)} U(t - t_i) , \tag{25}$$

where $V_i$ is the increment in DO level at top of $i$th subreach due to the aeration augmentation, $t_i$ the distance (time units) of top of $i$th subreach from top of reach and

$$\begin{aligned} U(t - t_i) &= 1 \quad \text{if } t \geq t_i \\ &= 0 \quad \text{if } t < t_i . \end{aligned}$$

Since budgets are often limited and stream standards often arbitrary, the stream managers would find it desirable to have a set of optimal aerator allocations corresponding to a spectrum of budgets and standards. This is accomplished in the following manner.

The objective function is defined as a weighted sum of "costs" denoting deviation from a nominal DO standard and the "costs" of aeration to achieve the resultant DO level. That is, the objective function is

$$\text{minimize } J = \sum_{i=1}^{N} \{w_1 (D_\ell - D_i)^2 + w_2 V_i^2\} \quad , \tag{26}$$

where $w_1$, $w_2 \geq 0$ and $w_1 + w_2 = 1$; $D_\ell$ is the deficit of nominal DO standard with reference to the DO saturation value $C_s$; $D_i$ the maximum DO deficit for $i$th subreach (assumed to occur at end of subreach) and $N$ the number of aerators.

Any other cost functions that may be more realistic can, of course, be specified in place of the above.

If $Y$ represents the total aeration capacity available, the system constraints will be

$$\sum_{i=1}^{N} V_i \leq Y \tag{27}$$

$$0 \leq D_i \leq C_s \qquad i = 1, 2, \ldots, N$$

$$0 \leq V_i \leq C_s \qquad i = 1, 2, \ldots, N \tag{28}$$

and the DO deficit at the end of each subreach,

$$D_i = (D_{i-1} - V_i)\, e^{-k_2 \Delta t} + \frac{k_1 L_0}{k_2 - k_1}\, e^{-(i-1) k_1 \Delta t} \times (e^{-k_1 \Delta t} - e^{-k_2 \Delta t}) \quad , \tag{29}$$

$$i = 1, 2, \ldots, N$$

where $\Delta t$ is the length of subreach in time units (all subreaches assumed of equal length).

Constraint (27) is eliminated by the introduction of a Lagrange multiplier $\lambda$, which can be parameterized provided that the constraint

$$\sum_{i=1}^{N} V_i(\lambda) = Y \tag{30}$$

is substituted. The objective function then becomes

$$\text{minimize} \sum_{i=1}^{N} \{w_1(D_\ell - D_i)^2 + w_2 V_i^2 + \lambda V_i\} \quad . \tag{31}$$

Consequently, various values of $w_1$ and $w_2$ will provide varying emphases on achieving the stream standard or adherence to a limited budget, while varying $\lambda$ will correspond to variation of the allowable budget.

Dynamic programming (DP) has been used to find the set of optimal allocation policies. The $N$ subreaches form $N$ stages, $D_i$ is the state variable and $V_i$ the decision variable at stage $i$. The $D_i$ for a specified value of $V_i$ is given by Eq. (10). The DP recursion equation is

$$\begin{aligned} f_i(D_i, \lambda) = \min_{V_i} \{ & w_1(D_\ell - D_i)^2 + w_2 V_i^2 + \lambda V_i \\ & + f_{i-1}(D_{i-1}, \lambda)\} \ , \end{aligned}$$

where $\lambda$ is a specified value, and $f_0(\ ) = 0$. The constraint (30) need not actually be used. By sweeping through a set of $\lambda$s, a corresponding set of $Y$s are immediately noted by summing the optimal $V_i$s.

Some total aeration capacities in milligrams per liter that were determined for values of $k_1 = 0.16 \text{ day}^{-1}$, $k_2 = 0.66 \text{ day}^{-1}$, $\Delta t = 0.25$ days, $C_s = 8.0$ mg/ℓ, $D_\ell = 3.0$ mg/ℓ, $L_0 = 30$ mg/ℓ,

$N = 24$ are shown in Table 2 (the $D_i$ were not given in the paper).

*Table 2*

*Required Aeration Capacities*

| $\lambda$ | $w_1 = 0.9$ | $w_1 = 0.5$ | $w_1 = 0.1$ |
|---|---|---|---|
| 100 | 5.76 | 5.76 | 5.76 |
| 10 | 5.76 | 5.76 | 5.76 |
| 1 | 6.73 | 6.26 | 5.78 |
| 0.1 | 7.35 | 6.99 | 5.79 |
| 0.01 | 7.35 | 7.04 | 5.83 |
| 0 | 7.35 | 7.06 | 5.83 |

Note that when $\lambda = 0$, there is no limitation on total available capacity. The so-called gap of the Lagrange multiplier approach [Everett [8]) in the resource allocation problem is demonstrated by the table, an optimal allocation for a resource less than 5.76 mg/ℓ not being determinable by this approach. The states $D_i$ were discretized into 11 parts from 0 to $C_s$ at each stage, and each 360/91 computer run took less than 2 sec.

A typical optimal policy and corresponding DO sag profile is shown in Fig. 3. The dashed line gives the DO sag profile with no instream aeration.

*C. MODEL STRUCTURE AND IDENTIFICATION*

The DO and BOD interactions in a river may involve numerous chemical and physical processes, some of which may be significant and some may have only marginal effect. The structuring of a suitable water quality model is important for prediction and control. An inadequate model will result in a distorted

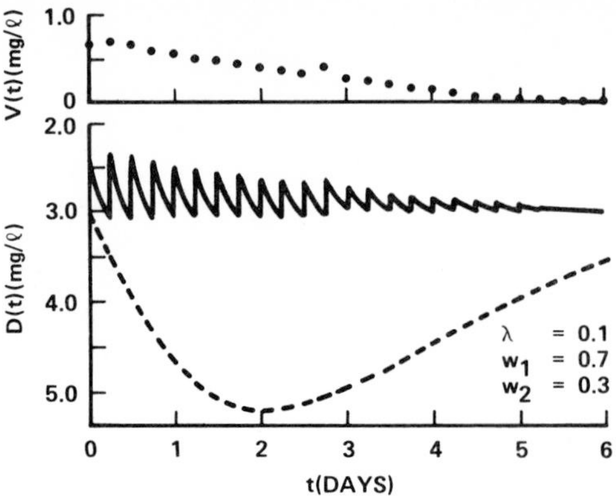


*FIG. 3. Typical optimal aeration capacity allocation policy and corresponding DO sag profile.*

estimation of the model parameters and consequent failure in pollution management. One method of addressing this problem is given by Beck and Young [9], who demonstrate the use of the extended Kalman filter (EKF) in determining whether or not an assumed model is statistically adquate and in indicating the direction of the possible model fixes.

The proposed model structures are based on an idealization of a river reach as shown in Fig. 4, where $u_1(t)$ is the DO concentration at top of reach (mg/ℓ), $u_2(t)$ the BOD concentration at top of reach (mg/ℓ), $x_1(t)$ the DO concentration at

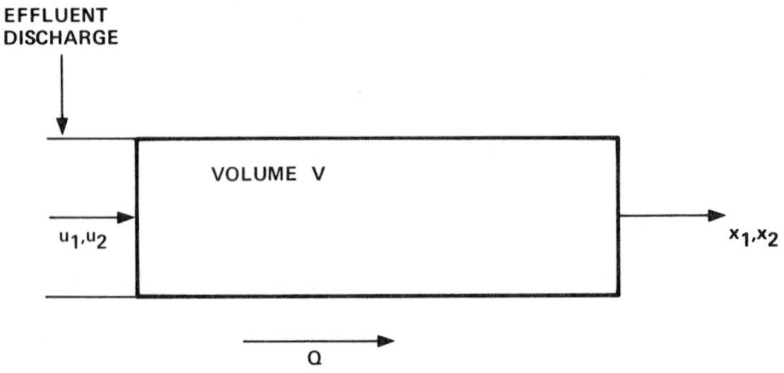


*FIG. 4. Schematic representation of river reach. (After Beck and Young [9].)*

bottom of reach (mg/ℓ), $x_2(t)$ the BOD concentration at bottom of reach (mg/ℓ), $V$ the mean volumetric holdup in reach (cu. ft), and $Q(t)$ the flow rate in reach (cu. ft/day).

Model I assumed that

$$\begin{aligned} \dot{x}_1 &= -[k_2 + \frac{Q}{V}]x_1 - k_1 x_2 + \frac{Q}{V} u_1 + k_2 C_S + \S_1(t) \\ \dot{x}_2 &= -[k_1 + \frac{Q}{V}]x_2 + \frac{Q}{V} u_2 + \S_2(t) \end{aligned} \tag{32}$$

where the $k$s and $C_s$ have the meanings previously defined ($k_1' = k_1$) and §s represent stochastic disturbances of the DO and BOD dynamics, respectively.

The parameters $k_1$ and $k_2$ are lumped with the state variables $x_1$ and $x_2$ and the EKF method is utilized recursively, the state-parameter estimations being updated serially with serial consideration of probably noisy, sampled observations of the system states. If the DO and BOD observations are taken at times $t_k$ and are denoted by $y_{1k}$, $y_{2k}$, respectively, then

$$\begin{aligned} y_{1k} &= x_{1k} + \eta_{1k} \quad , \\ y_{2k} &= x_{2k} + \eta_{2k} \quad , \end{aligned} \tag{33}$$

where the $\eta$s are random measurement errors on the output DO and BOD values.

The EKF method is described in detail by Gelb *et al.* [10]. Since the parameters multiply the state variables in Eq. (32), the EKF algorithm requires the linearization of the equation about current estimates at each step. It can be considered as a predictioncorrection procedure with the correction equation,

$$\hat{x}_k = \hat{x}_{k/k-1} + K_k(y_k - \hat{y}_{k/k-1}) \quad , \tag{34}$$

where $\hat{x}_k$ is the estimate of $x_k$ at $t_k$, $\hat{x}_{k/k-1}$ the *a priori* estimate of $x_k$ at $t_k^-$ given the value $x_{k-1}$ but prior to observation, $y_k$, $\hat{y}_{k/k-1}$ the *a priori* estimate of $y_k$ based on $\hat{x}_{k/k-1}$, and $K$ the matrix with time-varying elements based on the statistical matrices of the §s and ηs in accordance with Kalman filter theory.

Two criteria can be used for deciding the merits of an assumed model structure with the EKF method. Criterion 1 examines the behavior of relatively time-invariant parameters such as $k_1$ and $k_2$, and if the recursive estimates of these parameters indicate approximately stationary qualities, the models are judged adequate. Criterion 2 assumes that § and η are characteristically white noise and the model is judged adequate if the quantities $e_k = y_k - \hat{y}_{k/k-1}$, of the equivalent linear Kalman filter are likewise white noise sequences.

Model I identification results show a failure of criterion 1, as evidenced by Fig. 5. Model inadequacy is clear and $k_2$ even takes on negative values that are physically meaningless.

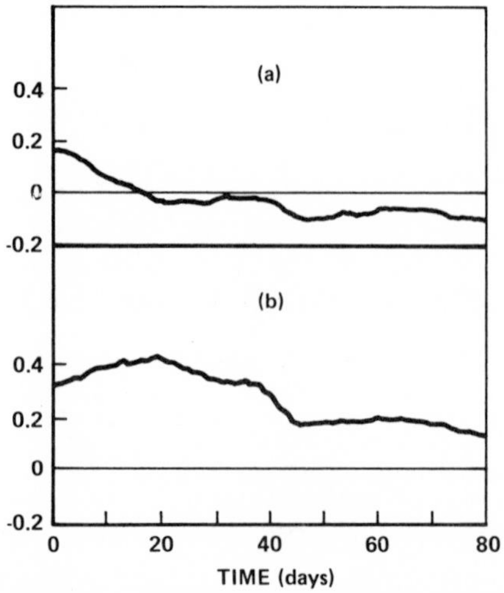


*FIG. 5. Model I identification, recursive parameter estimates: (a) $k_2$ in days$^{-1}$; (b) $k_1$ in days$^{-1}$. (After Beck and Young [9].)*

Model II is more realistic with the inclusion of two additional parameters. $D_B$, which is the net rate of DO addition due to photosynthetic-respiration of plants and algae and decomposition of mud deposits (mg/ℓ), is added to the first equation of (32). $L_A$, which is the rate of increase of BOD due to local runoff, is added to the second equation of (13). Model II parameter estimate results are shown in Fig. 6. The estimates of $k_1$ and $k_2$ show much greater constancy than those for model I. The parameters $D_B$ and $L_A$ have been estimated as random walks; however, these parameters seem now to have assumed the burden of model discrepancies.

Model III examines the actions of the algae in greater detail. The effects of hours of sunlight and river water

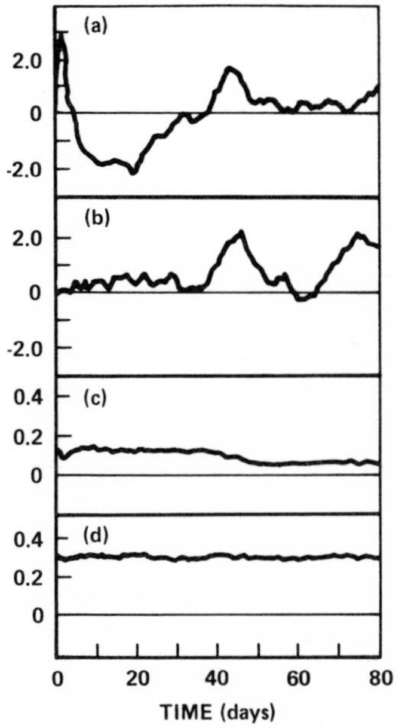


*FIG. 6. Model II identification, recursive parameter estimates: (a) $D_B$, in milligrams per liter per day; (b) $L_A$, in milligrams per liter per day; (c) $k_2$, in days$^{-1}$; (d) $k_1$, in days$^{-1}$. (After Beck and Young [9].)*

temperature are added to the DO equation (for details see Beck and Young [9]), and $D_B$ then represents only DO addition by decomposition of mud deposits so that it would be expected to be somewhat negative. Model III parameter estimate results are given in Fig. 7. Each of the estimate trajectories have been significantly improved, although $L_A$ exhibits disturbing variations in the 70 to 80 day period. Criterion 2 satisfaction is indicated by Fig. 8 in which models I and III residual errors $e_k$ are compared. Model I errors are clearly nonwhite, whereas model III results are, for the most part, nearly so.

As a matter of interest and as a further check on model adequacy, deterministic evaluations of models II and III are compared against each other and against observed data and are shown by Fig. 9.

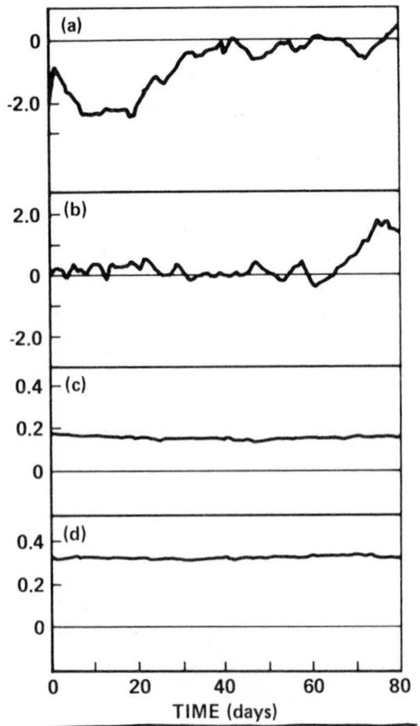


*FIG. 7. Model III identification, recursive parameter estimates: (a) $D_B$, in milligrams per liter per day; (b) $L_A$, in milligrams per liter per day; (c) $k_2$, in days$^{-1}$; (d) $k_1$, in days$^{-1}$. (After Beck and Young [9].)*

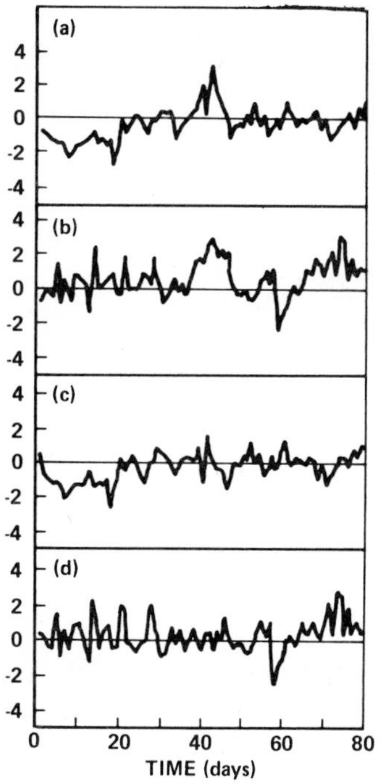


*FIG. 8. Innovation process residual errors $e_k$ for models I and III, in milligrams per liter: (a) model I, DO; (b) model I, BOD; (c) Model III, DO; (d) model III, BOD. (After Beck and Young [9].)*

## IV. APPLICATIONS TO INVERSE PROBLEM IN GROUNDWATER HYDROLOGY

### A. *AQUIFER PARAMETER IDENTIFICATION*

The partial differential equation describing groundwater flow depends, for its practical use, on the evaluation of two parameters, transmissivity (or hydraulic conductivity for an unconfined aquifer) and storage coefficient (or effective porosity for an unconfined aquifer). These parameters are not simply measurable from the physical point of view. However, considerable success has been achieved in the development of numerical models to simulate groundwater flow, provided that parameters are predetermined. Most models utilize either

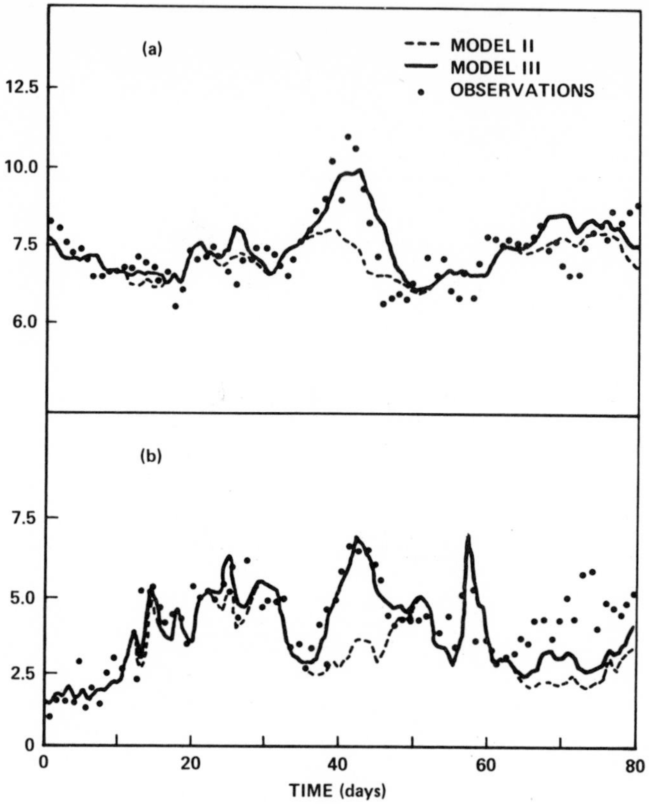


*FIG. 9. Deterministic simulation responses for models II and III, in milligrams per liter: (a) output DO $x_1$; (b) output BOD $x_2$. (After Beck and Young [9].)*

finite difference or finite element approximations to obtain solutions. The lack of a systematic and reliable method for aquifer parameter estimation makes the use of such models for field problems impractical.

It is only recently that attention has been directed toward developing analytical procedures for solving the parameter identification problem. Methods that have been proposed may be classified into two groups: the direct approach and the minimization approach. The direct approach treats the parameters as dependent variables in the form of a formal boundary-value problem. If the head variations are known over the entire flow region, the original governing equation becomes a linear first-order partial differential equation of hyperbolic type in terms of the unknown hydraulic conductivity function. With the aid of boundary conditions and flow data, a unique solution of the equation can be obtained. The minimization approach is an optimization procedure in which the algorithm starts from a set of initial estimates of the parameters and improves it in an iterative manner until the model response is sufficiently close to that of the field observations. The objective is usually the minimization of a norm of the differences between observed and calculated head variations only at the specified observation points. As contrast to the direct approach, the minimization method requires neither the observations of head variations nor its derivatives at every nodal point associated with a numerical discretization scheme. Various approaches to obtain the solution of the problem lead to different identification methods. Jacquard and Jain [11], and Thomas *et al.* [12] developed the methods in line with gradient searching procedures. Yeh and Tauxe [13] and Marino and Yeh [14] showed the use of quasilinearization allied with a finite difference scheme. Lin and Yeh [15] solved the inverse problem by the maximum principle and quasilinearization. Vermuri and Karplus [16] formulated the problem in terms of the optimal control and solved it by a gradient procedure utilizing a hybrid computer. Chen *et al.* [17] also treated the problem in an optimal control approach and solved it by both a steepest descent method and a conjugate gradient method. Slater and Durrer [18] presented the method based on least squares and linear programming. Yeh [19] presented numerical examples using five different approaches, quasilinearization, maximum principle, gradient method, influence coefficient method, and linear programming. Yeh [19] and Chang and Yeh [20] used a quadratic performance criterion subject to lower and upper bounds on parameters to be identified and solved the problem by quadratic programming. Other special considerations used in the minimization approaches are the use of subjective information in the estimation of parameters by Lovell *et al.* [21], the use of a sensitivity equation by Vemuri *et al.* [22], the decomposition and multilevel optimization by Haimes *et al.* [23], and the surrogate parameter approach by Labadie [24]. The method proposed by Coats *et al.* [25], in which the problem was formulated as a linear regression model and solved with the aid of least squares and linear programming technique, may also be considered to belong to this group. In the minimization approach, certain restrictions must be imposed on the function space. The parameter function is restricted to a certain class and the best solution is sought in the limited class. Customarily an aquifer is treated as homogeneously heterogeneous under the assumption that it can be divided into a number of subregions, each of which can be adequatele represented by a single parameter. More recently finite elements were used to represent the unknown parameter function parametrically in terms of nodal values over a suitable discretization of an aquifer (Distefano and Rath [26]; Yoon and Yeh [27]).

### 1. *Formulation of Inverse Problem*

A typical two-dimensional unsteady state groundwater flow in an unconfined aquifer can be represented by the following equation:

$$\frac{\partial}{\partial x}\left[K(x, y)h\frac{\partial h}{\partial x}\right] + \frac{\partial}{\partial y}\left[K(x, y)h\frac{\partial h}{\partial y}\right] = Q + S'\frac{\partial h}{\partial t}\ , \qquad (35)$$

subject to initial and boundary conditions of the following type:

$$h(x, y, 0) = h_0(x, y),\ x, y \in \Omega \ ;$$

$$h(x, y, t) = h_1(x, y, t),\ x, y \in \partial\Omega_1 \ ;$$

$$h\frac{\partial h}{\partial n} = h_n(x, y, t),\ x, y \in \partial\Omega_2 \ ;$$

in which $h(x, y, t)$ is the head in the aquifer, $K(x, y)$ the hydraulic conductivity, $Q$ the source or sink function, $x$, $y$ the horizontal space variables, $t$ the time variable, $S'(x, y)$ the effective porosity or specific storage, $\Omega$ the flow region, and $\partial\Omega$ the boundary of flow region $(\partial\Omega_1 U \partial\Omega_2 = \partial\Omega)$, and $\partial/\partial n$ denotes the normal derivative.

The objective is to determine the hydraulic conductivity $K(x, y)$ and the effective porosity $S'(x, y)$ using time histories of head recorded from a number of observation wells scattered within the aquifer system. It is assumed that the boundary and initial conditions are given.

The performance criterion generally employed for minimization is the standard least squares where the functional to be minimized is

$$J(K, S') = \sum_{\ell \in L} \sum_{t \in T} w_{\ell,t}\,[h_{\ell,t} - h^*_{\ell,t}]^2$$

in which $h_{\ell,t}$ is the computed head at observation well $\ell$ and time $t$, $h^*_{\ell,t}$ the observed head at observation $\ell$ and time $t$, $L$ the index set of observation well, $T$ the index set of time when observations are recorded, and $w_{\ell,t}$ the weighting factor, $0 \leq w_{\ell,t} \leq 1$.

### B. *QUASILINEARIZATION, GRADIENT PROCEDURE, MAXIMUM PRINCIPLE, AND LINEAR AND NONLINEAR ALGORITHMS FOR PARAMETER IDENTIFICATION*

In order not to obscure the trend of thought in a jungle of equations, a one-dimensional problem which is a simplified case based upon Eq. (35), has been analyzed by Yeh [28] by different methods. The basic concepts of some of those methods are developed in the fields of optimal control and operations research. Figure 10 shows the flow configurations.

The governing equation can be expressed as

$$\frac{\partial h}{\partial t} = D \frac{\partial}{\partial x}\left(h \frac{\partial h}{\partial x}\right) \tag{36}$$

subject to the following initial and boundary conditions:

$$h = h(x), \quad 0 \le x \le L, \quad t = 0; \quad h = h_0(t), \quad x = 0, \quad t > 0;$$

$$\frac{\partial h}{\partial x} = 0, \quad x = L, \quad t > 0 \quad ; \tag{37}$$

in which $h$ is the head in the aquifer, $x$ the space variable, $t$ the time variable, and $D$ the diffusivity of the aquifer $(= K/S')$.

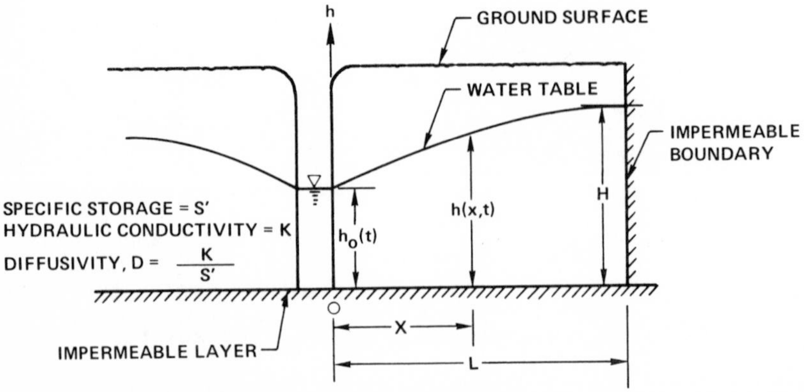


*FIG. 10. System configuration.*

Without losing generality, it has been assumed that the aquifer is homogeneous, isotropic, and overlays on a horizontal impermeable bed. The terms $h_0(t)$ and $\partial h/\partial x$ correspond to the known boundary conditions. To make the dependent variable dimensionless, the following changes in variables are introduced:

$$\Theta = h/H \quad , \quad y = x/L \quad , \quad \text{and} \quad \tau = Ht\ L^2 \tag{38}$$

in which $H$ is the maximum height of the water table above the impermeable layer, a known constant, and $L$ the distance from the river to the water divide, also a known constant. After substitution, Eq. (36) becomes

$$\frac{\partial \Theta}{\partial \tau} = D \frac{\partial}{\partial y}\left(\Theta \frac{\partial \Theta}{\partial y}\right) \tag{39}$$

subject to $\Theta = h(x)/H$, $0 \leq y \leq 1$, $\tau = 0$;

$$\Theta = h_0(t)/H, \quad y = 0, \quad \tau > 0;$$

$$\frac{\partial \Theta}{\partial y} = 0, \quad y = 1, \quad \tau > 0 \quad . \tag{40}$$

The parameter chosen for identification is the diffusivity $D$. It is assumed that observations on $\Theta$ are available at an observation well within the system. The objective is to uncover this unknown parameter, based upon some specified criterion function, along with the given observations and appropriate initial and boundary conditions.

Quasilinearization (Bellman and Kalaba [29]) and maximum principle (Sage [30]) were originally developed in the field of optimal control for systems governed by ordinary differential equations. Therefore, it is necessary to reduce Eq. (39) to a system of ordinary differential equations by means of finite-difference approximations. The space variable $y$ of Eq. (39) is discretized using the central difference scheme, while the time variable $\tau$ is being kept continuous. After simplification, it reduces to

$$\dot{\Theta}_i = D \frac{1}{2(\Delta y)^2} [\Theta_{i+1}^2 - 2\Theta_i^2 + \Theta_{i-1}^2]; \quad i = 1, 2, \ldots, (n-1) \tag{41}$$

in which $\dot{\Theta}_i = d\Theta_i/d\tau$.

To give a simple illustration, the following initial and boundary conditions are assumed:

$$\Theta = 1.0, \; 0 \leq y \leq 1, \quad \tau = 0; \quad \Theta = 0.5, \quad y = 0, \quad \tau > 0;$$

$$\Theta_n = \Theta_{n-1}, \quad y = 1, \quad \tau > 0 \quad . \tag{42}$$

The criterion function is the minimization of the following expression

$$J_0(D) = \sum_{i=1}^{T} [\Theta_m(\tau_i) - \Theta_m^*(\tau_i)]^2 \quad , \tag{43}$$

in which $\Theta_m^*(\tau_i)$ are the observations made from time 0 to $T$ at the $m$th discretized point corresponding to time $\tau_i$, $i = 1, 2, \ldots, T$. A uniform weighting function has been assumed in the least-squares criterion. The minimization is over the proper choice of the parameter $D$ and is subject to the imposed constraints [Eqs. (41) and (42)]. If more than one observation well exists, $J_0(D)$ would contain additional terms of the sum of the squares of errors associated with the available observations. Observation wells can also be weighted by introducing weighting factors to each term of the sum of the squares of errors.

### 1. *Quasilinearization*

The technique of quasilinearization involves the following steps:

(1) Equation (41) is linearized based upon some initial

estimate of the parameter (referred to as $D^o$).

$$\dot{\Theta}_i = D^o \frac{1}{2(\Delta y)^2} (\Theta^{o^2}_{i+1} - 2\Theta^{o^2}_i + \Theta^{o^2}_{i-1})$$
$$+ (\Theta^1_i - \Theta^o_i) [ \frac{D^o}{2(\Delta y)^2} (-4\Theta^o_i)] + (\Theta^1_{i+1} - \Theta^o_{i+1})$$
$$\times [ \frac{D^o}{2(\Delta y)^2} (2\Theta^o_{i+1})] + (\Theta^1_{i-1} - \Theta^o_{i-1}) [ \frac{D^o}{2(\Delta y)^2} (2\Theta^o_{i-1})]\delta$$
$$+ (D^1 - D^o) [ \frac{1}{2(\Delta y)^2} (\Theta^{o^2}_{i+1} - 2\Theta^{o^2}_i + \Theta^{o^2}_{i-1})] \quad ; \tag{44}$$

$i = 1, 2, \ldots, (n-1)$, $\delta = 0$ for $i = 1$,
$\delta = 1$ for $i \neq 1$

in which the superscript 1 represents the new approximation and the subscript 0 denotes the initial approximation.

In Eq. (44), the $\Theta^o$ values are obtained from the solution of Eq. (41) subject to condition (42) using the estimated parameter $D^o$ for the unknown $D$.

(2) The general solution to Eq. (44) is obtained by a linear combination of particular solutions $p$ and $q$

$$\Theta^1_i = D^1 p_i + q_i \; ; \quad i = 1, 2, \ldots, (n-1) \tag{45}$$

in which

$$\dot{p}_i = p_i [ \frac{D^o}{2(\Delta y)^2} (-4\Theta_i)] + p_{i+1} [ \frac{D^o}{2(\Delta y)^2} (2\Theta^o_{i+1})]$$
$$+ p_{i-1} [ \frac{D^o}{2(\Delta y)^2} 2\Theta^o_{i-1}]\delta + \frac{1}{2(\Delta y)^2} (\Theta^{o^2}_{i+1} - 2\Theta^{o^2}_i + \Theta^{o^2}_{i-1}) \; ;$$

$i = 1, 2, \ldots, (n-1)$, $\delta = 0$ for $i = 1$, and (46)
$\delta = 1$ for $i \neq 1$

subject to $p = 0$, $0 \leq y \leq 1$, $\tau = 0$; $p = 0$, $y = 0$, $\tau > 0$;

$$p_n = p_{n-1}, \quad y = 1, \quad \tau > 0 \tag{47}$$

and

$$\dot{q}_i = (q_i - \Theta_i^{\circ})\left[\frac{D^{\circ}}{2(\Delta y)^2}(-4\Theta_i^{\circ})\right] + (q_{i+1} - \Theta_{i+1}^{\circ})$$

$$\times \left[\frac{D^{\circ}}{2(\Delta y)^2}(2\Theta_{i+1}^{\circ})\right] + (q_{i-1} - \Theta_{i-1}^{\circ})\left[\frac{D^{\circ}}{2(\Delta y)^2}(2\Theta_{i-1}^{\circ})\right]\delta;$$

$$i = 1, 2, \ldots, (n-1), \quad \delta = 0 \quad \text{for} \quad i = 1, \text{ and} \tag{48}$$

$$\delta = 1 \quad \text{for} \quad i \neq 1$$

subject to $q = h(x)/H$, $0 \le y \le 1$, $\tau = 0$;

$$q = \frac{h_0(t)}{H}, \quad y = 0, \quad \tau > 0;$$

$$q_n = q_{n-1}, \quad y = 1, \quad \tau > 0 \quad . \tag{49}$$

The general solution at the $m$th discretized point, where the observations are made, is

$$\Theta_m^1 = D^1 \, p_m + q_m \quad . \tag{50}$$

The new estimate $D^1$ is still unknown. Imposing the necessary condition and requiring the function

$$J_1(D) = \sum_{j=1}^{T} [D^1 \, p_m + q_m) - \Theta_m^*(\tau_j)]^2 \tag{51}$$

to be minimum, yields

$$D^1 = \sum_{j=1}^{T} [\Theta_m^*(\tau_j) \, p_m - q_m p_m] / \sum_{j=1}^{T} p_m^2 \quad . \tag{52}$$

Equation (51) is actually the linearized form of Eq. (43) using the linearized general solution for $\Theta_m$. It serves as a means of identifying the unknown. Equations (44)-(52) describe a complete cycle and a new iteration is ready to start with the initial estimate replaced by the new estimate.

### 2. *Maximum Principle*

Since observations are made only at one observation well,

$$J(D) = \int_0^T (\Theta - \Theta^*)^2 \, d\tau \tag{53}$$

can only be minimized over time $\tau$ between 0 and $T$. Thus the Hamiltonian is defined as

$$H^* = (\Theta - \Theta^*)^2 + \lambda(\tau) \left[D \frac{\partial}{\partial y} (\Theta \frac{\partial \Theta}{\partial y})\right] + \mu(\tau)\ 0 \tag{54}$$

in which $\lambda(\tau)$ and $\mu(\tau)$ are the Lagrange multipliers associated with $\Theta$ and $D$. Applying the maximum principle (Lin and Yen [15]), the Hamilton canonic partial differential equations and boundary conditions are obtained as

$$\frac{\partial \Theta}{\partial \tau} = D \frac{\partial}{\partial y} (\Theta \frac{\partial \Theta}{\partial y}) \tag{55}$$

with $\Theta(y, 0) = 1$, $\Theta(0, \tau) = 0.5$, and $\partial\Theta/\partial y = 0$ at $y = 1$.

$$\dot{D} = 0, \quad \mu(0) = 0 \quad ; \tag{56}$$

$$\frac{\partial \lambda}{\partial \tau} = -2(\Theta - \Theta^*), \quad \lambda(T) = 0 \quad ; \tag{57}$$

$$\frac{\partial \mu}{\partial \tau} = - \frac{\partial}{\partial y} (\Theta \frac{\partial \Theta}{\partial y}), \quad \mu(T) = 0 \quad . \tag{58}$$

The finite difference approximations, using a central difference scheme for $y$ only and keeping $\tau$ continuous, for Eqs. (55)-(58)

$$\dot{\Theta}_i = \frac{D}{2(\Delta y)^2} (\Theta_{i+1}^2 - 2\Theta_i^2 + \Theta_{i-1}^2) \tag{59}$$

with $\Theta_i = 1$, $\tau = 0$, $\Theta_0 = 0.5$, $\Theta > 0$, and $\Theta_n = \Theta_{n-1}$.

$$\dot{D} = 0, \quad \mu(0) = 0 \quad ; \tag{60}$$

$$\dot{\lambda} = -2(\Theta_i - \Theta_i^*), \quad \lambda(T) = 0 \quad ; \tag{61}$$

$$\dot{\mu} = - \frac{\lambda}{2(\Delta y)^2} (\Theta_{i+1}^2 - 2\Theta_{i-1}^2 + \Theta_{i-1}^2), \quad \mu(T) = 0 \ ;$$

$$i = 1, 2, \ldots, (n-1) \quad . \tag{62}$$

Equations (59)-(62) can be easily solved by the technique

of quasilinearization. After linearizing Eqs. (59)-(62) (not repeated here), the general solution can be represented by a linear combination of homogeneous and particular solutions. Thus

$$\Theta_i^1 = p_i^\Theta + D^1 q_i^\Theta \quad , \tag{63}$$

$$\lambda^1 = p^\lambda + b^1 u^\lambda + D^1 q^\lambda \quad , \tag{64}$$

and

$$\mu^1 = p^\mu + D^1 a^\mu \quad , \tag{65}$$

in which $p$ and $q$ are particular solutions corresponding to the appropriate variables represented by the superscripts; $u$ is the homogeneous solution; and $b$ the missing initial conditions. In order to satisfy $\dot{D}^1 = 0$ and $\mu^1(T) = 0$, it is easily shown that

$$D^1 = \frac{p^\mu(T)}{q^\mu(T)} \tag{66}$$

To start the procedure, the required initial estimates are $D^0$ and its corresponding solution $\Theta_i^0$ and the estimated missing initial condition for the homogeneous equation. The homogeneous and particular equations are numerically integrated. The new estimate $D^1$, is calculated by Eq. (66). The second iteration is ready to start with initial estimates replaced by the new estimates. Quadratic convergence should be observed, if the algorithm converges (Bellman and Kalaba [29]).

The gradient procedure, the influence coefficient, and the linear programming formulation have the advantage of utilizing an implicit finite difference solution of the governing equation. Discretizing Eq. (39) both in the space and time directions, using central difference and backward difference, respectively, gives

$$\frac{\Theta_{i,k+1} - \Theta_{i,k}}{\Delta\tau} = D \frac{1}{2(\Delta y)^2} (\Theta^2_{i+1,k+1} - 2\Theta^2_{i,k+1} + \Theta^2_{i-1,k+1}) \ ;$$

$$i = 1, 2, \ldots, (n-1) \qquad (67)$$

in which $k$ denotes the time line in terms of $\Delta\tau$ units. Equation (67) is subject to the original initial and boundary conditions given by Eq. (42).

*3. Gradient Method*

Rewriting the criterion function, we have

$$J_0(D) = \sum_{j=1}^{T} [\Theta_m(\tau_j) - \Theta_m^*(\theta_j)]^2 \qquad (68)$$

in which the measurement times $\tau_j$ are assumed to be commensurate with $k\Delta\tau$.

In order to follow the direction of steepest descent, the gradient vector must be calculated. For the nonlinear problem as presented, it is not possible to evaluate theoretically the instantaneous gradient. However, for easier computer implementation, the finite difference approximation is used to compute the gradient

$$\nabla J_0(D) = \frac{J_0(D + \Delta D) - J_0(D - \Delta D)}{2\Delta D} \qquad (69)$$

in which $\Delta D$ is a small fraction of $D$, say 5% of $D$. Equation (69) represents the rate of change of the objective function with respect to a small change of the decision variable.

Specifically, the following steps are involved:

(1) Select an initial estimate for $D$, $D^o$, and solve Eq. (67) three times, corresponding to $D^o$, $D^o + \Delta D^o$, and $D^o - \Delta D$. Solutions are stored at the observation station for all observation times, i.e., $\Theta_m(\tau_j)$.

(2) Using values obtained in step 1 and Eq. (68), the gradient of the objective function with respect to $D$ is determined

by Eq. (69).

(3) Determine the direction $d$. In Cauchy's steepest descent method, the direction is the negative of the gradient. On the other hand, any modified gradient procedure can be used.

(4) Find a scalar $\beta^*$ such that $\beta^*$ minimizes $J_0(D^o + \beta d)$ by performing a one-dimensional search over $\beta$.

(5) Let $D^1 = D^o + \beta^* d$ and repeat, if necessary.

### 4. *The Influence Coefficient Method*

This method is mathematically related to both quasilinearization and gradient methods. The following basic steps are involved:

(1) Initial estimate of the parameter (referred to as $D^o$) is used to obtain solutions of Eq. (67) at the observation station $\Theta_m(\tau_j)$. These solutions are commensurate with $\tau_j$ and the consequent error $\varepsilon_j^o$

$$\varepsilon_j^o = \Theta_m^o(\tau_j) - \Theta_m^*(\tau_j) \tag{70}$$

is determined for all $j$.

(2) Independent unit change is made in $D^o$, say $D^o + \Delta D$, and the following influence coefficient matrix calculated. (In this example, there is only one parameter to be identified. Therefore it would be an array instead of a matrix.)

$$\begin{array}{c|cccc} & \varepsilon_1 & \varepsilon_2 & \cdots & \varepsilon_T \\ \hline D & A_{11} & A_{12} & \cdots & A_{1T} \end{array} \tag{71}$$

in which the elements are the ratios of changes of error with change in the parameter.

(3) Change in the initially estimated value of the parameter for the next iteration is sought

$$D^1 = D^o + \alpha^o \tag{72}$$

in which $\alpha^o$ is to be determined. New assessments of the errors are then given by

$$\varepsilon_j^1 = \varepsilon_j^o + A_{1j}\ \alpha^o \tag{73}$$

in which $A_{1j}$ are given by (71).

(4) Substituting Eq. (73) into the criterion function Eq. (43), imposing the necessary condition for a minimum that the gradient vanish with respect to the unknown $\alpha^o$, and solving for $\alpha^o$ give

$$\alpha^o = \sum_j \varepsilon_j\ A_{1j} / \sum_j\ A_{1j}^2 \quad . \tag{74}$$

Now the new estimate of $D$, $D^1$, can be computed using Eq. (72) and a new iteration is ready to start. Steps (1)-(4) can be repeated until satisfactory convergence is achieved.

### 5. *The Minimax and Linear Programming Approach*

If a uniform weighting function is assumed, the minimax criterion function is

$$\min_D \max_j |\varepsilon_j|; \ j = 1, 2, \ldots, T \tag{75}$$

in which $\varepsilon_j = \Theta_m(\tau_j) - \Theta_m^*(\tau_j)$. Steps (1)-(3) are the same as the influence coefficient method.

To determine $\alpha^o$, Eq. (73) is substituted into the new criterion function, Eq. (75), and manipulated to yield the following linear programming formulation: min $\varepsilon$, subject to

$$\pm\ [\varepsilon_j + \mathrm{A}_{1j}\alpha^o] \le \varepsilon_j; \quad \varepsilon_j \le \varepsilon; \quad \varepsilon_j,\ \varepsilon,\ \alpha \ge 0;$$

$$\text{and} \quad j = 1, 2, \ldots, \ T \ . \tag{76}$$

The preceding standard linear programming representation

has at least 3 constraints. The variables are $\varepsilon_j$, $\alpha^o$, and $\varepsilon$, and the quantities $\varepsilon_j^o$ and $A_{1j}^o$, are computationally known. The solutions that can be obtained from the linear programming algorithm, using the Simplex method, include $\alpha^o$. Thus, the new estimate of the parameter $D$ can be computed using Eq. (72), i.e., $D^1 = D^o + \alpha^o$. A new iteration is ready to enter.

Table 3 shows the generated observations at the fifth discretized point using a value of 1 for parameter $D$. These observations were obtained by a direct numerical integration of Eq. (41) subject to Eq. (42) using the Runge-Kutts method furnished by the IBM scientific subroutine. The space and time increments used were $\Delta y = 0.1$ and $\Delta\tau = 0.002$. Identical results were also observed when Eq. (67) was solved by the Newton-Raphson method for each time line using the same space and time increments.

*Table 3*

*Observed Values of Dimensionless Head at Fifth Discretized Point, D = 1.0*

| $j$ (1) | $\tau_j$ (2) | $\Theta_5^*(\tau_j)$ (3) |
|---|---|---|
| 1 | 0 | 1.000 |
| 2 | 0.1 | 0.905 |
| 3 | 0.2 | 0.838 |
| 4 | 0.3 | 0.772 |
| 5 | 0.4 | 0.728 |
| 6 | 0.5 | 0.692 |
| 7 | 0.6 | 0.663 |
| 8 | 0.7 | 0.639 |
| 9 | 0.8 | 0.618 |
| 10 | 0.9 | 0.601 |
| 11 | 1.0 | 0.587 |

The generated observations and the corresponding parameter used are referred to as the true values. The task now is to uncover the true value of $D$ that was used to generate these observations. An initial estimate of 0.1 was used for parameter *D*. Note that this initial guess is actually 10 times smaller than the true value. Table 4 indicates the results of successive approximations using different methods. In all cases, convergence was obtained within five iterations.

*Table 4*

*Results for D of Successive Approximations by Different Methods*

| Method (1) | Zero (2) | First (3) | Second (4) | Third (5) | Fourth (6) | True D (7) |
|---|---|---|---|---|---|---|
| A[a] | 0.1 | 0.482 | 0.923 | 0.977 | 1.001 | 1.000 |
| B[b] | 0.1 | 0.300 | 0.594 | 0.994 | -- | 1.000 |
| C[c] | 0.1 | 0.493 | 0.864 | 0.990 | 1.000 | 1.000 |
| D[d] | 0.1 | 0.530 | 0.930 | 1.010 | 1.000 | 1.000 |
| E[e] | 0.1 | 0.590 | 0.950 | 1.010 | -- | 1.000 |

[a]*Quasilinearization.*
[b]*Maximum principle.*
[c]*Gradient method (Fletcher-Powell method).*
[d]*The influence coefficient method.*
[e]*The minimax and linear programming approach.*

## A. INVARIANT IMBEDDING AND NONLINEAR BOUNDARY VALUE PROBLEM IN GROUNDWATER

A large class of groundwater flow problems involve solving differential equations that are of two-point boundary-value type. Most problems encountered are nonlinear and beset by various difficulties. Numerically, there exists no convenient technique for estimating the missing initial conditions, and a trial-and-error method is generally used. The concept of invariant imbedding developed in the field of optimal control (Bellman and Kalaba [29]; Lee [32]) is especially suited for estimating missing initial conditions. It is usually allied with the technique of quasilinearization in obtaining solutions for such problems.

Bellman *et al.* [31] have recently proposed to use the invariant imbedding concept for the mathematical modeling, identification, and analysis of groundwater aquifer systems. It has been shown by Bellman *et al.* [31] that under certain conditions Eq. (1) can be approximated by a system of ordinary differential equations in the following form:

$$dX/dt = f(\underset{\sim}{x}, t) \quad , \tag{77}$$

where $\underset{\sim}{x}$ and $f$ are $(n + m)$-dimensional vectors with components $y_1, y_2, \ldots, y_n, p_1, p_2, \ldots, p_m$, and $f_1, f_2, \ldots, f_n$, 0, 0, ..., 0, respectively. The $y$s and $p$s correspond to the unknown hydraulic conductivities and the unknown parameters in the system that may include the unknown sink or source functions. It will be assumed that not all the state variables can be measured and some of the $y$s can be measured only in certain combinations with other variables. Let

$$\underset{\sim}{z}(t) = \underset{\sim}{h}(\underset{\sim}{x}, t) + \text{(measurement errors)} \tag{78}$$

with $0 \leq t \leq t_f$. The vectors $\underset{\sim}{z}$ and $\underset{\sim}{h}$ are $q$-dimensional vectors. The number $q$ represents the number of measurable quantities. $\underset{\sim}{h}$ represents the true value of the measurable quantities and $\underset{\sim}{z}$ represents the measured value with noise.

Our problem is: on the basis of the measurements of observations $z(t)$, $0 \leq t \leq t_f$, estimate the $(n + m)$ values

$$\underset{\sim}{x}(t) = \underset{\sim}{c} \tag{79}$$

for Eq. (77), such that the least-squares expression

$$J = \int_0^t \sum_{j=0}^{n+m} (z_j(t) - h_j(\underset{\sim}{x}, t))^2 \, dt \tag{80}$$

is minimized. Instead of the simple least-squares expression, the following weighted criterion also can be used:

$$J = \int_0^t [\underset{\sim}{z}(t) - \underset{\sim}{h}(\underset{\sim}{x}, t)]^T \underset{\sim}{Q}(t)[\underset{\sim}{z}(t) - \underset{\sim}{h}(\underset{\sim}{x}, t)]dt \quad , \tag{81}$$

where $[\underset{\sim}{z} - \underset{\sim}{h}]^T \underset{\sim}{Q}(\underset{\sim}{z} - \underset{\sim}{h})$ represents the quadratic form associated with the matrix $Q(t)$, which is a symmetric $q \times q$ matrix. The expansion of this quadratic form leads to a weighted sum of squares of the elements of $[\underset{\sim}{z} - \underset{\sim}{h}]$, with the weighting determined by the elements of $Q$. Notice that if $Q(t)$ is a unit matrix, Eq. (81) reduces to Eq. (80). By choosing the elements in $Q$ suitably, the observations on any variable can be made more important than observations on any other variable.

By the use of invariant imbedding, the estimator equations for the above problem can be obtained easily (Bellman *et al.* [31]):

$$\frac{d\underset{\sim}{e}}{da} = \underset{\sim}{f}(\underset{\sim}{e}, a) + \underset{\sim}{q}(a)[\underset{\sim}{h}_{\underset{\sim}{e}}(\underset{\sim}{e}, a)]^T \underset{\sim}{Q}(a)[\underset{\sim}{z}(a) - \underset{\sim}{h}(\underset{\sim}{e}, a)] \tag{82}$$

$$\begin{aligned}\frac{d\underset{\sim}{q}}{da} &= \underset{\sim}{f}_{\underset{\sim}{e}}(\underset{\sim}{e}, a)\underset{\sim}{q}(a) + q(a)[f_e(e, a)]^T \\ &\quad + \underset{\sim}{q}(a)\underset{\sim}{h}_{\underset{\sim}{e}\underset{\sim}{e}}(\underset{\sim}{e}, a)\underset{\sim}{Q}(a)[\underset{\sim}{z}(a) - \underset{\sim}{h}(\underset{\sim}{e}, a)]\underset{\sim}{q}(a) \\ &\quad - \underset{\sim}{q}(a)[\underset{\sim}{h}_{\underset{\sim}{e}}(\underset{\sim}{e}, a)]^T \underset{\sim}{Q}(a)\underset{\sim}{h}_{\underset{\sim}{e}}(\underset{\sim}{e}, a)\underset{\sim}{q}(a) \quad ,\end{aligned} \tag{83}$$

where $\underset{\sim}{e}$ is the optimal estimate for $\underset{\sim}{x}$ at $t = a$ and $\underset{\sim}{q}$ is a weighting vector. Thus, by the simultaneous solution of Eqs. (82) and (83), the optimal estimates for $x$ can be obtained. Notice that Eqs. (82) and (83) are simple ordinary differential equations of the initial-value type. Because of their fast convergence rate, they form powerful forecasting equations.

The above is only a simple illustration of how the proposed technique works by the use of invariant imbedding. Notice that the parameters and the variables are estimated easily by solving a system of ordinary differential equations. The solution is straightforward, no iteration or trial-and-error is involved. The invariant imbedding approach can also be used to perform sensitivity analysis as well as the identification of the mathematical structure of the groundwater system.

REFERENCES

1. L. BECKER and W. W-G. YEH. "Optimization of Real Time Operation of a Multiple Reservoir System," *Water Resources Res. 10*, No. 6, 1107-1112 (1974).
2. R. H. HICKS *et al.* "Large Scale Nonlinear Optimization of Energy Capability for the Pacific Northwest Hydroelectric System." Presented at the IEEE PES Winter Meeting, New York City, January 1974.
3. R. FLETCHER and C. M. REEVES. "Function Minimization by Conjugate Gradients," *Computer J.* 149-154 (1964).
4. W. J. TROTT and W. W-G. YEH. "Optimization of Multiple Reservoir System," *J. Hydr. Div. ASCE 99(HY10)*, 1865-1884 (1973).
5. R. E. BELLMAN and S. E. DREYFUS. "Applied Dynamic Programming," Princeton University Press, Princeton, New Jersey, 1962.
6. H. W. STREETER and E. B. PHELPS. U.S. Public Health Bull. 146 (1925).
7. S. CHANG and W. W-G YEH. "Optimal Allocation of Artificial Aeration Along a Polluted Stream Using Dynamic Programming," *Water Resources Bull. 9*, No. 4, 985-997 (1973).
8. H. EVERETT. "Generalized Lagrange Method for Solving Problems of Optimum Allocation of Resources," *Operations Res. J. 11*, 399-417 (1963).
9. B. BECK and P. YOUNG. "Systematic Identification of DO-BOD Model Structure," *J. Environ. Eng. Div. ASCE 102*, No. EE5, 909-927 (1976).
10. A. GELB *et al.* "Applied Optimal Estimation," pp. 180-191. MIT Press, Cambridge, Massachusetts, 1974.
11. P. JACKQUARD and C. JAIN. "Permeability Distribution from Field Pressure Data," *Soc. Pet. Eng. J. 5(4)*, 281-294 (1965).
12. L. K. THOMAS, L. J. HELLUMS, and G. M. REHEIS. "A Nonlinear Automatic History Matching Technique for Reservoir Simulation Models," *Soc. Pet. Eng. J. 12(6)*, 508-514 (1972).
13. W. W-G YEH and G. W. TAUXE. "Optimal Identification of Aquifer Diffusivity Using Quasilinearization," *Water Resources Res. 7(4)*, 955-962 (1971).
14. M. A. MARINO and W. W-G YEH. "Identification of Parameters in Finite Leaky Aquifer Systems," *J. Hydr. Div. ASCE 99(HY2)*, 319-336 (1973).
15. A. C. LIN and W. W-G YEH. "Identification of Parameters

in an Inhomogeneous Aquifer by Use of the Maximum Principle of Optimal Control and Quasilinearization," *Water Resources Res. 10(4)*, 829-838 (1974).

16. V. VEMURI and W. J. KARPLUS. "Identification of Nonlinear Parameters of Groundwater Basin by Hybrid Computation," *Water Resources Res. 5(1)*, 172-185 (1969).
17. W. H. CHEN, G. R. GAVALAS, J. H. SEINFELD, and M. L. WASSERMAN. "A New Algorithm for Automatic Historic Matching," *Soc. Pet. Eng. J. 14(6)*, 593-608 (1974).
18. G. E. SLATER and E. J. DURRER. "Adjustment of Reservoir Simulation Models to Match Field Performance," *Soc. Pet. Eng. J. 11(3)*, 295-305 (1971).
19. W. W-G YEH. "Aquifer Parameter Identification," *J. Hydr. Div. ASCE 101(HY9)*, 1197-1209 (1975).
20. S. CHANG and W. W-G YEH. "Invariant Imbedding and Unsteady Groundwater Flow," *J. Hydr. Div. ASCE 100(HY10)*, 1343-1352 (1974).
21. R. E. LOVEL, L. DUCKSTEIN, and C. C. KISIEL. "Use of Subjective Information in Estimation of Aquifer Parameters," *Water Resources Res. 8(3)*, 680-690 (1972).
22. V. VEMURI, J. A. DRACUP, R. C. ERDMANN, and N. VERMURI. "Sensitivity Analysis Method of System Identification and its Potential in Hydrologic Research," *Water Resources Res. 5(2)*, 341-349 (1969).
23. Y. Y. HAIMES, R. L. PERRINE, and D. A. WISMER. "Identification of Aquifer Parameters by Decomposition and Multilevel Optimization," *Water Resources Center Contribution No. 123*, University of California Los Angeles, 1968.
24. J. W. LABADIE. "Decomposition of a Large Scale Nonconvex Parameter Identification Problem in Geohydrology," Report No. ORC 72-73, Operations Research Center, University of California, Berkeley, 1972.
25. K. H. COATS, J. R. DEMPSEY, and J. H. HENDERSON. "A New Technique for Determining Reservoir Description from Field Performance Data," *Soc. Pet. Eng. J. 10(1)*, 66-74 (1970).
26. N. DISTEFANO and A. RATH, "An Identification Approach to Subsurface Hydrological Systems," *Water Resources Res. 11(6)*, 1005-1012 (1975).
27. Y. S. YOON and W. W-G. YEH. "Parameter Identification in an Inhomogeneous Medium with Finite Element Method," *Soc. Pet. Eng. J.* 217-226 (1976).
28. W. W-G. YEH. "Optimal Identification of Parameters in an Inhomogeneous Medium with Quadratic Programming," *Soc. Pet. Eng. J. 15(5)*, 371-375 (1975).
29. R. BELLMAN and R. KALABA. "Dynamic Programming Invariant Imbedding and Quasilinearization: Comparison and Interconnections," RM-4038-PR, RAND Corp., Santa Monica, California, March 1964.
30. A. P. SAGE. "Optimal Systems Control," Prentice-Hall, Englewood Cliffs, New Jersey, 1968.
31. R. BELLMAN, S. E. LEE, and W. W-G. YEH. "Invariant Imbedding and Nonlinear Boundary Value Problems in Groundwater Modeling," Proposal submitted to National Science Foundation, November 1976.
32. E. S. LEE. "Quasilinearization and Invariant Imbedding," Academic Press, New York, 1968.

*ADDITIONAL REFERENCES*

Y. BARD. "Nonlinear Parameter Estimation," Academic Press, New York, 1974.

R. BELLMAN, H. H. KAJIWADA, and R. E. KALABA. "Quasilinearization, Boundary-Value Problems, and Linear Programming," *IEEE Trans. Auto. Control AC-10*, No. 2, 199 (1965).

R. E. BELLMAN and R. E. KALABA. "Quasilinearization and Nonlinear Boundary-Value Problems," Elsevier Publishing Co., Amsterdam, The Netherlands, 1965.

S. CHANG and W. W-G. YEH. "A Proposed Algorithm for the Solution of the Large Scale Inverse Problem in Ground Water," *Water Resourses Res. 12(3)*, 365-374 (1976).

Y. EMSELLEM and G. deMARSILY. "An Automatic Solution for the Inverse Problem," *Water Resources Res. 7(5)*, 1264-1283 (1971).

E. O. FRIEND and G. F. PINDER. "Galerkin Solution of the Inverse Problem for Aquifer Transmissivity," *Water Resources Res. 9(5)*, 1397-1410 (1973).

H. P. JAHNS. "A Rapid Method for Obtaining a Two-Dimensional Reservoir Description from Well Response Data," *Soc. Pet. Eng. J. 6(4)*, 315-327 (1966).

H. H. KAGIWADA. "System Identification: Methods and Applications," Addison-Wesley, Reading, Massachusetts, 1974.

D. KLEINECKE. "Use of Linear Programming for Estimating Geohydrologic Parameters of Groundwater Basins," *Water Resources Res. 7(2)*, 367-375 (1971).

G. W. LABADIE. "A Surrogate-Parameter Approach to Modeling Groundwater Basins," *Water Resources Bull. 11(1)*, 97-114 (1975).

D. McLAUGHLIN. "Investigation of Alternative Procedures for Estimating Ground-Water Basin Parameters," Prepared for OWRT, USDI, Resources Engineers, Walnut Creek, California, 1975.

R. W. NELSON. "In Place Determination of Permeability Distribution for Heterogeneous Porous Media through Analysis of Energy Dissipation," *Soc. Pet. Eng. J. 8(1)*, 33-42 (1968).

R. W. NELSON and W. L. McCOLLUM. "Transient Energy Dissipation Methods of Measuring Permeability Distributions in Heterogeneous Porous Materials," Rep. CSC 691229, Water Resour. Div., USGS, Washington, D.C. 1969.

R. W. NELSON. "Conditions for Determining a Real Permeability Distribution by Calculation," *Soc. Pet. Eng. J. 2(3)*, 223-224 (1962).

S. P. NEUMAN. "Calibration of Distributed Parameter Groundwater Flow Models Viewed as a Multiple Objective Decision Process Under Uncertainty," *Water Resources Res. 9(4)*, 1006-1021 (1973).

D. A. NUTBROWN. "Identification of Parameters in a Linear Equation of Groundwater Flow," *Water Resources Res. 11(4)*, 581-588 (1975).

B. SUGAR, S. YAKOWITZ, and L. DUCKSTEIN. "A Direct Method for the Identification of the Parameters of Dynamic Nonhomogeneous Aquifers," *Water Resources Res. 11(4)*, 563-570 (1975).

R. W. VEATCH, Jr. AND G. W. THOMAS. "A Direct Approach for History Matching," Paper SPE 3515 Presented at SPE-AIME 46th Annual Fall Meeting, New Orleans, Louisiana, October 3-6, 1971.

# Sensitivity Analysis and Optimization of Large Scale Structures

J. S. ARORA AND E. J. HAUG, JR.

*Division of Materials Engineering*
*College of Engineering*
*The University of Iowa*
*Iowa City, Iowa*






ISBN 0-12-012715-6

## I. INTRODUCTION

The purpose of this chapter is to present an optimal structural design technique for large-scale structures that integrates the concept of substructuring into its formulation. While substructuring is commonly used for large-scale structural analysis, no general method of optimization that incorporates substructuring has yet been presented in the literature. This is partly due to the fact that design sensitivity analysis with substructuring, which is required in any optimal design algorithm, involves a large number of calculations. No efficient method for such calculations has been available in the past. Recently, however, an efficient method for design sensitivity analysis with substructuring has been developed by Govil and Arora [1]. The method was developed by using adjoint variables and ideas presented by Noor and Lowder [2] for calculating displacement sensitivity vectors with substructuring.

The displacement method of structural analysis, linear expansions of various expressions, the design sensitivity analysis of Govil and Arora [1], and Kuhn-Tucker necessary conditions of nonlinear programming are employed to develop the technique. Incorporation of the substructuring concept in analysis essentially means partitioning of the nodal coordinates into two subsets: (1) the boundary coordinates, and (2) the interior coordinates. The equilibrium equations for the structure are also partitioned accordingly. The present method exploits these partitioned equations both in the structural analysis and in the optimal design procedures. The boundary and interior coordinate concept is effectively integrated to develop the design sensitivity vectors for optimization. The method is reduced to an iterative algorithm.

To illustrate applicability of the method, a computer program is developed to optimize general trusses. The program is used to optimize a 25-member transmission tower and a 200-member plane truss that are subjected to multiple loading conditions. The results obtained with substructuring are compared with the results obtained without substructuring.

During the past 15 yr, considerable research has been done in the area of optimal design of structures, using primarily nonlinear programming methods. Recent books by Pope and Schmit [3], and Gallagher and Zienkiewicz [4] summarize the state of the art quite nicely. The algorithm presented in this chapter is motivated by the optimal control techniques of Bryson and his co-workers [5], where a clear distinction is made between the state variables (which describe state of the system), and the design variables (which describe the system) of the problem. An algorithm based on these concepts [6] has been applied successfully to a wide range of mechanical and structural system optimal design problems [7]-[10].

Kirsch *et al.* [11] have also considered the problem of optimum design of structures by partitioning it into a number of substructures. Their procedure is to sequentially optimize each substructure, with respect to its own design variables, while keeping other design variables constant. The iterative procedure is continued until no further improvements are possible. This procedure is inefficient and can lead to nonoptimal solutions, as pointed out by the authors. In contrast, the procedure presented in this chapter considers all substructures simultaneously, and allows all design variables to vary at the same time. Thus the procedure is quite general and efficient.

## II. THE DESIGN PROBLEM

A general problem of optimal design of structures may be defined as follows: find a vector $b$ that minimizes a cost function

$$\bar{J} = \bar{J}(b, z) \tag{1}$$

satisfying the equilibrium equation (state equation)

$$K(b)z = S(b) \tag{2}$$

and subject to the constraints

$$\phi^s(b, z) \leq 0 \tag{3}$$

and

$$\phi^d(b) \leq 0 \quad , \tag{4}$$

where $b$ is a vector of $m$ design variables, such as cross-sectional areas, moments of inertia, thickness, etc.; $z$ is a state variable vector of $n$ nodal displacements; $n$ is the number of degrees of freedom of the structure; $K(b)$ is a $(n,n)$-structural stiffness matrix; $S(b)$ is a vector of loads on the structure; and $\phi^s(b, z) \varepsilon R^{\alpha_1}$ and $\phi^d(b) \varepsilon R^{\alpha_2}$ are vector functions. Therefore, the total number of constraints is $\alpha = \alpha_1 + \alpha_2$. The inequalities in Eqs. (3) and (4) apply to each component of the vector functions. Multiple loading conditions for the structure are treated by simply expanding the state variable vector $z$.

The cost function of Eq. (1) is quite general and may represent weight of the structure, displacements of critical points of the structure, and certain critical member forces. The cost function depends only on the design variables, if it represents weight of the structure. The inequality (3) represents constraints that depend upon state and design variables, which may include member stress and the nodal displacement constraints. The inequality (4) represents constraints that depend only on design variables, which are generally explicit bounds on the design variables or relationships between them.

A hindrance in both structural analysis and optimization is the dimension of the state variable and matrix $K$ in Eq. (2). To alleviate this problem, one may employ the substructuring technique, in which vector $z$ (and hence, $S$) is partitioned into a boundary vector $z_B$ and an interior vector $z_I$. The state equation Eq. (2) is then partitioned as

$$\begin{bmatrix} K_{BB} & K_{BI} \\ K_{IB} & K_{II} \end{bmatrix} \begin{bmatrix} z_B \\ z_I \end{bmatrix} = \begin{bmatrix} S_B \\ S_I \end{bmatrix} , \tag{5}$$

where the subscripts B and I refer to boundary and interior quantities for all substructures $z_B$ and $S_B \varepsilon R^N$; $z_I$ and $S_I \varepsilon R^{(n-N)}$; and $N$ is the number of boundary degrees of freedom. The matrices $K_{BB}$, $K_{BI}$, $K_{IB}$, and $K_{II}$ have the required compatible dimension and these matrices, along with vectors $S_B$ and $S_I$, are understood to be functions of the design vector.

Following the substructuring approach, the interior displacements $z_I$ are eliminated from the first line of Eq. (5) using the second line and the following reduced equation is obtained:

$$K_B z_B = F_B \tag{6}$$

where

$$K_B = K_{BB} + K_{BI} Q \tag{7}$$

$$F_B = S_B + Q^T S_I \tag{8}$$

$$Q = - K_{II}^{-1} K_{IB} \quad . \tag{9}$$

Once the boundary displacements $z_B$ have been computed from Eq. (6), the interior displacements $z_I$ can be computed. One first computes the boundary stiffness matrix $K_B$ and the force vector $F_B$ by considering contributions from each substructure, as follows [12]:

$$K_B = \sum_{r=1}^{L} \beta^{(r)^T} K_B^{(r)} \beta^{(r)} \quad ; \tag{10}$$

$$F_B = S_B + \sum_{r=1}^{L} \beta^{(r)^T} Q^{(r)^T} S_I^{(r)} \quad ; \tag{11}$$

where $\beta^{(r)}$ is a Boolean transformation matrix, the superscript $r$ represents the $r$th substructure, the matrices $K_B^{(r)}$ and $Q^{(r)}$ have expressions similar to those given in Eqs. (7)-(9), and $L$ is the total number of substructures. The interior displacements may then be computed by considering each substructure separately, as follows:

$$z_I^{(r)} = \left[ K_{II}^{(r)} \right]^{-1} \left[ S_I^{(r)} \quad -K_{IB}^{(r)} \quad z_B^{(r)} \right] \quad . \tag{12}$$

The forces in various members of the structure are computed from the nodal displacements, using well-known procedures [12].

In terms of the vectors $z_B$ and $z_I$, the optimal design problem may now be defined as follows: find a design vector $b$ that minimizes the cost function

$$J = J(b, z_B, z_I) \tag{13}$$

satisfying the equilibrium equation (5), and subject to the constraints

$$\phi^s(b, z_B, z_I) \leq 0 \tag{14}$$

and the constraints of Eq. (4).

## III. DESIGN SENSITIVITY ANALYSIS

In preparation for an iterative design procedure, it is assumed that an engineering estimate of the design variable $b$ has been made and the associated values of $z_B$ and $z_I$ determined from Eqs. (6) and (12). In order to find an improved design, one must first determine the effect, or sensitivity, due to a small change $\delta b$ in design. Due to the well-posed nature of the structural analysis problem, the state variables $z_B$ and $z_I$ will be changed by small amounts. All equations of the problem can now be linearized using a Taylor series expansion.

In the following development, let a ~ over the constraint functions represent inclusion of only the 'ε-active' constraints [7]. A constraint $\phi_i \leq 0$ is said to be ε-active if it satisfies the inequality $\phi_i + \varepsilon \geq 0$, where ε is a small positive number. Linear expansions of the cost function of Eq. (13) and the constraints of Eqs. (14) and (4) yield the following expressions:

$$\delta J = \frac{\partial J}{\partial b}\,\delta b + \frac{\partial J}{\partial z_B}\,\delta z_B + \frac{\partial J}{\partial z_I}\,\delta z_I \quad , \tag{15}$$

$$\delta\tilde{\phi}^s = \frac{\partial\tilde{\phi}^s}{\partial b}\,\delta b + \frac{\partial\tilde{\phi}^s}{\partial z_B}\,\delta z_B + \frac{\partial\tilde{\phi}^s}{\partial z_B}\,\delta z_I \quad , \tag{16}$$

$$\delta\tilde{\phi}^d = \frac{\partial\tilde{\phi}^d}{\partial b}\,\delta b \quad , \tag{17}$$

where all derivatives are evaluated at the known values of $b$, $z_B$, and $z_I$. Now in order to determine design sensitivity, one would like to express terms containing $\delta z_B$ and $\delta z_I$ as functions of $\delta b$. For this purpose, let us write first-order Taylor's expansion of Eq. (5):

$$K_{BB}\ \delta z_B + K_{BI}\ \delta z_I = C_1\ \delta b \quad , \tag{18}$$

$$K_{IB}\ \delta z_B + K_{II}\ \delta z_I = C_2\ \delta b \quad , \tag{19}$$

where

$$C_1 = \frac{\partial S_B}{\partial b} - \frac{\partial}{\partial b}\,(K_{BB} z_B) - \frac{\partial}{\partial b}\,(K_{BI} z_I) \tag{20}$$

and

$$C_2 = \frac{\partial S_I}{\partial b} - \frac{\partial}{\partial b}\,(K_{IB} z_B) - \frac{\partial}{\partial b}\,(K_{II} z_I) \quad . \tag{21}$$

Using Eq. (19) and the matrices $Q$ and $K_B$ calculated during structural analysis, $\delta z_I$ is eliminated from Eq. (18) to obtain

$$K_B\ \delta z_B = C\ \delta b \quad , \tag{22}$$

where

$$C = C_1 + Q^T C_2 \quad . \tag{23}$$

Equation (22) represents a linearized version of the state equation (6).

Now, a vector $\lambda_I$ and a matrix $\bar{\lambda}_I$ are defined to be solutions of the following equations:

$$K_{II}\lambda_I = \frac{\partial J}{\partial z_I}^T \tag{24}$$

$$K_{II}\bar{\lambda}_I = \frac{\partial \tilde{\phi}^s}{\partial z_I}^T \quad , \tag{25}$$

which may be readily solved since $K_{II}$ was decomposed during structural analysis.

Taking the transpose of Eqs. (24) and (25), postmultiplying by $\delta z_I$, and substituting from Eq. (19), one obtains

$$\lambda_I^T\ [C_2\ \delta b - K_{IB}\ \delta z_B] = \frac{\partial J}{\partial z_I}\ \delta z_I \quad , \tag{26}$$

and

$$\bar{\lambda}_I^T\ [C_2\ \delta b - K_{IB}\ \delta z_B] = \frac{\partial \tilde{\phi}^s}{\partial z_I}\ \delta z_I \quad . \tag{27}$$

The right-hand side of Eqs. (26) and (27) are exactly the terms needed in Eqs. (15) and (16) to eliminate $\delta z_I$. Substituting Eqs. (26) and (27) into Eqs. (15) and (16), one obtains

$$\delta J = \left[ \frac{\partial J}{\partial b} + \lambda_I^T C_2 \right] \delta b + \left[ \frac{\partial J}{\partial z_B} - \lambda_I^T K_{IB} \right] \delta z_B \tag{28}$$

and

$$\delta \tilde{\phi}^s = \left[ \frac{\partial \tilde{\phi}^s}{\partial b} + \bar{\lambda}_I^T C_2 \right] \delta b + \left[ \frac{\partial \tilde{\phi}^s}{\partial z_B} - \bar{\lambda}_I^T K_{IB} \right] \delta z_B \quad . \tag{29}$$

Similarly, a vector $\lambda_B$ and a matrix $\bar{\lambda}_B$ are defined to be solutions of the following equations:

$$K_B \lambda_B = (\partial J^T / \partial z_B) - K_{BI} \lambda_I \quad ; \tag{30}$$

$$K_B \bar{\lambda}_B = (\partial \tilde{\phi}^{s^T} / \partial z_B) - K_{BI} \bar{\lambda}_I \quad . \tag{31}$$

One may substitute for $\lambda_I$ and $\bar{\lambda}_I$ from Eqs. (24) and (25) into Eqs. (30) and (31), respectively, to obtain

$$K_B \lambda_B = \frac{\partial J^T}{\partial z_B} + Q^T \frac{\partial J^T}{\partial z_I} \quad , \tag{32}$$

and

$$K_B \bar{\lambda}_B = \frac{\partial \tilde{\phi}^{s^T}}{\partial z_B} + Q^T \frac{\partial \tilde{\phi}^{s^T}}{\partial z_I} \quad . \tag{33}$$

Since the matrix $K_B$ was decomposed during structural analysis, these equations are efficiently solved.

Taking the transpose of Eqs. (30) and (31), postmultiplying by $\delta z_B$, and substituting from Eq. (22), one obtains

$$\lambda_B^T C \, \delta b = \left[ \frac{\partial J}{\partial z_B} - \lambda_I^T K_{IB} \right] \delta z_B \quad , \tag{34}$$

$$\bar{\lambda}_B C \, \delta b = \left[ \frac{\partial \tilde{\phi}^s}{\partial z_B} - \bar{\lambda}_I^T K_{IB} \right] \delta z_B \quad . \tag{35}$$

The right-hand sides of Eqs. (34) and (35) appear in Eqs. (28) and (29). Making this substitution, Eqs. (28) and (29) reduce to

$$\delta J = \Lambda^{J^{\mathrm{T}}} \delta b \quad , \tag{36}$$

$$\delta \tilde{\phi}^{s} = \Lambda^{s^{\mathrm{T}}} \delta b \quad , \tag{37}$$

where

$$\Lambda^{J} = \frac{\partial J^{\mathrm{T}}}{\partial b} + C_2^{\mathrm{T}} \lambda_{\mathrm{I}} + C^{\mathrm{T}} \lambda_{\mathrm{B}} \tag{38}$$

and

$$\Lambda^{s} = \frac{\partial \tilde{\phi}^{s^{\mathrm{T}}}}{\partial b} + C_2^{\mathrm{T}} \overline{\lambda}_{\mathrm{I}} + C^{\mathrm{T}} \overline{\lambda}_{\mathrm{B}} \quad . \tag{39}$$

Equations (36) and (37) represent the desired sensitivity information needed for structural design. These equations tell the designer what the effect a candidate design change will have on his cost function and constraint functions. Perhaps more important, they provide a method for rational optimization. Virtually any optimization method to be applied requires this derivative information. A steepest-descent method preferred by the authors is used in the following sections to solve illustrative examples.

Before proceeding to the optimization phase, it is important to note that the design sensitivity data contained in the matrices of Eqs. (38) and (39) are quite efficiently calculated, since matrix equations previously solved in the finite element analysis phase have been utilized, with only a change in the nonhomogeneous right-hand side. The sensitivity analysis is thus carried out with a minimum of computational effort. Details of this computation are presented in Govil and Arora [1].

## IV. DESIGN OPTIMIZATION

With the sensitivity data (or design derivatives) of the preceding section, one is now in a position to undertake design optimization. A steepest-descent method is employed here [6]-[10]. It is a "design improvement" oriented algorithm that simultaneously seeks to correct constraint errors, while moving in a constrained steepest-descent direction.

A restriction is placed on the linearized constraint functions by requiring that the design change $\delta b$ be computed in such a manner that it corrects or reduces error in all the violated constraints. These requirements on Eqs. (17) and (37) can be stated by the following inequalities:

$$\Lambda^{s^T} \delta b \leq \Delta\phi^s \quad ; \tag{40}$$

$$\Lambda^{d^T} \delta b \leq \Delta\phi^d \quad ; \tag{41}$$

where

$$\Lambda^d = \frac{\partial \tilde{\phi}^{d^T}}{\partial b} \tag{42}$$

and $\Delta\phi^s$ and $\Delta\phi^d$ are desired corrections in constraint violations. If a constraint $\phi_i \leq 0$ is $\varepsilon$-active, i.e., $\phi_i + \varepsilon \geq 0$, then $\Delta\phi_i = -\phi_i$. The constraints of Eqs. (40) and (41) can be written in compact form as

$$\Lambda^T \delta b \leq \Delta\phi \tag{43}$$

where

$$\Lambda = \left[\Lambda^s \ \Lambda^d\right] \quad , \tag{44}$$

and

$$\Delta\phi^T = \left[\Delta\phi^{s^T} \ \Delta\phi^{d^T}\right] \quad . \tag{45}$$

The reduced problem of computing an optimum design change $\delta b$ can now be stated as follows: Find $\delta b$ to minimize the cost function of Eq. (36), subject to the constraint of Eq. (43) and the step-size constraint

$$\delta b^{\mathrm{T}} W \, \delta b \leq \xi^2 \quad , \tag{46}$$

where $W$ is a positive definite weighting matrix and $\xi$ is a small number. This is exactly the problem defined in refs. [6] and [7]. An application of the Kuhn-Tucker conditions of nonlinear programming gives the following solution [6,7]:

$$\delta b = -\eta \, \delta b^1 + \delta b^2 \quad ; \tag{47}$$

$$\delta b^1 = W^{-1} \, [\Lambda^J + \Lambda\mu^1] \quad ; \tag{48}$$

$$\delta b^2 = -W^{-1} \, \Lambda\mu^2 \quad ; \tag{49}$$

$$B[\mu^1 \vdots \mu^2] = [(-\Lambda^{\mathrm{T}} W^{-1} \Lambda^J) \vdots -\Delta\phi] \quad ; \tag{50}$$

$$B = \Lambda^{\mathrm{T}} W^{-1} \Lambda \quad ; \tag{51}$$

$$\mu = \mu^1 + (1/\eta) \, \mu^2 \quad ; \tag{52}$$

where $\eta$ is a step size to be chosen by the designer [6], [7]. This method can now be described by the following step-by-step algorithm:

*Step 1* At the current design point $b^{(j)}$, solve Eqs. (6) and (12) for $z_{\mathrm{B}}^{(j)}$ and $z_{\mathrm{I}}^{(j)}$, respectively. Here $j$ denotes the iteration number.

*Step 2* Compute the vectors $\lambda_{\mathrm{I}}$ and $\lambda_{\mathrm{B}}$ from Eqs. (24) and (32), respectively. Assemble the matrix $\Lambda^J$ of Eq. (38).

*Step 3* Check the constraints of Eq. (14) and form the vector $\tilde{\phi}^s$. Compute matrices $\bar{\lambda}_{\mathrm{I}}$ and $\bar{\lambda}_{\mathrm{B}}$ from Eqs. (25) and (33), respectively. Assemble the matrix $\Lambda^s$ of Eq. (39).

*Step 4* Check the constraints of Eq. (4) and form a vector $\tilde{\phi}^d$. Compute the matrix $\Lambda^d$ of Eq. (42).

*Step 5* Assemble the matrix $\Lambda$ and the vector $\Delta\phi$ of Eqs. (44) and (45), respectively.

*Step 6* Compute vectors $\mu^1$ and $\mu^2$ from Eq. (50). Choose a step size $\eta$ and compute the Lagrange multiplier from Eq. (52).

*Step 7* Check the sign of each component of $\mu$. If any components of $\mu$ are negative, remove the corresponding rows from $\Lambda^T$ and $\Delta\phi$, and return to Step 5.

*Step 8* Compute $\delta b^1$, $\delta b^2$, and $\delta b$ from Eqs. (48), (49), and (47), respectively. Put $b^{(j+1)} = b^{(j)} + \delta b$.

*Step 9* If all constraints are satisfied and $||\delta b^1||$ is sufficiently small [6], terminate the process. Otherwise, return to Step 1 with $b^{(j+1)}$ as the best available design.

## V. AN APPLICATION OF THE METHOD TO TRUSSES

In order to test efficiency of the method, optimal design of general trusses is considered here. The method requires a considerable amount of calculation. Therefore, these calculations should be carefully planned and complete advantage of special properties of any matrices should be taken to effect computational efficiency. This paragraph discusses calculation of various sensitivity matrices, using substructuring.

The problem of optimal design of trusses is defined as follows: Find the cross-sectional area of each member of the truss such that the weight of truss is minimized and stress, displacement, and member size constraints remain satisfied. Since the weight of the structure is to be minimized, the cost function of Eq. (13) depends only on the design variables. This eliminates some of the calculation of the preceding paragraph. For example, $\partial J/\partial z_B$ and $\partial J/\partial z_I$ are null vectors. Therefore, the vectors $\lambda_I$ and $\lambda_B$ of Eqs. (24) and (30) are also null. Thus, the vector $\Lambda^J$ of Eq. (38) is simply $\partial J^T/\partial b$, which is quite easy to calculate. A general flow chart that was used to develop a computer program for truss optimization [1] is

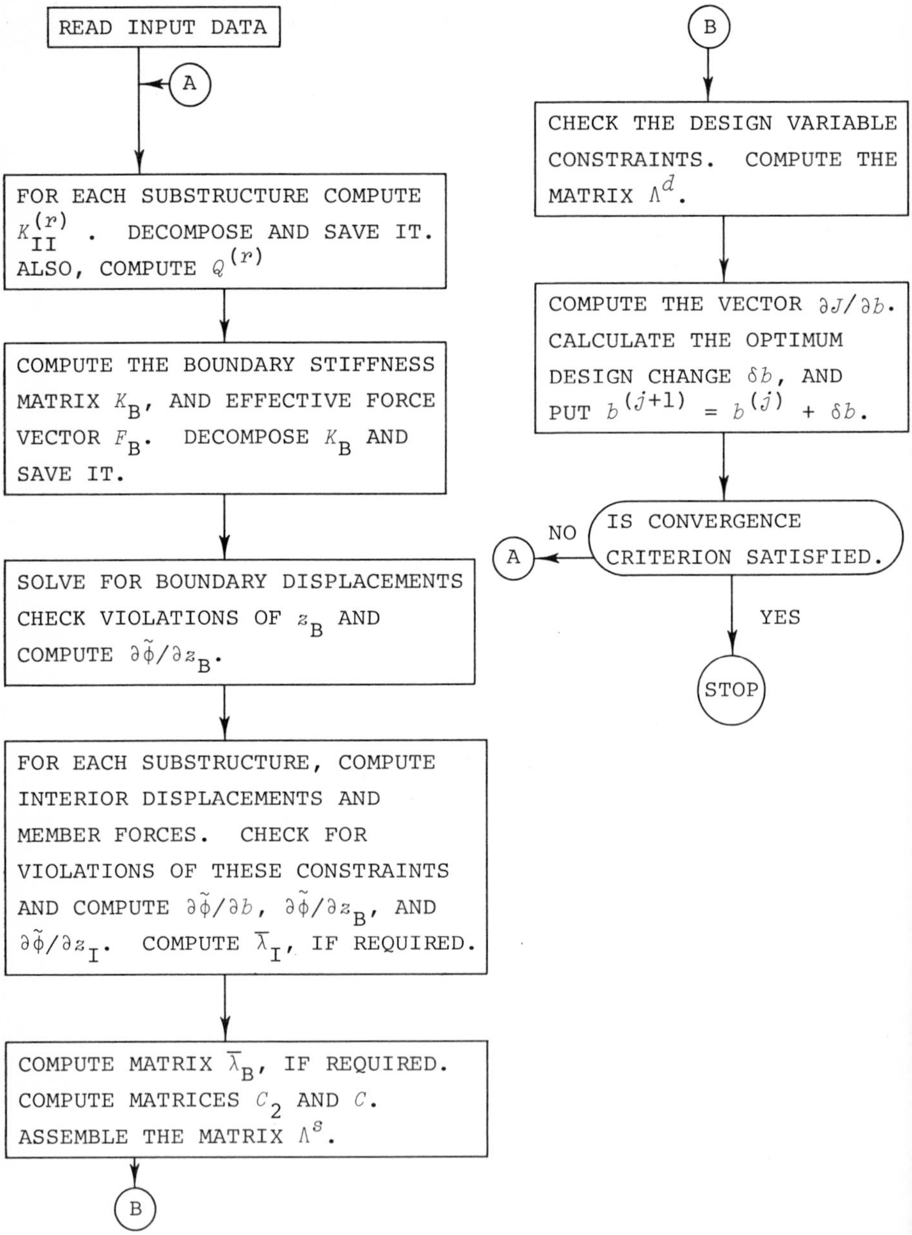


*FIG. 1. Flow diagram for truss optimization computer program.*

shown in Fig. 1.

Computation of the sensitivity matrices $\partial\tilde{\phi}^s/\partial b$, $\partial\tilde{\phi}^s/\partial z_B$, $\partial\tilde{\phi}^s/\partial z_I$, and $\partial\tilde{\phi}^d/\partial b$ is also quite easy once the form of constraint functions $\tilde{\phi}^s$ and $\tilde{\phi}^d$ has been decided. It should be noted that some components of $\tilde{\phi}^s$ may not depend on all of the three variables $b$, $z_B$, and $z_I$. Some components may depend on $b$ and $z_B$ or on $b$ and $z_I$, and others may depend only on $z_B$ or $z_I$. For example, the stress constraint depends on $b$ and $z_I$ for members connected only to interior nodes, and depends on $b$ and $z_B$ for members connected only to boundary nodes. The stress constraint depends on all the variables $b$, $z_B$, and $z_I$ if a member is connected to both interior and boundary nodes. Also, a displacement constraint at an interior node will depend only on $z_I$ and at a boundary node it will depend only on $z_B$. In numerical calculations, some advantage can be taken of these simplifications, to enhance computational efficiency of the algorithm [1]. For the present, however, all components of $\tilde{\phi}^s$ will be assumed to depend on all the variables $b$, $z_B$, and $z_I$.

Calculation of the matrix $\Lambda^s$ of Eq. (39) requires computation of the matrices $C_2$, $C$, $\bar{\lambda}_I$, and $\bar{\lambda}_B$. For these calculations, partitioning of the structure into various substructures can be utilized quite effectively [1]. In computing the matrix $\bar{\lambda}_I$ from Eq. (25), the matrix $K_{II}$ for the entire structure need not be considered at one time. Once the substructures have been defined, the stiffness matrices corresponding to the interior degrees of freedom for each substructure, $K_{II}^{(r)}$, are completely uncoupled. Therefore, the matrix $K_{II}$ has the form

$$K_{II} = \begin{bmatrix} K_{II}^{(1)} & & 0 \\ & \ddots & \\ 0 & & K_{II}^{(L)} \end{bmatrix} . \tag{53}$$

Also, all the components of $\tilde{\phi}^s$ will not depend upon the interior coordinates $z_I^{(r)}$ of the $r$th substructure. Therefore, the matrix $\partial\tilde{\phi}^s/\partial z_I$ is also partitioned as

$$\frac{\partial\tilde{\phi}^s}{\partial z_I} = \begin{bmatrix} \dfrac{\partial\tilde{\phi}^s}{\partial z_I^{(1)}} & & & 0 \\ & \cdot & & \\ & & \cdot & \\ & & & \cdot \\ 0 & & & \dfrac{\partial\tilde{\phi}}{\partial z_I^{(L)}} \end{bmatrix}. \tag{54}$$

The matrix $\bar{\lambda}_I$ is also partitioned accordingly and each matrix $\bar{\lambda}_I^{(r)}$ is now computed from the equation

$$K_{II}^{(r)}\ \bar{\lambda}_I^{(r)} = \frac{\partial\tilde{\phi}^{s^T}}{\partial z_I^{(r)}} \quad .$$

It can be easily shown that calculations of the matrix $C_2^T\ \bar{\lambda}_I$ can proceed substructure-wise.

In computing the matrix $\bar{\lambda}_B$ from Eq. (33), a matrix $(\partial\tilde{\phi}^s/\partial z_I)$ is required. The calculation of this matrix can also proceed substructure-wise, because the matrix $Q$ can be arranged in a form similar to Eq. (54). Similarly, computation of the matrix $Q^T C_2$ in Eq. (23) can be carried out by considering one substructure at a time.

The externally applied loads are taken to be independent of the design variables for the truss problem. Therefore, the terms $\partial S_B/\partial b$ and $\partial S_I/\partial b$ drop out of Eqs. (20) and (21). In order to calculate the matrices $C_1^{(r)}$ and $C_2^{(r)}$ from Eqs. (20) and (21), $(\partial/\partial b)(K_{BB}^{(r)}\ z_B^{(r)})$, $(\partial/\partial b)(K_{BI}^{(r)}\ z_I^{(r)})$, etc., are required. These matrices can be computed quite conveniently by considering one member of a substructure at a time.

A computer program SOS1 (Structural Optimization by Substructuring) has been developed for optimal design of trusses and is automatic, in the sense that it requires the pertinent design information as input data only in the beginning [1]. Some other computational aspects for improving efficiency of the algorithm, which were discussed previously [7], are also incorporated in the present program. The program developed has been used to optimize two trusses of fixed geometry subjected to multiple loading conditions, with stress, displacement, and design variable constraints. These results were obtained by starting from a uniform initial design of 1.0 $in^2$ for all members. The computational times reported are for an IBM 360-65(H) computer.

### A. *EXAMPLE 1. TWENTY-FIVE-MEMBER TRANSMISSION TOWER*

Figure 2 shows geometry and dimensions of a 25-member transmission tower. Design data and loading conditions for the structure are given in Table I. The tower is divided into two substructures by partitioning it at joints 1, 2, 5, and 6. Members 1 through 13 belong to substructure 1 and the remaining members, 14 to 25, belong to substructure 2.

Table II gives the optimum solution for two cases: (1) stress constraint, and (2) stress and displacement constraints. It also gives a comparison of computational times between TRUSSOPT3 [7] and SOS1 [1] computer programs. A reduction of approximately 6 to 9% in computing time was achieved with present computer program.

### B. *EXAMPLE 2. TWO-HUNDRED-MEMBER PLANE TRUSS*

The second structure optimized by the present technique is

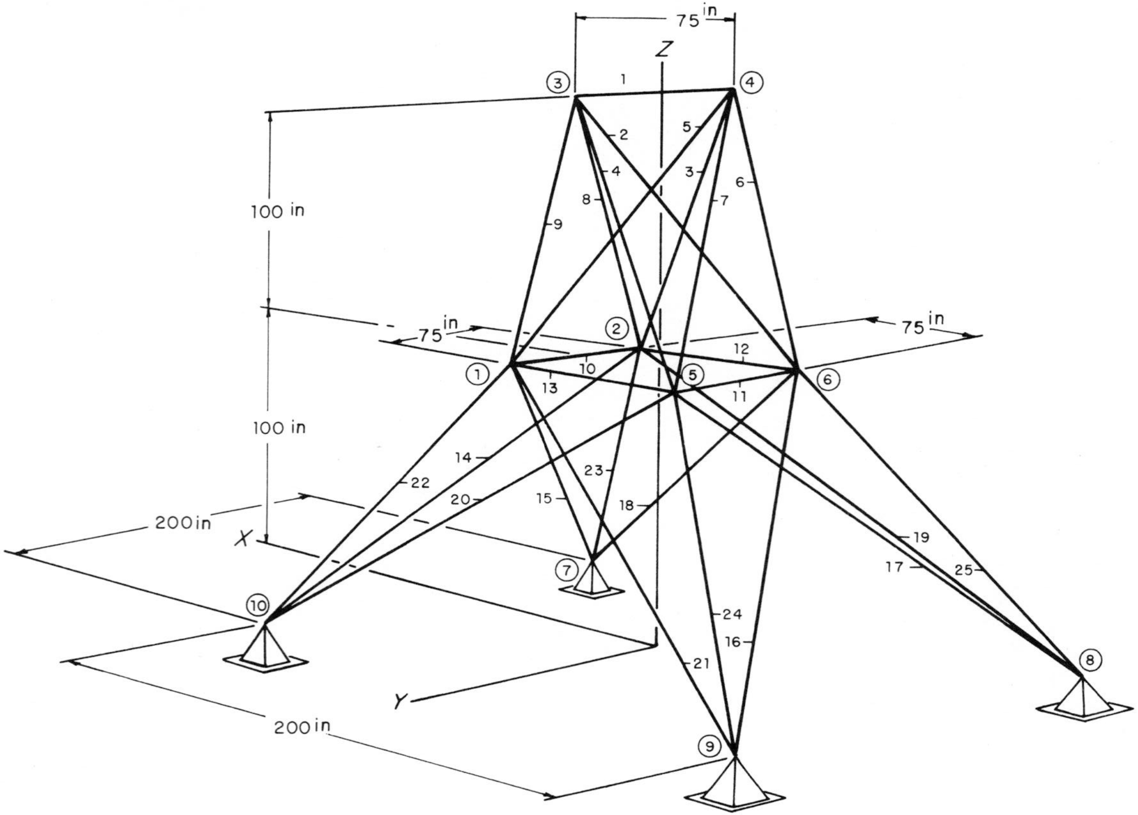


*FIG. 2. Twenty-five-member transmission tower.*

*TABLE I*

*Design Data for 25-Member Transmission Tower*

| | |
|---|---|
| *Modulus of elasticity* | = $10^4$ ksi |
| *Material density* | = 0.10 lb/in$^3$ |
| *Stress limits* | = ±40 ksi |
| *Lower limit on cross-sectional areas* | = 0.10 in$^2$ (stress constraints only) |
| | = 0.01 in$^2$ (stress and displacement constraints) |
| *Upper limit on cross-sectional areas* | = none |
| *Displacement limits at all nodes and in all directions* | = 0.35 in |
| *Number of loading conditions* | = two (all loads in kips) |

| Load condition | Node | Direction of load | | |
|---|---|---|---|---|
| | | x | y | z |
| 1 | 1 | -0.5 | 0 | 0 |
| | 2 | -0.5 | 0 | 0 |
| | 3 | -1.0 | -10.0 | -5.0 |
| | 4 | 0 | -10.0 | -5.0 |
| 2 | 3 | 0 | -20.0 | -5.0 |
| | 4 | 0 | 20.0 | -5.0 |

a 200-member plane truss, shown in Fig. 3. The design data for the truss are given in Table III. For the purpose of obtaining computational times with different number of substructures, the following three cases are considered for this example:

*Case 1:* The truss is divided into two substructures by partitioning it at joints 34 to 42. Members 1 to 93 are in substructure 1, and 94 to 200 in substructure 2.

*Case 2:* The truss is divided into three substructures by partitioning it at joints 29 to 33 and 57 to 61. Members 1 to

*TABLE II*

*Optimum 25-Member Transmission Tower*

| | | Final area ($in^2$) | |
|---|---|---|---|
| No. | Member numbers | With stress constraints only | With stress and displacement constraints only |
| 1 | 1 | 0.1000 | 0.0100 |
| 2 | 2-5 | 0.3759 | 2.0476 |
| 3 | 6-9 | 0.4716 | 2.9965 |
| 4 | 10-13 | 0.1000 | 0.0100 |
| 5 | 14-17 | 0.1000 | 0.6853 |
| 6 | 18-21 | 0.2778 | 1.6217 |
| 7 | 22-25 | 0.3799 | 2.6712 |
| Weight (lbs) | | 91.18 | 545.04 |
| Number of tight constraints at optimum | | 7 | 5 |
| Computation time per iteration in seconds | | | |
| By SOS1 [1] | | 0.62 | 0.49 |
| By TRUSSOPT3 [7] | | 0.66 | 0.54 |

76, 77 to 152, and 153 to 200 are in substructures 1, 2, and 3, respectively.

*Case 3:* The truss is divided into five substructures by partitioning it at joints 15 - 19, 29 - 33, 43 - 47, and 57 - 61. Members 1 - 38, 39 - 76, 77 - 114, 115 - 152, and the rest (153 - 200) are in substructures 1, 2, 3, 4, and 5, respectively.

In cases 2 and 3, nodes 1 - 5 were considered as boundary nodes. It may be pointed out that in all the cases, the nodes may be renumbered in order to provide a minimum upper bandwidth for the matrices $K_B$, $K_{BB}^{(r)}$, and $K_{II}^{(r)}$.

Table IV gives optimum results for this problem, under two separate constraint conditions, along with the value

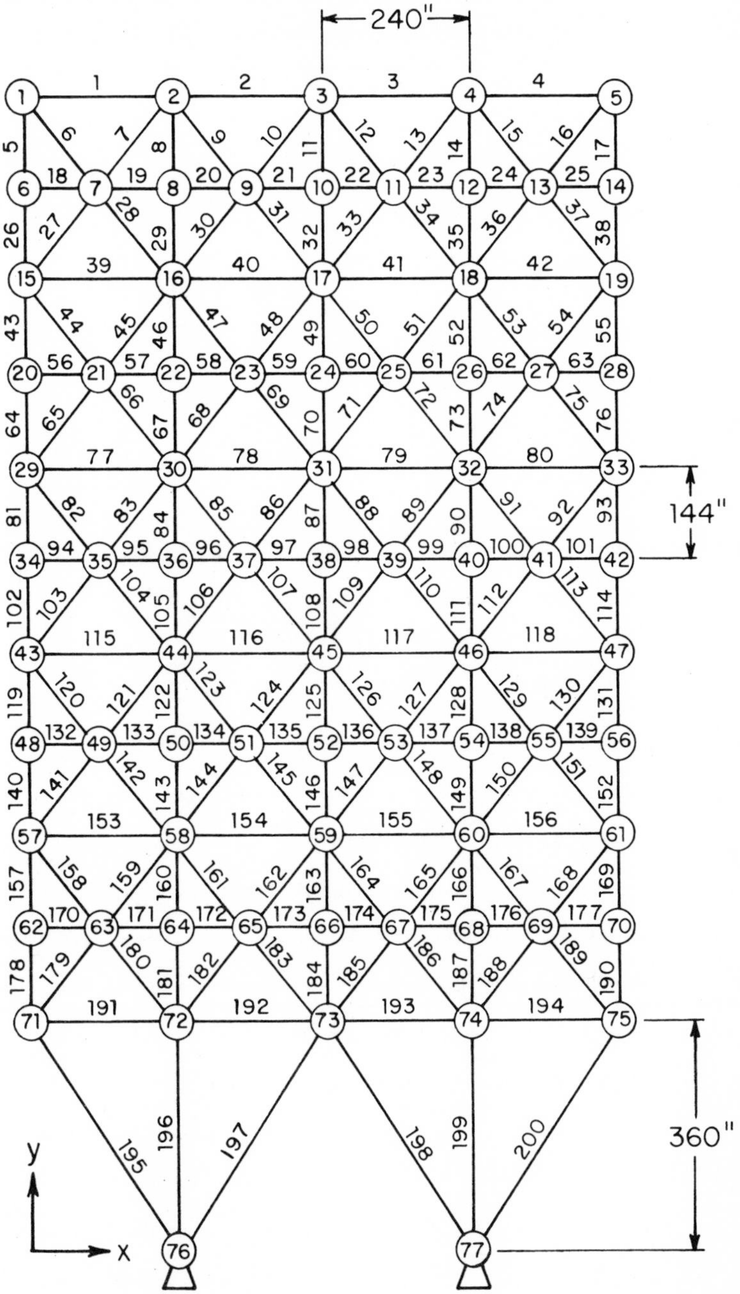


*FIG. 3. Two-hundred-member plane truss.*

*TABLE III*

*Design Data for 200-Member Truss*

| | | |
|---|---|---|
| *Modulus of Elasticity* | = | *30,000 ksi* |
| *Material density* | = | *0.283* $lb/in^3$ |
| *Lower limit on cross-sectional areas* | = | *0.10* $in^2$ |
| *Upper limit on cross-sectional areas* | = | *none* |
| *Stress limits* | = | *±30 ksi* |
| *Displacement limits* | = | *±0.5 in* |

*There are three loading conditions and they are as follows:*

*Loading condition 1. One kip acting in positive x direction at node points 1, 6, 15, 20, 19, 34, 43, 48, 57, 62, 71.*

*Loading condition 2. 10 kips acting in negative y direction at node points 1, 2, 3, 4, 5, 6, 8, 10, 12, 14, 15, 16, 17, 18, 19, 20, 22, 24, ..., 71, 72, 73, 74, 75.*

*Loading condition 3. Loading conditions 1 and 2 acting together.*

of cost function, which remained the same in all cases. Table V shows a comparison of computational times between TRUSSOPT3 and SOS1 computer programs. It also shows the number of tight constraints at the optimum. The total computational time for each iteration depends upon many factors, such as the number of constraint violations, the number of loading conditions, and the number of substructures under which violations occur. The computational times given in Table V are for the number of violations near the optimum point. Also, for the solution with only stress constraints, violations occurred in all substructures and in all loading conditions for all three cases. For the stress and displacement constraint case, violations occurred in all substrucutres and in all loading conditions

TABLE IV

Optimum 200-Member Truss

| No. | Member numbers | Final area ($in^2$) With stress constraints only | With stress and displacement constraints |
|---|---|---|---|
| 1 | 1,4 | 0.1000 | 0.1878 |
| 2 | 2,3 | 0.1000 | 0.1000 |
| 3 | 5,17 | 0.2469 | 4.7832 |
| 4 | 6,16 | 0.1185 | 0.1703 |
| 5 | 7,15 | 0.1000 | 0.1000 |
| 6 | 8,14 | 0.3675 | 2.3462 |
| 7 | 9,13 | 0.1000 | 0.1876 |
| 8 | 10,12 | 0.1000 | 0.1000 |
| 9 | 11 | 0.2454 | 2.8809 |
| 10 | 18,25,56,63, 94,101,132,139, 170,177 | 0.1000 | 0.1000 |
| 11 | 19,20,23,24 | 0.1000 | 0.1000 |
| 12 | 21,22 | 0.1000 | 0.1000 |
| 13 | 26,38 | 0.5815 | 6.7767 |
| 14 | 27,37 | 0.1000 | 0.1000 |
| 15 | 28,36 | 0.1736 | 0.2361 |
| 16 | 29,35 | 0.7008 | 3.3133 |
| 17 | 30,34 | 0.1000 | 0.1732 |
| 18 | 31,33 | 0.1000 | 0.2227 |
| 19 | 32 | 0.5792 | 4.1473 |
| 20 | 39,42 | 0.1035 | 0.1000 |
| 21 | 40,41 | 0.1000 | 0.1000 |
| 22 | 43,55 | 0.6760 | 8.1292 |
| 23 | 44,54 | 0.2489 | 0.2476 |
| 24 | 45,53 | 0.1000 | 0.1000 |
| 25 | 46,52 | 1.2622 | 4.4206 |
| 26 | 47,51 | 0.1000 | 0.2802 |
| 27 | 48,50 | 0.1000 | 0.2673 |
| 28 | 49 | 0.8402 | 4.7929 |
| 29 | 57,58,61,62 | 0.1000 | 0.1000 |
| 30 | 59,60 | 0.1000 | 0.1002 |
| 31 | 64,76 | 1.0080 | 9.3889 |
| 32 | 65,75 | 0.1000 | 0.1000 |
| 33 | 66,74 | 0.3103 | 0.3362 |
| 34 | 67,73 | 1.5955 | 5.0733 |
| 35 | 68,72 | 0.1000 | 0.3008 |
| 36 | 69,71 | 0.1000 | 0.3096 |
| 37 | 70 | 1.1731 | 5.5744 |
| 38 | 77,80 | 0.1892 | 0.4967 |
| 39 | 78,79 | 0.1000 | 0.3865 |
| 40 | 81,93 | 1.0402 | 9.5196 |
| 41 | 82,92 | 0.4108 | 0.9366 |
| 42 | 83,91 | 0.1000 | 0.1000 |

*TABLE IV. continued*

*Optimum 200-Member Truss*

| No. | Members numbers | Final area ($in^2$) With stress constraints only | Final area ($in^2$) With stress and displacement constraints |
|---|---|---|---|
| 43 | 84,90 | 2.2576 | 6.2617 |
| 44 | 85,89 | 0.1000 | 0.3508 |
| 45 | 86,88 | 0.1000 | 0.4835 |
| 46 | 87 | 1.4567 | 5.8679 |
| 47 | 95,96,99,100 | 0.1000 | 0.1000 |
| 48 | 97,98 | 0.1000 | 0.1000 |
| 49 | 102,114 | 1.3481 | 10.4800 |
| 50 | 103,113 | 0.1111 | 0.1108 |
| 51 | 104,112 | 0.4795 | 1.0313 |
| 52 | 105,111 | 2.5909 | 6.8203 |
| 53 | 106,110 | 0.1007 | 0.5012 |
| 54 | 107,109 | 0.1000 | 0.3754 |
| 55 | 108 | 1.7922 | 6.4768 |
| 56 | 115,118 | 0.2932 | 1.9807 |
| 57 | 116,117 | 0.1000 | 1.4784 |
| 58 | 119,131 | 1.2625 | 9.1546 |
| 59 | 120,130 | 0.6050 | 3.1979 |
| 60 | 121,129 | 0.1000 | 0.1000 |
| 61 | 122,128 | 3.3587 | 9.0271 |
| 62 | 123,127 | 0.1000 | 0.2074 |
| 63 | 124,126 | 0.1069 | 0.9717 |
| 64 | 125 | 2.0771 | 6.5338 |
| 65 | 133,134,137,138 | 0.1000 | 0.1000 |
| 66 | 135,136 | 0.1002 | 0.1219 |
| 67 | 140,152 | 1.5748 | 9.9624 |
| 68 | 141,151 | 0.1325 | 0.1341 |
| 69 | 142,150 | 0.6817 | 3.3000 |
| 70 | 143,149 | 3.6920 | 9.5771 |
| 71 | 144,148 | 0.1092 | 0.9814 |
| 72 | 145,147 | 0.1000 | 0.2269 |
| 73 | 146 | 2.4116 | 7.0561 |
| 74 | 153,156 | 0.4250 | 2.5500 |
| 75 | 154,155 | 0.1000 | 0.6074 |
| 76 | 157,169 | 1.3251 | 7.5376 |
| 77 | 158,168 | 0.8462 | 4.1216 |
| 78 | 159,167 | 0.1000 | 0.1000 |
| 79 | 160,166 | 4.5880 | 13.3290 |
| 80 | 161,165 | 0.1000 | 1.8691 |
| 81 | 162.164 | 0.1276 | 0.3045 |
| 82 | 163 | 2.6772 | 7.4246 |
| 83 | 171,172,175,176 | 0.1000 | 0.1000 |
| 84 | 173,174 | 0.1002 | 0.1000 |
| 85 | 178,190 | 1.6597 | 8.2183 |

*TABLE IV. continued*

*Optimum 200-Member Truss*

| | | Final area ($in^2$) | |
|---|---|---|---|
| No. | Members numbers | With stress constraints only | With stress and displacement constraints |
| 86 | 179,189 | 0.1156 | 0.1000 |
| 87 | 180,188 | 0.9181 | 4.1916 |
| 88 | 181,187 | 4.9213 | 13.8330 |
| 89 | 182,186 | 0.1339 | 0.3354 |
| 90 | 183,185 | 0.1002 | 1.9082 |
| 91 | 184 | 3.0100 | 7.8840 |
| 92 | 191,194 | 1.2391 | 5.8649 |
| 93 | 192,193 | 0.8521 | 3.4248 |
| 94 | 195,200 | 2.3257 | 10.6560 |
| 95 | 196,199 | 5.9232 | 17.7770 |
| 96 | 197,198 | 2.5718 | 7.7140 |
| Optimum weight in pounds | | 7,488 | 28,963 |

in some initial iterations. However, near the optimum point only displacement violations occurred under the second and the third loading conditions and in the substructure number 1 for all the cases.

A comparison of structural analysis times indicates that as the number of substructures is increased from 2 to 5, the analysis time is reduced by a factor of approximately 2. This reduction in analysis time is due to the reduced bandwidth of the matrix $K_{II}^{(r)}$. In comparison with TRUSSOPT3 results, structural analysis time was more by 50% with two substructures and was less by 24%, with five substructures. The total computational time was also reduced in all the cases, with the greatest reduction occurring in case 3. This indicates that there is an optimum number of substructures for minimum computing time. When the number of substructures is below or above this optimum number, efficiency of the substructuring technique is reduced.

TABLE V

Comparison of Computational Times for 200-Member Truss[a]

| | | With stress constraints only | | | | With stress and displacement constraints | | | |
|---|---|---|---|---|---|---|---|---|---|
| | | SOS1 [1] program | | | TRUSSOPT3 [7] Program | SOS1 [1] program | | | TRUSSOPT3 [7] Program |
| | | Case 1 | Case 2 | Case 3 | | Case 1 | Case 2 | Case 3 | |
| Computational time/iteration for analysis only[b] | | 4.54 | 3.06 | 2.37 | 3.11 | - | - | - | - |
| Total computational time/iteration | | 20.57 | 16.88 | 15.19 | 22.00 | 7.06 | 5.20 | 4.43 | 13.00 |
| Percentage reduction in computational time from TRUSSOPT3 [7] | | 6.5 | 23.3 | 31.0 | - | 45.7 | 60.0 | 65.9 | - |
| Number of tight constraints at optimum | Stress and/or displacement | 50 | 49 | 50 | 50 | 5 | 4 | 4 | 5 |
| | Design variable | 40 | 40 | 41 | 40 | 23 | 23 | 25 | 23 |

[a]All times are in seconds.

[b]Excluding initial setup time.

It may be observed from Table IV that many design variables are at their lower bounds at the optimum solution. A side computer run was made by keeping these variables constant for case 1 with only stress constraints in order to see the reduction in computing time. For this run, computing time per cycle was 12.49 sec, as compared with 20.57 sec. This is a significant reduction of approximately 40%. In many practical situations it is known *a priori* that certain members of the structure must have assigned values, minimum or maximum areas. The above calculation indicates that the designer should take full advantage of such knowledge to achieve maximum efficiency.

## VI. DISCUSSION AND CONCLUSIONS

This chapter has presented a general method for optimal design of structures that incorporates substructuring in its formulation in a profitable manner. The design sensitivity analysis for the problem is discussed and the method is applied to general truss structures. Results for two example problems are obtained with a computer program that is based on the present algorithm. The first example problem is a 25-member transmission tower. For this relatively small-scale problem, computing time is reduced by only 6 to 9%, as compared to an algorithm without substructuring. The second example is a moderate size 200-member plane truss. For this example, reduction in computing times is more significant, up to 66%, and the computer storage requirement is reduced by approximatley 15%. For large-scale problems, reduction in computing time is expected to be greater than that achieved for the present examples.

There are two major phases of any optimal design algorithm:

(1) structural analysis and (2) optimization. Thus, reduction in the computing time can be obtained in either of the two phases. The present examples indicate that with substructuring there can be a significant reduction in computing times in the structural analysis phase. However, most of the efficiency in the algorithm was achieved in the optimization phase. This is due to the fact that substructuring is effectively utilized in the design sensitivity analysis of the problem. For example, if all the constraint violations take place in one substructure, then the adjoint matrix $\bar{\lambda}_I^{(r)}$ is computed for only that substructure. For case 1 of the 200-member plane truss example, this means that in design sensitivity analysis one has to solve only a $66 \times 66$ system of equations, rather than a $150 \times 150$ system of equations, each having a half bandwidth of 20. Thus, the number of calculations is reduced considerably, even with only two substructures. The system of equations is further reduced when the number of substructures is increased. In general, design sensitivity calculations with substructures involve manipulation of matrices having smaller dimensions. This contributes to a reduction in computational time, as well as computer storage requirements.

The results presented herein indicate that the present method is more efficient than a similar method without substructuring (compare with TRUSSOPT3 results). In ref. [7], TRUSSOPT3 results were shown to be quite efficient as compared to other available results. Therefore, it may be concluded that substructuring can be incorporated quite profitably into an optimal design algorithm.

The optimal structural design method with substructuring has recently been extended [13] to include (a) constraints and cost functions that depend on structural eigenvalues, and (b) fail-safe constraints under probable damage conditions.

REFERENCES


1. A. K. GOVIL and J. S. ARORA. "A Technique for Optimal Structural Design with Substructuring," Technical Report No. 27, Division of Materials Engineering, University of Iowa, Iowa City, Iowa 52242, October 1976.
2. A. K. NOOR and H. E. LOWDER. "Approximate Reanalysis Techniques with Substructuring," *J. Struc. Div. ASCE 101*, No. ST8, 1687-1698 (1975).
3. G. G. POPE and L. A. SCHMIT (eds.). "Structural Design Applications of Mathematical Programming Techniques," AGARDograph No. 149, National Technical Information Service, Springfield, Virginia, February 1972.
4. R. H. GALLAGHER and O. C. ZIENKIEWICZ (eds.). "Optimum Structural Design: Theory and Applications," Wiley, New York, 1973.
5. A. E. BRYSON, Jr. and Y. C. HO. "Applied Optimal Control," Ginn and Co., Waltham, Massachusetts, 1969.
6. E. J. HAUG, Jr., K. C. PAN, and T. D. STREETER. "A Computational Method for Optimal Structural Design: I Piecewise Uniform Structures," *Int. J. Numer. Meth. Eng. 5*, 171-184 (1972).
7. J. S. ARORA and E. J. HAUG, Jr. "Efficient Optimal Design of Structures by Generalized Steepest Descent Programming," *Int. J. Numer. Meth. Eng. 10*, No. 4, 746-766 (1976).
8. J. S. ARORA, E. J. HAUG, Jr., AND K. RIM. "Optimal Design of Plane Frames," *J. Struc. Div. ASCE 101*, No. ST10 (1975).
9. E. J. HAUG, Jr., and J. S. ARORA. "Optimal Design Techniques Based on Optimal Control Methods," *Proc. 1st ASME Design Technol. Transfer Conf., New York*, 65-74 (1974).
10. J. S. ARORA. "On Improving Efficiency of an Algorithm for Structural Optimization and a User's Manual for Program TRUSSOPT3," Technical Report No. 12, Department of Mechanics and Hydraulics, University of Iowa, Iowa City, September 1976, 112 pages.
11. U. KIRSCH, M. REISS, and U. SHAMIR. "Optimum Design by Partitioning into Substructures," *J. Struc. Div. ASCE 98*, No. ST1, 249-267 (1972).
12. J. S. PRZEMIENIECKI. "Theory of Matrix Structural Analysis," McGraw-Hill, New York, 1968.
13. D. T. NGUYEN, A. K. GOVIL, J. S. ARORA, and E. J. HAUG, Jr. "Fail-Safe Optimal Design of Structures with Substructuring," Technical Report No. 45, Division of Materials Engineering, University of Iowa, Iowa City, Iowa 52242, August 1978.

# Advances in Adaptive Filtering

LEONARD CHIN

*Communication Navigation Technology Directorate*
*Naval Air Development Center**
*Warminster, Pennsylvania*



*Views or conclusions contained in this chapter should not be interpreted as representing the official opinion or policy of the United States Navy.*




ISBN 0-12-012715-6

## I. INTRODUCTION

### A. *ADAPTIVE FILTERING AND PURPOSE*

The advance of adaptive filtering is a natural evolution of the fundamental Kalman-Bucy filter, which requires both the linear dynamic system and linear measurements of the state vector to be known and disturbed only by white Gaussian noise with known first- and second-order statistics. Such requirements are hardly ever fulfilled in practical applications, because complete knowledge for the design of the optimal filter is usually not available. For example, the system model (e.g., the order of the differential equation) may not be adequate and/or the noise statistics may not be accurate (e.g., parameters generated based on insufficient data points). In order to compensate for these and other inadequacies, adaptive techniques must be used. An adaptive filter therefore, may be viewed as a Kalman-Bucy filter plus an added learning feature such that the integrated system can successfully achieve its optimal performance even if the environment under which the system operates is only partially known. If the system can adapt to the environment, it must be able to perceive the current state and the change of the state of the environment, and then make the necessary decision to adjust its behavior based on the latest observations of the environment. This point of view can be translated into a schematic diagram shown in Fig. 1, which illustrates how the adaptive filter works in a very broad sense. Obviously specific situations require appropriate adaptive schemes, which can be best for the given set of conditions. In general, all existing adaptive schemes can be classified according to the nature of

- the mathematical model that represents the physical system, e.g., linear or nonlinear differential/difference equations;
- uncertainties associating with the model, e.g., parametric or structural (the order of the differential/difference equation); time-varying or time-invariant;
- the constraints imposed on the adaptation process.

In this chapter, different classes of linear problems will be discussed; nonlinear problems will be treated separately in another volume of the Control and Dynamic Systems.

One of the most popular applications of the adaptive filter is found in optimal control systems, in which the purpose of the filter is to learn the state-of-knowledge concerning the statistical data, dynamic system structure, and performance index, which are utilized to generate the optimal control law. For example, consider a control system described by the following linear continuous-time vector differential equations:

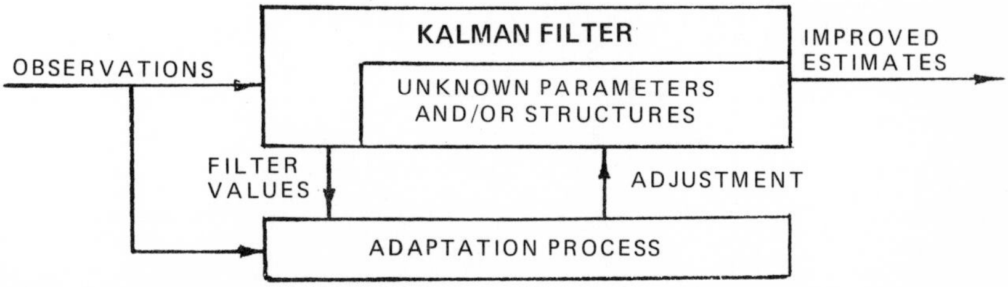


*FIG. 1. Adaptive filter diagram.*

$$\dot{x} = Fx + GU + \Gamma W$$

$$y = Hx$$

$$z = y + V$$

where $x$, $y$, and $z$ are *state*, *output*, and *observed output* vectors, respectively; $U$ is the control vector; $W$ and $V$ are plant noise and measurement noise, respectively; $F$, $G$, $\Gamma$, and $H$ are time-varying rectangular matrices with appropriate dimensions.

If the control system is to be adaptive, two types of dynamic processes must take place:

(a) The state variable ($x$) of the control object are estimated using measurements, and corresponding control action is taken in accordance with the control law existing at that time.

(b) The statistical information concerning $V$, $W$, and elements of matrices $F$, $G$, $\Gamma$, and $H$, and performance index are learned by the filter, and corresponding adjustments are made to the optimal control law from time to time. This example is schematically shown in Fig. 2. Other applications of the adaptive filter are frequently found in communication systems, pattern recognition, biomedical systems, nuclear reactors,

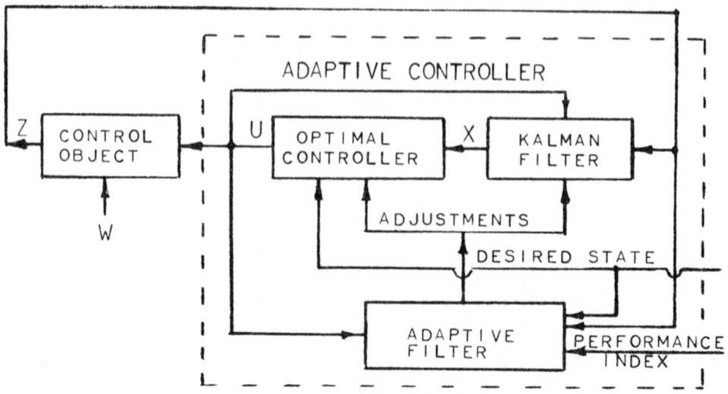


*FIG. 2. Adaptive control system diagram.*

electric power systems, and many others.

### *B. ORGANIZATION*

The material is organized into five sections. This section provides an introduction to the concept of adaptive filtering via definitions and purposes. It also provides a brief treatment of general adaptive control systems to permit discussion of adaptive filtering within a common frame of reference. Section II provides a discussion on various analysis techniques for adaptive filters within the realm of linear systems; mathematical models are defined to limit the boundary that covers significant topics such as covariance analysis, model testing, and implementation considerations. Section III covers two important topics of synthesis: adaptive filtering and error bounding techniques. The discussion on adaptive techniques is divided into three parts: noise parameter adaptation, system parameter adaptation, and system structure adaptation. The discussion on error bounding techniques is divided into two categories: computer constraints and divergence problems. Section IV examines a practical application of the adaptive filtering theory to improve a certain inertial navigation system accuracies. The problem is treated based on realistic operational conditions. Simulation results for selected cases are obtained to demonstrate the advantages of the adaptive filter vice the nonadaptive filter.

## II. ANALYSIS

### *A. INTRODUCTION*

The origin of the estimation theory can be traced back to

the earliest times when the Babylonians used a form of mathematics that was similar to the Fourier series to interpret observations and make decisions. However, the foundation of the estimation theory (in a sense, of minimizing various errors) must be attributed to Galileo Galilei, founder of experimental physics and astronomy.

During the seventeenth and eighteenth centuries, prominent mathematicians, including Euler, Lagrange, Laplace, and Bernoulli, had contributed to the advancement of the estimation theory in many ways. Readers who were interested in the historical developments should consult Neugebauer [1] and Sorenson [2].

During the early nineteenth century, the solution to a class of estimation problems in finding the best value of an unknown parameter corrupted by an additive noise is given by Gauss and Legendre, who formulated the method of least-squares. They were interested in the unknown parameters in astronomical studies, in which motions of heavenly bodies were studied using the telescope, and values of the unknown parameters were inferred from the measurement data. Hence the concepts of reduncant measurement data, observability, dynamic modeling, etc., were introduced. These concepts of Gauss and Legendre, together with many available mathematical tools in probability and statistics, have provided the foundation for many new developments in the past 150 years.

During the nineteenth century and up to the 1910s, the advancement of the least-squares estimation theory was limited to the computational aspects, including the use of orthogonal polynomials. Among the principal contributors were Cauchy, Schmidt, Gram, Chebyshev, and Gauss, himself.

During the 1920s and 1930s, Fisher [3] was considered to be one of the principal investigators in the development of statistical estimation methods. Among other contributions, Fisher was well known for his study of estimator properties such as sufficiency, efficiency, and consistency.

During the 1940s Kolmogorov [4] and Wiener [5] were responsible for the successful development of the linear, minimum, mean-square-error estimator (for the case where the measurement and signal are scalar, continuous-time, and stationary random process). The main contribution of Kolmogorov and Wiener was the establishment of the Wiener-Hopf equation, which can only be solved explicitly under special conditions. For this reason, filter design using the Wiener-Hopf equation approach had only limited practical applications.

During the 1950s, Swerling [6], Carlton and Follin [7] had successfully circumvented the difficulty of solving the integral (Wiener-Hopf) equation for the filtering problem, and the concept of obtaining the least-squares estimate in the recursive manner was formulated. This new development is considered to be the starting point of the modern estimation theory, which is subsequently referred to as "Kalman filtering."

During the early 1960s, Kalman and Bucy [8,9] generalized the results of the modern estimation theory. Their main contribution was the transformation of the Wiener-Hopf integral equation into an equivalent nonlinear differential equation, whose solution yields the covariance matrix of the minimum filtering error, which in turn contains all necessary information for the design of the optimal filter. Hence, by demanding a numerical rather than an analytical solution for the Wiener-Hopf equation, Kalman has succeeded in developing a simple recursive filter that can be conveniently synthesized on a digital computer. However, the Kalman filter is not optimal unless the dynamical system is linear and a precise knowledge of the *a priori* Gaussian statistics of initial conditions and of the model (process noise and measurement noise) are available. In practical applications, neither the dynamical system linear nor the *a priori* statistics are known, and in some cases, the structure (order of the differential or difference equations) of the dynamic system is at most partially known (see references [10] and [48]). Furthermore, it is convenient to treat this brief review from two viewpoints: analysis and synthesis. Analysis refers to the study of filter efficiency and consistency by evaluating the actual error covariance matrix and its sensitivity and stability properties, as well as considering implementation problems. Synthesis refers to the improvement of Kalman filter design in an uncertain environment through adaptive techniques as well as error bounding methods.

### *B. MATHEMATICAL MODELS*

The survey to be conducted here is limited to optimal filter analysis and synthesis of the following linear models:

$$\dot{x}(t) = F(t)\,x(t) + G(t)\,u(t) \quad , \tag{1}$$

$$z(t) = H(t)\,x(t) + v(t) \quad , \tag{2}$$

for continuous-time system, and

$$x(k+1) = \Phi(k)\,x(k) + \Gamma(k)\,u(k) \quad , \tag{3}$$

$$z(k) = H(k)\,x(k) + v(k) \quad , \tag{4}$$

for the discrete-time system where

$x(\cdot)$ : $n \times 1$ state vector

$\Phi(\cdot)/F(\cdot)$ : $n \times n$ transition fundamental matrix

$\Gamma(\cdot)/G(\cdot)$ : $n \times q$ input matrix

$z(\cdot)$ : $r \times 1$ measurement matrix

$H(\cdot)$ : $r \times n$ output matrix

$u(\cdot)$ : $q \times 1$, $v(\cdot)$ : $r \times 1$ — correlated Gaussian white noise process/sequence with means and covariances as follows:

$$E[u(\cdot)] = \mu_u \quad ,$$

$$E[v(\cdot)] = \mu_v \quad ,$$

$$E[u(\cdot)u(\cdot)^T] = Q\delta(\cdot) \quad ,$$

$$E[v(\cdot)v(\cdot)^T] = R\delta(\cdot) \quad ,$$

$$E[u(\cdot)v(\cdot)^T] = C\delta(\cdot) \quad ,$$

where $E[\cdot]$ is the expected value, $\delta(\cdot)$ the delta function, and $C^T$ the transpose of matrix or vector $C$.

The initial value $x(0)$ is assumed to be a random vector (not necessarily with zero mean) and the initial value $P(0)$ is assumed to be a known covariance matrix. The problem is to find the optimal estimate $x(k)$ denoted by $\hat{x}(k|k)$, which is a linear function of the measurement data minimizing the quadratic function

$$J = E\{[x(k) - \hat{x}(k|k)]^T w[x(k) - \hat{x}(k|k)]\} \quad , \tag{5}$$

where $w$ is a symmetric positive definite matrix. If it is assumed that dimensions of all vectors and matrix, elements in all matrices, and the first- and second-order statistics of the Gaussian noises are known, then the Kalman-Bucy filter for the continuous-time system is described by the following equations:

$$\dot{\hat{x}}(t) = F(t)\hat{x}(t) + G(t)\mu_u + K(t)[z(t) - H(t)\hat{x}(t) - \mu_v] \tag{6}$$

$$K(t) = [P(t)H^{\mathrm{T}}(t) + G(t)C(t)]R^{-1}(t) \tag{7}$$

$$\dot{P}(t) = F(t)P(t) + P(t)F^{\mathrm{T}}(t) + G(t)Q(t)G^{\mathrm{T}}(t) - K(t)R(t)K^{\mathrm{T}}(t) \tag{8}$$

The Kalman-Bucy filter for the discrete-time system is described by the following equations:

$$\hat{x}(k|k) = \hat{x}(k|k-1) + K(k|k-1)[z(k) - H(k)\hat{x}(k|k-1) - \mu_v(k)]; \tag{9}$$

$$\hat{x}(k|k-1) = \Phi(k,k-1)\hat{x}(k-1|k-1) + \Gamma(k-1)\mu_u(k-1) + K_p(k-1)[z(k-1) - H(k-1)\hat{x}(k-1|k-1) - \mu_v(k-1)]; \tag{10}$$

$$K_p(k) = \Gamma(k)C(k)R^{-1}(k); \tag{11}$$

$$K(k|k-1) = P(k|k-1)H^{\mathrm{T}}(k)[H(k)P(k|k-1)H^{\mathrm{T}}(k) + R(k)]^{-1}; \tag{12}$$

$$P(k|k) = P(k|k-1) - K(k|k-1)H(k)P(k|k-1); \tag{13}$$

$$P(k|k-1) = [\Phi(k,k-1) - K_p(k-1)H(k-1)]P(k-1|k-1) \times [\Phi(k,k-1) - K_p(k-1)H(k-1)]^{\mathrm{T}} + \Gamma(k-1)Q(k-1)\Gamma^{\mathrm{T}}(k-1) - K_p(k-1)R(k-1)K_p^{\mathrm{T}}(k-1). \tag{14}$$

Note that $\hat{x}(k|k)$ is the optimal estimate of $x(k)$ using measurements up to and including $z(k)$ and

$$P(k|k) \triangleq \mathrm{E}\{[x(k) - \hat{x}(k|k)]^{\mathrm{T}}[x(k) - \hat{x}(k|k)]/z(1),\ldots,z(k)\}.$$

### *C. ANALYSIS TECHNIQUES*

During the second half of the 1960s, a great number of authors investigated errors in the Kalman filter when it is implemented using various approximations. For example, Heffes [11] studied the actual error covariance [$P$ matrix] for the discrete-time case by considering uncertainties in $P_0$, $Q$, and $R$; Nishimura [12] extended this study to include the continuous-time case. Neal [13] studied the effect of uncertainties in the $F$ matrix and a recursive equation for $P$ was derived. Griffin and Sage [14] have shown, for continuous and discrete filters, as well as continuous-smoother, the performance degradation due to uncertainties in $F$, $Q$, $R$, and $P_0$. Price [15] derived recursive equations for the suboptimal estimation error covariance matrix in the discrete-time case. Huddle and Wismer [16] studied the performance degradation when a reduced-order filter is used. Problems concerning error bounds and divergence were investigated by Sorenson [17], Fitzgerald [18], and Seblee *et al.* [19]. Problems concerning model testing were discussed by Berkovec [20] and Mehra [21]. Problems concerning practical implementations of the Kalman filter were reported by Schmidt [22], Huddle [23] and Batin and Levine [24]. In general, all research work done in these areas can be classified into three groups: covariance analysis, model testing, and implementation errors, which are summarized below.

### 1. *Covariance Analysis*

The basic approach of the covariance analysis is to determine the theoretical behavior of estimation errors in a suboptimal linear filter, suboptimal in the sense that the filter does not contain the exact model of the system dynamics and/or measurement process. In order to simplify this summary, further classification of the covariance analysis is necessary.

Filtering error. Filtering error is defined as the behavior of the error covariance when correct values of $F$ and $H$ are used in implementing the filter, but approximations are made in computing the filter gain. For example, consider the continuous-time case in which $Q$ and $R$ are uncorrelated. The error covariance for the filter with the same structure as the Kalman filter, but with an arbitrary gain matrix, is given by

$$\dot{P} = (F - KH)P + P(F - KH)^{T} + GQG^{T} + KRK^{T} \quad , \tag{15}$$

in which the time dependent notation $(t)$ for each symbol is omitted for clarity. The error covariance for the optimal filter in which the optimal gain $K$ does not appear explicitly is given by

$$\dot{P} = FP + PF^{T} - PH^{T}R^{-1}HP + GQG^{T} \quad . \tag{16}$$

Using Eq. (16) and design values of $F$, $H$, $G$, $Q$, $R$, and $P(o)$, compute the error covariance time function, which is used to compute $K$ in Eq. (15) (the filter gain would be optimal if the design values are correct). Next, using the correct values of $F$, $H$, $G$, $Q$, $R$, and $P(o)$, insert $K$ into Eq. (15) to compute the actual error covariance history. Since Eq. (15) is based only on the structure of the Kalman filter and not on any assumption that the optimal $K$ is employed, it can be used to analyze the filter error covariance for any set of filter gains.

For the discrete-time case, a pair of equations that are equivalent to Eq. (15) are

$$P_k(+) = (I - K_kH_k)P_k(-)(I - K_kH_k)^{T} + K_kR_kK_k^{T} \quad , \tag{17}$$

$$P_{k+1}(-) = \Phi_kP_k(+)\Phi_k^{T} + \Gamma_kQ_k\Gamma_k^{T} \quad . \tag{18}$$

For clarity, $P_k(+)$ and $P_{k+1}(-)$ are used to denote $P(k|k)$ and $P(k+1|k)$, respectively; also, subscript $k$ is used to denote $(k|k)$. Similarly, a pair of equations that are equivalent to Eq. (16) are

$$P_k(+) = P_k(-) - P_k(-)H_k^{T}(H_kP_k(-)H_k^{T} + R_k)^{-1} H_kP_k(-) \quad , \tag{19}$$

$$P_{k+1}(-) = \Phi_kP_k(+)\Phi_k^{T} + \Gamma_kQ_k\Gamma_k^{T} \quad . \tag{20}$$

<u>Filter sensitivity</u>. Filter sensitivity is defined as the behavior of error covariance when incorrect values of matrices are used in implementing the filter; optimal or other methods are used in computing the filter gain. The incorrect matrices may be attributed to

(1) When the number of states to be estimated is a subset of the total states.

Let $F^*$ and $H^*$ be the incorrect matrices and $K^*$ be the filter gain computed using the incorrect $F$ and $H$, i.e., $F \neq F^*$ and $H \neq H^*$.

From the fundamental theory of Kalman filtering, equations for the estimated state $\hat{x}$, is given by

$$\dot{\hat{x}}(k) = F^* \hat{x} + K^* [z - H^* \hat{x}] \quad . \tag{21}$$

Define

$$\tilde{x} = \hat{x} - x \quad . \tag{22}$$

From Eq. (21),

$$\dot{\tilde{x}} = (F^* - K^* H^*)\hat{x} - (F - K^* H)x - Gu + K^* v \quad . \tag{23}$$

Define

$$\Delta F = F^* - F \qquad \Delta H = H^* - H \quad .$$

Equation (23) can be written in terms of $F$ and $H$:

$$\dot{\tilde{x}} = (F^* - K^* H^*)\tilde{x} + (\Delta F - K^* \Delta H)x - Gu + K^* v \quad . \tag{24}$$

In order to obtain a set of desirable error sensitivity equations, define a new vector $x'$:

$$x' = \begin{bmatrix} x \\ \cdots \\ x \end{bmatrix} \quad . \tag{25}$$

Therefore the differential equation for $x'$ can be written as

$$\dot{x}' = \begin{bmatrix} F^* - K^*H^* & \vdots & \Delta F - K^*\Delta H \\ \cdots\cdots\cdots & \vdots & \cdots\cdots\cdots \\ 0 & \vdots & F \end{bmatrix} \begin{bmatrix} \tilde{x} \\ \\ x \end{bmatrix} + \begin{bmatrix} K^*v - Gu \\ \\ Gu \end{bmatrix} . \tag{26}$$

Let

$$E[x' \; x'^T] = \begin{bmatrix} P & Y^T \\ Y & W \end{bmatrix} , \tag{27}$$

where

$$P = E[\tilde{x} \; \tilde{x}^T] \; ; \quad Y = E[x \; \tilde{x}^T] \; ; \quad W = E[x \; x^T] \quad .$$

To obtain differential equations for $P$, $Y$, and $W$, take

$$\frac{d}{dt} (E[x' \; x'^T]) = \begin{bmatrix} \dot{P} & \dot{Y}^T \\ \dot{Y} & \dot{W} \end{bmatrix} \tag{28}$$

and yield

$$\dot{P} = (F^* - K^*H^*)P + P(F^* - K^*H^*)^T + (\Delta F - K^*\Delta H)Y$$
$$+ Y^T(\Delta F - K^*\Delta H)^T + GQG^T + K^*RK^{*T} \tag{29}$$

$$\dot{Y} = FY + Y(F^* - K^*H^*)^T + W(\Delta F - K^*\Delta H)^T - GQG^T \quad , \tag{30}$$

$$\dot{W} = FW + WF^T + GQG^T \quad , \tag{31}$$

since the initial uncertainty in the estimate is the same as the uncertainty in the state,

$$E[x(0)x(0)^T] = P(0) = Y(0) = W(0) \quad .$$

(2) When the number of states to be implemented is a subset of the total states.

Let $F^*$ and $H^*$ be the incorrect matrices, which do not have the same dimensions as the true matrices $F$ and $H$. To facilitate the analysis, it is desirable to form $F$ and $H$, and this can be done through an appropriate transformation, i.e.,

$$\Delta F = T^{\mathrm{T}} F^{*} T - F \quad , \tag{32}$$

where $T$ is a nonsquare matrice of proper dimensions. Similarly,

$$\Delta H = H^{*} T - H \quad , \tag{33}$$

since

$$X_{\text{implemented}} = T X_{\text{true}} \quad . \tag{34}$$

Repeating the same operations from Eqs. (24) to (28), the following sensitivity equations can be obtained:

$$\begin{aligned} \dot{P} = {} & T^{\mathrm{T}}(F^{*} - K^{*} H^{*}) T P + P T^{\mathrm{T}}(F^{*} - K^{*} H^{*})^{\mathrm{T}} T \\ & + (\Delta F - T^{\mathrm{T}} K^{*}\, \Delta H) Y + Y^{\mathrm{T}}(\Delta F - T^{\mathrm{T}} K^{*}\, \Delta H)^{\mathrm{T}} \\ & + G Q G^{\mathrm{T}} + T^{\mathrm{T}} K^{*} R K^{*\mathrm{T}} T \quad ; \end{aligned} \tag{35}$$

$$\dot{Y} = F Y + Y T^{\mathrm{T}}(F^{*} - K^{*} H^{*})^{\mathrm{T}} T + W(\Delta F - T^{\mathrm{T}} K^{*}\, \Delta H)^{\mathrm{T}} - G Q G^{\mathrm{T}} \quad ; \tag{36}$$

$$\dot{W} = F W + W F^{\mathrm{T}} + G Q G^{\mathrm{T}} \quad . \tag{37}$$

A similar set of sensitivity equations for the discrete filter was given by Gelb [25]. A complex set of equations for the case in which the filter state is a linear combination of the true states is given by Nash *et al.* [26].

### 2. *Model Testing*

The covariance analysis discussed above is based on certain assumed system model from which error equations are derived. However, it does not test the validity of the system model. In order to do that, actual system data must be used, and Mehra [21] has suggested an algorithm to test the optimality of the assumed system model. The concepts of correlation function and innovation process are used. The innovation process is defined as

$$\tilde{z} = z - \hat{z} \tag{38}$$

and the filter can be expressed in terms of $\tilde{z}$:

$$\dot{\hat{x}} = F\hat{x} + K\tilde{z} \quad , \tag{39}$$

in which $K$ is the optimal gain. By defining $\tilde{x} = \hat{x} - x$, the innovation process can also be written as

$$\tilde{z} = - H\tilde{x} + v \quad . \tag{40}$$

Thus, for $t_2 > t_1$, the correlation function is

$$\begin{aligned} \mathrm{E}[\tilde{z}(t_2)\tilde{z}(t_1)^{\mathrm{T}}] = {} & H(t_2)\ \mathrm{E}[\tilde{x}(t_2)\tilde{x}^{\mathrm{T}}(t_1)]H^{\mathrm{T}}(t_1) \\ & - H(t_2)\ \mathrm{E}[\tilde{x}(t_2)v^{\mathrm{T}}(t_1)] \\ & + R(t_1)\ \delta(t_2 - t_1) \end{aligned} \tag{41}$$

From Eqs. (4), (5), (38), and (39), it can be shown that $\tilde{x}$ satisfies the differential equation:

$$\dot{\tilde{x}} = (F - KH)\tilde{x} - Gu + Kv \quad . \tag{42}$$

The solution to Eq. (42) is

$$\begin{aligned} \tilde{x}(t_2) = {} & \Phi(t_2, t_1)\tilde{x}(t_1) - \int_{t_1}^{t_2} \Phi(t_2, \tau)[G(\tau)\ u(\tau) \\ & - K(\tau)v(\tau)]d\tau \end{aligned} \tag{43}$$

where $\Phi(t_2, t_1)$ is the transition matrix corresponding to $(F - KH)$. Using Eq. (43), the following two equations can be obtained:

$$\mathrm{E}[\tilde{x}(t_2)\tilde{x}^{\mathrm{T}}(t_1)] = \Phi(t_2, t_1)P(t_1) \quad , \tag{44}$$

$$\mathrm{E}[\tilde{x}(t_2)v^{\mathrm{T}}(t_1)] = \Phi(t_2, t_1)K(t_1)R(t_1) \quad . \tag{45}$$

Substituting Eqs. (44) and (45) into Eq. (41):

$$\begin{aligned} \mathrm{E}[\tilde{z}(t_2)\tilde{z}^{\mathrm{T}}(t_1)] = {} & H(t_2)\Phi(t_2, t_1)[P(t_1)H^{\mathrm{T}}(t_1) - K(t_1)R(t_1)] \\ & + R(t_1)\ \delta(t_2 - t_1) \quad . \end{aligned} \tag{46}$$

If the gain is optimal, i.e.,

$$K(t_1) = P(t_1)\, H^{T}(t_1) R^{-1}(t_1) \quad , \tag{47}$$

then Eq. (46) becomes

$$E[\tilde{z}(t_2)\tilde{z}^{T}(t_1)] = R(t_1)\, \delta(t_2 - t_1) \quad , \tag{48}$$

which is called the "innovation property." That is to say, when $K$ is the optimal gain, the correlation function of the innovation process is zero for $t_1 \neq t_2$. In other words, $\tilde{z}(t)$ is a white-noise process and there is no information left in it when $K$ is optimal. From the filter-testing point of view, the innovation property can be used as a criterion to test for optimality. From the filter-design point of view, Eq. (46) can be used to derive the optimal gain by "whitening" the innovation process.

In Mehra [21], a discrete, time-invariant steady-state system was considered. A correlation matrix of the last $N$ samples was constructed and the method of Jenkins and Watts was used to test whether the innovation sequence possessed the white noise property. The concept of whitening is an effective method to identify unknowns $Q$ and $R$ for known $F$, $G$, and $H$. However, if $F$, $G$, and $H$ are also unknown, the whiteness of the innovation sequence is not a sufficient condition for system identification. When this is the case, other methods (Section III) must be used.

In practice, an important consideration is placed on the real-time processing of the correlation matrix. The method for model testing discussed here may not be efficient from the computational point of view. This plus other computational problems relative to the implementation of the Kalman filter will be examined next.

### 3. *Implementation Errors*

The analysis of implementation errors deals with real-time applications of the Kalman filter, which is programmed in an on-line operational digital computer. Filtering errors occur when a simplified system model is used to represent the true systems model as a simplified algorithm is used to perform the required computation. These simplifications are necessary because programming the theoretical filter equation is impossible due to limitations of the digital computer. In this section, some practical considerations relative to modeling problems, computer problems, and special computing algorithms designed to minimize filtering errors will be reviewed.

Modeling problems. Performance of the Kalman filter based on assumed models operating in a real-time computer is usually degraded from theoretical predictions. The discrepancy between the predicted value and computed value is referred to as "divergence," which can be classified into two groups: "apparent" and "true". Apparent divergence occurs when modeling errors cause the implemented filter to be suboptimal. In this case, estimation errors approach values that are larger than those predicted by the theory. (This type of behavior is similar to those cases discussed in the covariance analysis section.) True divergence occurs when the filter is unstable or when there are unmodeled states causing the error to grow without bound. Figure 3 illustrates the behavior of these two types of divergence. Different methods in filter design dealing with divergence problems will be reviewed in Section III.

Computer constraints. The use of a digital computer to

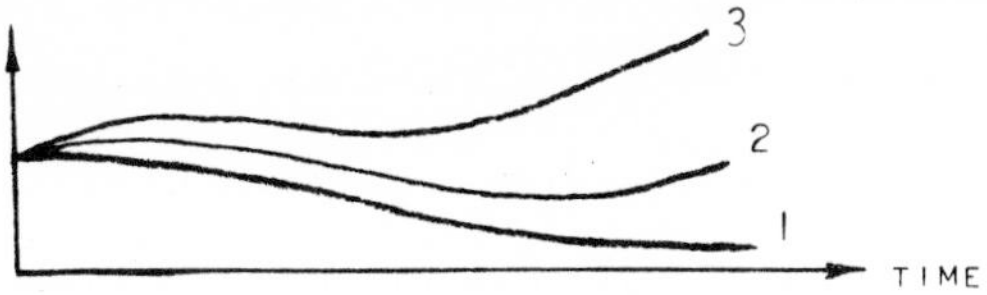


*FIG. 3. Filter errors versus time: (1) theoretically predicted errors; (2) apparent divergence; (3) true divergence.*

implement the Kalman filter is usually constrained by the complexity of the filtering equations that consequently place the burden on the computer speed and storage memory. In addition, because of the inherent nature of the digital computer, which cannot solve analytic equations accurately, truncation error occurs due to approximate mathematical operations and roundoff error occurs due to finite word length of the computer. In order to compromise between the computer constraint and the algorithm complexity, it is necessary to form a memory and time requirement estimate so that a cost-effectiveness tradeoff can be made. In practical situations the computer is often selected without tradeoff considerations and the filter is designed to fit the selected computer. In the case when the selected computer is inadequate, then simplification of the algorithm and reduction in number of computations are necessary. Available techniques to accomplish these include reducing the number of states, decoupling equations, or prefiltering. In addition, sensitivity to word length and roundoff errors can be minimized using special computing algorithms, which will be discussed next.

Special computing algorithms. Due to inherent finiteness of the digital computer, computational errors are unavoidable. However, specially designed algorithms can be devised to minimize them. This section examines a few of the many developed techniques.

*(1) Algorithm for integration.* For real-time applications, the Euler's algorithm for integration is found to be unsatisfactory often due to nonconvergence of the solution. The modified Euler algorithm, which considers the first three terms of the Taylor series, is found to be satisfactory. The Runge-Kutta method, which includes the fourth derivative, increases the accuracy of the approximation. However the storage memory, execution time, and roundoff errors are also increased. Hence a tradeoff must be made in deciding which method is to be used.

*(2) Algorithm for transition matrix.* The calculation of the transition matrix is a special case of the general integration problem, i.e., the solution of

$$\Phi = F\Phi \tag{49}$$

for constant $F$ is

$$\Phi(t_2, t_1) = e^{\Delta t F} = \sum_{n=0}^{\infty} \frac{(\Delta t)^n F^n}{n!} , \tag{50}$$

where

$$\Delta t = t_2 - t_1 .$$

When $\Delta t$ is less than the dominant time constants in the system, it can be computed with an accuracy in the order of six to eight decimal places for $n$ in the range of 12 to 15. If it is very much smaller than the dominant time constants, $n = 2$ or 3 may be sufficient.

For time-varying $F$, solution to Eq. (49) is

$$\Phi(t_2, t_1) = \sum_{n=0}^{\infty} \frac{(t_2 - t_1)^n}{n!} F^n(t_1) . \tag{51}$$

when $t_2 - t_1$ is much less than the time required for significant changes in $F$, the fundamental property of the transition matrix can be used:

$$\Phi(t_3, t_1) = \Phi(t_3, t_2)\Phi(t_2, t_1) \quad . \tag{52}$$

*(3) Algorithm for Q.* In practical applications, a time-continuous system is always being discretized to fit the digital computation since $Q$ and $\Phi$ are related by

$$Q_k = \int_{t_k}^{t_k+1} \Phi(t_{k+1}, \tau)\; Q(\tau)\Phi^T(t_{k+1}, \tau)d\tau \quad . \tag{53}$$

Using Leibnitz' formula, differentiation of Eq. (53) yields

$$\dot{Q}_k = FQ_k + Q_kF^T + Q \quad , \tag{54}$$

where $Q$ is constant and $Q_k(0) = 0$. Let the computation of $Q_k$ be divided into $N$ equal intervals, i.e.,

$$Q_k = Q(N\ \Delta t) \quad ; \tag{55}$$

hence for $N = 1$:

$$\begin{aligned} Q(\Delta t) &= Q_k(0) + [F_0Q_k(0) + Q_k(0)F_0^T + Q]\Delta t \\ &= Q\Delta t \quad ; \end{aligned} \tag{56}$$

for $N = 2$:

$$Q(2\Delta t) = 2Q\Delta t + (F_1Q + QF_1^T)\Delta t^2 \quad ; \tag{57}$$

in general,

$$\begin{aligned} Q(N\Delta t) &= NQ\Delta t \\ &\quad + \left(\sum_{i=1}^{N-1} iF_iQ + Q\sum_{i=1}^{N-1} iF_i^T\right)\Delta t^2 + Q\Delta t^3 \quad . \end{aligned} \tag{58}$$

*(4) Algorithm for avoiding matrix inversion.* Experiences indicate that matrix inversion requires large executive time and memory and introduces computation errors. Our purpose here

is to show that matrix inversion in the Kalman filter can be avoided. This can be done by considering the simultaneous measurement components that occur serially over a zero-time span, because the covariance update equation

$$P_k(+) = P_k(-) - P_k(-)\ H_k^{\mathrm{T}}[H_k P_k(-) H_k^{\mathrm{T}} + R_k]^{-1}\ H_k P_k(-) \tag{59}$$

contains an inverse of the dimension of the measurement vector, whose dimension can be taken to equal one.

*(5) "Joseph" algorithm for P.* The conventional equation for updating the covariance matrix is

$$P(+) = (I - KH)P(-) \quad . \tag{60}$$

The "Joseph" equation equivalent to (60) is

$$P(+) = (I - KH)P(-)(I - KH)^{\mathrm{T}} + KRK^{\mathrm{T}} \quad , \tag{61}$$

where

$$K = P(-)\ H^{\mathrm{T}}[HP(-)H^{\mathrm{T}} + R]^{-1} \quad . \tag{62}$$

Note that although Eq. (60) is simpler, it is computationally inferior to Eq. (61), as discussed in detail by Bucy and Joseph [27]. It can be shown that for a small change of $\delta K$, the equivalent change of Eq. (60) is

$$\delta P(+) = -\ \delta K\ HP(-) \quad , \tag{63}$$

and the change of Eq. (61) is

$$\delta P(+) = \delta K[RK^{\mathrm{T}} - HP(-)(I - KH)^{\mathrm{T}}] + [KR - (I - KH)P(-)H^{\mathrm{T}}]\ \delta K^{\mathrm{T}} \quad . \tag{64}$$

Substituting Eq. (62) into Eq. (64) yields $\delta P(+) = 0$ to the first-order approximation. It is apparent that the Joseph algorithm gives a better accuracy. However, it requires considerably more computation time than the conventional algorithm.

*(6) The square root algorithm.* In order to assure that the covariance matrix is always positive definite, a matrix $A$ is introduced, such that $P = AA^{T}$. The advantage of this so-called square root algorithm is that it gives the same accuracy in single precision as does the conventional method in double precision. The disadvantage is that matrix $A$ is not unique, and this leads to many different versions of the same theory. Contributors in this area, including Andrews [28], Bellantoni and Dodge [29], Dyer and McReynolds [30], and Kaminski *et al.* [31], have compiled a bibliography on this subject.

Another disadvantage of the square root algorithm is that, in general, extra calculations are required. Recently, Carlson [32] has devised an algorithm using a lower triangle form for matrix $A$ to improve computational speed. It was demonstrated that this algorithm matches or exceeds the speed of all conventional methods depending on the dimension of $P$.

*(7) Chandrasekhar algorithm.* The solution to the Kalman filtering problem is usually obtained by solving an $n \times n$ matrix Riccati equation. In general, $n(n + 1)/2$ simultaneous nonlinear differential equations have to be solved. Recently, in an attempt to minimize the calculation burden of the computer, Kailath [33], Lindquist [34], and Morf *et al.* [35] have developed, based on the work of Chandrasekhar [36], a variation of the Kalman-Bucy filter that is computationally advantageous under certain conditions. The simplified algorithm, applicable only to linear time-invariant systems, is called the Chandrasekhar method because it is similar to the technique that he used to solve the Wiener-Hopf equation in finite-time intervals.

The distinct advantage of this algorithm is that it solves

only $2np$ or $n(m + p)$ simultaneous nonlinear differential equations, where $m$ is the dimension of $G$ and $p$ is the dimension of $H$. For example, Desalu *et al.* [37] have applied this algorithm to study a problem concerning air pollution, in which only 125,000 computations are required as compared to the conventional Riccati-based algorithm that requires 72 million computations.

## III. SYNTHESIS

### A. *INTRODUCTION*

In nearly two decades, the problem of filter synthesis has been studied concurrently with the problem of filter analysis. Similar to the analysis, synthesis also encompasses a large area of interest. For example, Magill [38] considered the Gaussian scalar Markov process and studied the optimal adaptive filter problem in which certain constant parameters of the system are unknown, but assumed that they can be initially selected at random from a finite set of values with known *a priori* probabilities of selection. Hilborn and Lainiotis [39] extended Magill's results to include the vector case, in which neither assumption in Gaussian process nor constant unknown parameters are necessary. Shellenbarger [40] used the maximum likelihood method to develop an algorithm for the suboptimal filter in which $Q$ and $R$ are the only unknown parameters. Because the computation requires a ratio relationship between the system and measurement states, the usefulness of this method is limited. Kailath [41] and Mehra [42] used the innovation sequence correlation technique to solve the same problem by determining $Q$ and $R$ first, then using the results to determine the filter gain $K$. Carew and Belanger [43] determined $K$ directly without the intermediate step of finding $Q$ and $R$. Friedland [44] considered that the mean values of the system noise and measurement noise as the only unknown parameters, which are viewed as additional state vector, can be estimated by decoupling the bias estimate from the state estimate. Lin and Sage [45] solved the same problem using the invariant imbedding technique. Sage and Husa [46] considered a more general case, in which $Q$, $R$, as well as the means are unknown; in addition, $P$ is not restricted to be a diagonal matrix. Godbole [47] solved the same problem by applying the innovation sequence correlation concept to show that the identification of the noise covariance can be achieved without the knowledge of noise means. Thus the estimation of Gaussian noise statistics can be treated separately in two stages: first the covariance, then the means. Chin [48] applied similar concepts to solve an inertial navigation system accuracy improvement problem, and computer simulation results were obtained for special cases. Boroson and Schwartz [49] derived an algorithm for simultaneous parameter identification and state observation of a linear, time-invariant system. Krause and Graupe [50] solved the problem of simultaneously identifying both transition matrix elements as well as $Q$ and $R$, for the case where no *a priori* information on any of these parameters existed. Lainiotis [51,52] treated the general problem concerning simultaneous adaptation of system structure and parameters as well as adaptation of system structure alone. Mostafa [53] studied the specific problem of identifying unknown dynamics of an electric power system.

Another important aspect of filter designing is the

computational consideration concerning the problem of divergence. For example, Schmidt [54] attempted to bound the estimated error by constraining the filter gain but including the effects of unknown parameters. Cosaert and Gottzein [55] developed an algorithm (decoupled shifted memory) to avoid divergence and reduce the sensitivity of performance due to unknown noise parameters. Jazwinski [56] solved the problem by approximating model errors with a Gaussian white noise input and estimated $Q$ and $R$ in an adaptive manner. Jazwinski [57] also considered a deterministic plant and designed an algorithm (limited memory filter) from a probabilistic viewpoint. Other similar error-bounding techniques have been developed by Tarn and Zaborsky [58], Miller [59], Sacks and Sorenson [60], Nahi and Schaefer [61], and Jazwinski [62].

In summary, the discussion of filter synthesis may be approached from two standpoints: adaptive techniques and error-bounding (divergence preventing) techniques.

## B. *ADAPTIVE TECHNIQUES*

In general adaptive problems can be divided into three classes: noise parameter adaptation, system parameter adaptation, and system structure adaptation.

### 1. *Noise Parameter Adaptation*

Many techniques are available for the adaptation of the first- and second-order statistics of the Gaussian noise that assumed to be system and measurement disturbances. The following four well-known approaches are presented:

a. <u>Bayesian approach</u>. The development of the Bayesian approach is found in ref. [38]. The usefulness of this technique is not limited to the identification of noise parameters, it is also applicable to identifying unknown parameters of $F$ or $\Phi$ matrix as well as $X(0)$. In this method, all unknown parameters are being represented by a vector $\alpha$. Assume that all possible values of $\alpha$ form a finite set of stochastic processes ($\{\alpha_i\}$, $i = 1, \ldots, N$) with known *a priori* probability of each process occurring $p(\alpha_i)$. Consider the scalar case of Eqs. (1) and (2), in which the system noise variance $R$ is known to have one of $N$ possible values $[R_i\colon\ i = 1, \ldots, N]$. The problem of finding the best estimate of $x$ (in the sense of minimizing a quadratic error function) can be formulated as $N$ different stochastic processes (generated by $N$ different possible parameters and noise covariances) in which the *a priori* probability of each process occurring is known. The solution to this problem is, of course, the conditional mean:

$$\hat{x} = \int_A x p(x|z_k)\, dx \quad , \tag{65}$$

where $\hat{x}$ is the best estimate of $x$, $A$ the state space of $x$, $p(x|z_k)$ the conditional probability of $x$ given measurement data $z_k$, which is a set consisting of $k$ elements denoted by

$$z_k = \{z^1, z^2, \ldots, z^k\} \quad . \tag{66}$$

In the case of parameter adaptation, it is required to obtain recursive equations for the *a posteriori* probability density of $x$, the unknown parameter $\alpha$ and the conditional probability function $p(\alpha|z_k)$. For the continuous-time system

$$\hat{x} = \int_A \hat{x}(\alpha)\ p(\alpha|z_k)\, d\alpha \quad . \tag{67}$$

For the discrete-time system:

$$\hat{x} = \sum_{i=1}^{N} \hat{x}(\alpha_i)\ p(\alpha_i|z_k) \quad . \tag{68}$$

Equations (67) and (68) are optimal Bayes estimators, which consist of a combination of estimators, each calculates a value for $\hat{x}(\alpha_i)$ and each $\hat{x}(\alpha_i)$ is being weighted by $p(\alpha_i|z_k)$.

If the stochastic process is assumed to be Gauss-Markov, an algorithm for the *adaptive* optimal estimator can be derived. Consider the following time-discrete system:

$$x(k+1;\alpha_i) = \Phi(k|\alpha_i)x(k;\alpha_i) + \Gamma(k|\alpha_i)u(k|\alpha_i) \quad ; \tag{69}$$

$$z(k;\alpha_i) = H(k|\alpha_i)x(k;\alpha_i) + v(k|\alpha_i) \quad . \tag{70}$$

Accordingly, the following notations will be used in the adaptive filter: $Q(k|\alpha_i)$, $R(k|\alpha_i)$, $K(k|\alpha_i)$, etc., to indicate that these matrices are functions of $\alpha_i$. Let $\hat{x}(k;\alpha_i)$ be the best estimate of $x(k;\alpha_i)$, i.e.,

$$\hat{x}(k;\alpha_i) = \Phi(k|\alpha_i)\hat{x}(k-1;\alpha_i) + K(k|\alpha_i)\tilde{z}(k-1;\alpha_i) \quad , \tag{71}$$

where

$$\tilde{z}(k;\alpha_i) = z(k) - \hat{z}(k;\alpha_i) \quad , \tag{72}$$

$$\hat{z}(k;\alpha_i) = H(k|\alpha_i)\Phi(k|\alpha_i)\hat{x}(k-1;\alpha_i) \quad , \tag{73}$$

$$K(k|\alpha_i) = P_x(k;\alpha_i)H^{\mathrm{T}}(k|\alpha_i)P_z^{-1}(k;\alpha_i) \quad , \tag{74}$$

$$P_x(k;\alpha_i) = P_x(k-1;\alpha_i) - K(k|\alpha_i)P_z(k;\alpha_i)K^{\mathrm{T}}(k|\alpha_i) \quad , \tag{75}$$

$$P_x^{+}(k-1;\alpha_i) = \Phi(k-1|\alpha_i)P^{-1}(k-1;\alpha_i)\Phi^{\mathrm{T}}(k-1|\alpha_i) + \Gamma(k-1|\alpha_i)Q(k-1|\alpha_i)\Gamma^{\mathrm{T}}(k-1|\alpha_i) \quad , \tag{76}$$

$$P_z(k;\alpha_i) = H(k|\alpha_i)P(k,\alpha_i)H^{\mathrm{T}}(k|\alpha_i) + R(k|\alpha_i) \quad , \tag{77}$$

note that

$$P_x(k;\alpha_i) = \mathrm{E}[\tilde{x}(k;\alpha_i)\tilde{x}^{\mathrm{T}}(k;\alpha_i)] \quad , \tag{78}$$

$$P_z(k;\alpha_i) = \mathrm{E}[\tilde{z}(k;\alpha_i)\tilde{z}^{\mathrm{T}}(k;\alpha_i)] \quad . \tag{79}$$

Since the adaptive filter is

$$\hat{x}(k|k) = \sum_{i=1}^{N} \hat{x}(k|k;\alpha_i)p(\alpha_i|z_k) \quad , \tag{80}$$

it remains to show how $p(\alpha_i|z_k)$, the nonlinear weighting coefficient, can be computed. Using Bayes rule:

$$p(\alpha_i|z_k) = \frac{p(z_k|\alpha_i)p(\alpha_i)}{\sum_{j=1}^{N} p(z_k|\alpha_j)p(\alpha_j)} \quad . \tag{81}$$

Since $p(\alpha_i)$ is given and $p(z_k|\alpha_i)$ is obtained from the system equations, then $p(\alpha_i|z_k)$ can be written as

$$p(\alpha_i|z_k) = \frac{|P_z(k|k-1;\alpha_i)|^{-1/2} \exp[-\frac{1}{2} L(k|\alpha_i)]p(\alpha_i|z_{k-1})}{\sum_{j=1}^{N} |P_z(k|k-1;\alpha_j)|^{-1/2} \exp[-\frac{1}{2} L(k|\alpha_i)]p(\alpha_i|z_{k-1})} \tag{82}$$

where

$$P_z(k|k-1;\alpha_i) = H(k|\alpha_i)P_x(k|k-1;\alpha_i)H^T(k|\alpha_i) + R(k|\alpha_i) \quad , \tag{83}$$

$$L(k|\alpha_i) = \tilde{z}^T(k|k-1;\alpha_i)P_z^{-1}(k|k-1;\alpha_i)\tilde{z}(k|k-1;\alpha_i) \quad , \tag{84}$$

$$\tilde{z}(k|k-1;\alpha_i) = z(k) - \hat{z}(k|k-1;\alpha_i) \quad , \tag{85}$$

$$z(k) = \{z_1, z_2, \ldots, z_k\} \quad , \tag{86}$$

$$\hat{z}(k|k-1;\alpha_i) = H(k|\alpha_i)\Phi(k|\alpha_i)\hat{x}(k|k-1;\alpha_i) \quad . \tag{87}$$

b. Maximum likelihood approach. When the probability density function of the uncertainty in the system is not available, the maximum likelihood method may be used to design an adaptive filter. This approach is based on the concept that the most likely value of the unknown is the value that yields the maximum probability of occurrence within a given set of measurements

$$Z_k = \{z_1, z_2, \ldots, z_k\} \quad . \tag{88}$$

Consider the parameter adaptation problem in which the state vector $x$ and the unknown parameter vector $\alpha$ are to be estimated simultaneously. Let $h(x_k, \alpha, Z_k)$ be a function to be maximized, i.e.,

$$h(\hat{x}_k, \hat{\alpha}, Z_k) = \max_{x_k, \alpha} h(x_k, \alpha, Z_k) \quad , \tag{89}$$

where $h(\cdot)$ is the likelihood function chosen to be the conditional probability density function

$$h(x_k, \alpha, Z_k) = p(x_k, \alpha | Z_k) \quad . \tag{90}$$

The maximizing values for $x_k$ and $\alpha$ can be obtained by equating the derivative of log $\{p(x_k, \alpha | Z_k)\}$ to zero, i.e., let

$$L(x_k, \alpha) = \log p(x_k, \alpha | Z_k) \tag{91}$$

and set $\partial L/\partial x_k = 0$ and $\partial L/\partial \alpha = 0$. Using Bayes' rule to expand the probability density function, it can be shown that

$$\begin{aligned} L(x_k, \alpha) = &-\frac{1}{2} \{\log|P_{k|k}(\alpha)| + ||x_k - \hat{x}_{k|k}(\alpha)||^2_{P^{-1}_{k|k(\alpha)}} \\ &- \frac{1}{2} \sum_{i=1}^{k} ||z_i - H_i \hat{x}_{i|i}(\alpha)||^2_{(H_i P_{i|i-1}(\alpha) H_i^{\mathrm{T}} + R)^{-1}} \\ &+ \log|H_i P_{i|i-1}(\alpha) H_i^{\mathrm{T}} + R|\} \\ &+ \log P(\alpha) + \text{const} \quad , \end{aligned} \tag{92}$$

taking the derivative of $L(x_k, \alpha)$ with respect to the $j$th component of $\alpha$,

$$\begin{aligned} \frac{\partial L(x_k, \alpha)}{\partial \alpha^j} = &-\frac{1}{2} \operatorname{Tr} \left(P^{-1}_{k|k} \frac{\partial P_{k|k}}{\partial \alpha^j}\right) + \frac{1}{2} \sum_{i=1}^{k} \{\operatorname{Tr}(B_i^{-1} - B_i^{-1} \\ &\times \nu_i \nu_i^{\mathrm{T}} B_i^{-1}) \frac{\partial B_i}{\partial \alpha^j} - 2B_i^{-1} \nu_i \frac{\partial \hat{x}^{\mathrm{T}}_{i|i-1}}{\partial \alpha^j} H_i^{\mathrm{T}}\} + \frac{\partial \log P(\alpha)}{\partial \alpha^j} \end{aligned} \tag{93}$$

where Tr(•) is the trace of a matrix and

$$B_k = H_k P_{k|k-1} H_k^{\mathrm{T}} + R \quad , \tag{94}$$

$$\nu_k = z_k - H_k \hat{x}_{k|k-1} \quad , \tag{95}$$

$$P_{k|k-1} = \mathrm{E}\{(x_k - \hat{x}_{k|k-1})(x_k - \hat{x}_{k|k-1})^{\mathrm{T}} | Z_k, \alpha\} \quad . \tag{96}$$

Note that if no *a priori* distribution of $\alpha$ is given, the terms $\log p(\alpha)$ in Eq. (92) and $\partial \log p(\alpha)/\partial\alpha^j$ in Eq. (93) do not exist. In other words, the joint probability function of the measurements, given $\alpha$, is maximized. Equation (93) must be solved simultaneously for $\alpha$. Using the Newton-Raphson approximation method together with the assumption that (a) the system is time-invariant, completely controllable and completely observable; (b) adaptation begins after the filter has reached a steady state; (c) the last term of Eq. (93) does not exist; (d) the first term of Eq. (93) is neglected since for large $k$, it is much smaller than the second term, and (e) $\alpha$ is chosen from elements of $[\,{}^{B}_{K}\,]$ or a transformation of it, ref. [42] has shown that the suboptimal maximum likelihood filter is given by

$$\hat{x}_{k+1|k} = \Phi[\hat{x}_{k|k-1} + \hat{K}_k \nu_k] \quad , \tag{97}$$

$$\hat{K}^{jm}_{k+1} = \hat{K}^{jm}_{k} + A^{-1}_{k+1} g_{k+1} \quad , \tag{98}$$

where $K^{jm}_k$ denotes the $j,m$ element of the $K$ matrix based on measurements up to $k$.

$$A_{k+1} = A_k + \mathrm{Tr}\,\left(H \frac{\partial \hat{x}_{k+1|k}}{\partial K^{jm}} \cdot \frac{\partial \hat{x}^{\mathrm{T}}_{k+1|k}}{\partial K^{jm}} H^{\mathrm{T}}\right) \quad , \tag{99}$$

$$g_{k+1} = g_k + \mathrm{Tr}\,\left(\nu_{k+1} \frac{\partial \hat{x}_{k+1|k}}{\partial K^{jm}} H^{\mathrm{T}}\right) \quad , \tag{100}$$

$$\frac{\partial \hat{x}_{k+1|k}}{\partial K^{jm}} = \Phi(I - \hat{K}_k H) \frac{\partial \hat{x}_{k|k-1}}{\partial K^{jm}} + I_{jm} \nu_k , \tag{101}$$

where $I_{jm}$ is a matrix of all zeros except one in the $j,m$ element.

Abramson [63] used Eq. (93) to develop an algorithm for the case in which the vector $\alpha$ consists of two unknown quantities, $Q$ and $R$. The following results are obtained:

$$\hat{Q}_k^{jj} = \frac{k-1}{k} \hat{Q}_{k-1}^{jj} + \frac{1}{k} [\Gamma_k^{-1}(\Delta x^* \Delta x^{*T} + P_{k|k}^* - U_k^*)(\Gamma_k^{-1})^T]^{jj}, \tag{102}$$

$$\hat{R}_k^{jj} = \frac{k-1}{k} \hat{R}_{k-1}^{jj} + \frac{1}{k} [\Delta z_k^* \Delta z_k^{*T} + H_k P_{k|k}^* H_k^T]^{jj} , \tag{103}$$

where

$$\Delta x_k = P_{k|k-1} H_k^T B_k^{-1} \Delta z_k , \tag{104}$$

$$\Delta z_k = z_k - H_k \hat{x}_{k|k-1} , \tag{105}$$

$$U_k = \Phi_k P_{k|k-1} \Phi_k^T . \tag{106}$$

In deriving Eqs. (102) and (103), the following approach is taken:

(1) $Q$, $R$, and $x_k$ are initially estimated independent of *a priori* information on the probability density functions of $Q$ and $R$. After the initial calculations are obtained, the *a priori* estimates and their covariances are considered in obtaining a combined estimate of $Q$ and $R$.

(2) In computing $Q$, $R$ is assumed to be known and vice versa. Since estimates of $Q$ and $R$ are dependent, it is expected that biased estimates in them are inseparable.

Two other types of maximum likelihood adaptive filter may be considered as special cases.

(1) Conditional mode estimate, in which the conditional probability density function $p(x_k|Z_k)$ is maximized with respect to $x$. This adaptive filter can be obtained using the Bayesian approach discussed previously.

(2) Marginal maximum likelihood, in which the marginal density function $p(\alpha|Z_k)$ is maximized with respect to $x$. This special case reduces the complexity of Eq. (92) to

$$L(\alpha) = -\frac{1}{2}\sum_{i=1}^{k}\{||z_i - H_i\hat{x}_{i|i-1}(\alpha)||^2_{(H_iP_{i|i-1}(\alpha)H_i^{\mathrm{T}}+R)^{-1}}$$
$$+ \log|H_i P_{i|i-1}(\alpha) H_i^{\mathrm{T}} + R|\}$$
$$+ \log p(\alpha) + \text{const} \quad . \tag{107}$$

c. <u>Covariance-matching method</u>. This method is applicable to the design of adaptive filter in which $Q$ and $R$ are the only unknowns. The basic concept of this method is to make the actual covariances (obtained from the filter) consistent with their theoretical values. As an example, consider the innovation sequence

$$\nu_k = -H\tilde{x}_k + v_k \quad , \tag{108}$$

which has a theoretical covariance

$$\mathrm{E}[\nu_k\nu_k^{\mathrm{T}}] = HP_{k|k-1}H^{\mathrm{T}} + R \quad , \tag{109}$$

where $P_k$ is the covariance of $\tilde{x}_k$ and is related to $Q$ by

$$P_{k|k-1} = \Phi P_{k-1|k-1}\Phi^{\mathrm{T}} + \Gamma_{k|k-1}Q\Gamma^{\mathrm{T}}_{k|k-1} \quad . \tag{110}$$

The actual covariance of $\nu_k$ can be approximated by taking the sample of

$$C = \frac{1}{m}\sum_{i=1}^{m}\nu_i\,\nu_i^{\mathrm{T}} \tag{111}$$

If the comparison between $C$ and $E[\nu_k \nu_k^T]$ shows that the actual covariance is much smaller than the theoretical value, then the value of $Q$ must be decreased in order to match the covariances. This is done through Eq. (110), which in turn decreases the value of the expression $[HP_kH^T + R]$. Since the true values of $Q$ and $R$ are unknown, the convergence of the matching technique is poor. However, in some applications when $Q$ is known and $R$ is the only unknown, it can be estimated by using

$$\hat{R}_k = \frac{1}{m} \sum_{j=1}^{m} \nu_{k-j} \nu_{k-j}^T - HP_{k|k-1} H^T \quad , \tag{112}$$

where $P_{k|k-1}$ is obtained from the filter. The convergence of this estimate is good.

Another example of applying the covariance matching technique can be seen in the mechanization of Eqs. (102) and (103). Initially the values of $P^*_{k|k}$, $U^*_k$, and $\Delta x^*$ are computed as functions of *a priori* estimates of $Q$ and $R$ or some past estimates. If the actual $Q$ and $R$ are significantly different from their theoretical values, a feedback scheme is used to adjust the differences. In ref. [63], a technique was developed to evaluate the filter performance and compare numerical results. In general, the convergence of this and other covariance-matching techniques are very doubtful.

d. Correlation method. This method is applicable to the design of adaptive filter in which the second-order statistics of the system and measurement noises are the only unknowns. The basic concept of this method is to correlate the output or a function related to the output of the system. From this operation a set of equations is generated relating the autocorrelation function to the unknown parameters, which can be solved algebraically. The correlation method has two variations (stochastic approximations are considered as special cases, not variations), depending on the manner in which the autocorrelation function is formed. If it is generated directly from the output, the method is referred to as output correlation, if it is generated from the innovation sequence, it is called innovation correlation. The output correlation method is applicable to systems in which $\Phi$ is stable or $z_k$ is a stationary sequence only, whereas the innovation correlation method is applicable to cases in which $\Phi$ is not stable. However, both methods require the system to be completely controllable and completely observable. In general, estimates obtained using the innovation correlation method are more efficient than those obtained from the output correlation method, and this is due to the fact that innovation sequence is less correlated than the output sequence.

*(1) Output correlation.* Assume that the output sequence $z_k$ is stationary so that the sutocorrelation is a function of lag only. Define

$$C_\ell = E\{z_k \, z_{k-\ell}^T\} \tag{113}$$

in which $\ell$ is the finite lag. From Eqs. (6) and (7):

$$C_\ell = \begin{cases} HSH^T + R \quad , & \ell = 0 \\ H\phi SH^T \quad , & \ell > 0 \end{cases} \tag{114}$$

where

$$S_k = E\{x_k \, x_k{}^T\} \quad , \tag{115}$$

$$S_k = \phi S_{k-1}\phi^T + \Gamma Q \Gamma^T \quad . \tag{116}$$

$Q$ and $R$ can be solved as follows:

First obtain a set of correlation functions for $\ell = 1, \ldots, m$, then write

$$\begin{bmatrix} C_1 \\ \cdot \\ \cdot \\ \cdot \\ C_m \end{bmatrix} = ASH^{\mathrm{T}} \quad , \tag{117}$$

where

$$A^{\mathrm{T}} = [\phi^{\mathrm{T}} H^{\mathrm{T}}, \ \ldots, \ (\phi^{\mathrm{T}})^m H^{\mathrm{T}}] \quad ; \tag{118}$$

assume that the system is completely observable and $\phi$ is non-singular, then rank $(A) = m$ and the inverse for $(A^{\mathrm{T}}A)$ exists, yielding

$$SH^{\mathrm{T}} = (A^{\mathrm{T}}A)^{-1}A^{\mathrm{T}} \begin{bmatrix} C_1 \\ \cdot \\ \cdot \\ \cdot \\ C_m \end{bmatrix} . \tag{119}$$

To obtain $R$, let $\ell = 0$ and from Eq. (114):

$$R = C_0 - H\ (SH^{\mathrm{T}}) \quad . \tag{120}$$

To obtain $Q$, Eqs. (119) and (116) must be solved simultaneously. In general a unique solution for $Q$ is not possible. However, a unique $K$, the optimal steady-state filter gain, can be obtained:

$$K = (S - D)\ H^{\mathrm{T}}\ (C_0 - HDH^{\mathrm{T}})^{-1} \quad , \tag{121}$$

where

$$D_k = \mathrm{E}\{\hat{x}_{k|k-1}\ \hat{x}^{\mathrm{T}}_{k|k-1}\} \quad , \tag{122}$$

which can be shown to be

$$D_k = \phi[D_{k-1} + KBK^{\mathrm{T}}]\phi^{\mathrm{T}} \quad , \tag{123}$$

where

$$B = C_0 - HDH^{\mathrm{T}} \quad . \tag{124}$$

Substitute Eqs. (124) and (121) into Eq. (123):

$$D_k = \phi[D_{k-1} + (S - D_{k-1})H^{\mathrm{T}}$$
$$\times (C_0 - HD_{k-1}H^{\mathrm{T}})^{-1}H(S - D_{k-1})]\phi^{\mathrm{T}} \quad . \tag{125}$$

$D_k$ can be solved using Eq. (125), since $SH^{\mathrm{T}}$ is computed using Eq. (119). Finally, the autocorrelation for the output sequence must be estimated.

Using the ergodic property of $z_k$:

$$\hat{C}_\ell = \frac{1}{N} \sum_{i=\ell}^{N} z_i z_{i-\ell}^{\mathrm{T}} \quad , \tag{126}$$

where $N$ is the sample size, Eq. (126) can be computed recursively for $\ell = 0, 1, 2, \ldots, n$:

$$\hat{C}_\ell^{N} = \hat{C}_\ell^{N-1} + \frac{1}{N} (z_N z_{N-\ell}^{\mathrm{T}} - \hat{C}_\ell^{N-1}) \quad . \tag{127}$$

*(2) Innovation correlation.* The innovation process and innovation sequence are defined in Eqs. (38) and (108), respectively. In the previous section it was shown that if the Kalman filter is optimal, then $\nu$ must be a zero-mean Gaussian white noise process. Conversely, if the filter is not optimal, the innovation process must be correlated. Consider a discrete-time system and define

$$C'_\ell = \mathrm{E}[\nu_k \nu_{k-\ell}^{\mathrm{T}}] \quad , \tag{128}$$

where $\nu_k$ is a stationary Gaussian random sequence. Substituting Eq. (108) into Eq. (128):

$$C'_\ell = \mathrm{E}[(H\tilde{x}_k + v_k)(H\tilde{x}_{k-\ell} + v_{k-\ell})^{\mathrm{T}}] \quad ; \tag{129}$$

the following results can be derived easily:

$$C'_\ell = \begin{cases} HPH^T + R & \text{for} \quad \ell = 0 \\ H[\phi(I - KH)]^{\ell-1} \phi[PH^T - KC'_0] & \text{for} \quad \ell > 0 \end{cases} \tag{130}$$

The autocorrelation for the innovation sequence can be estimated:

$$\hat{C}'_\ell = \frac{1}{N} \sum_{k=\ell}^{N} \nu_k \nu^T_{k-\ell} \quad . \tag{131}$$

Define

$$A = \begin{bmatrix} H\phi \\ H\phi(I - KH)\phi \\ \cdot \\ \cdot \\ \cdot \\ H[\phi(I - KH)]^{\ell-1} \phi \end{bmatrix} . \tag{132}$$

Following the same procedure developed previously for the output correlation method, then for $\ell > 0$, Eq. (130) becomes

$$PH^T = KC'_0 + A^\# \begin{bmatrix} \hat{C}'_1 \\ \cdot \\ \cdot \\ \cdot \\ \hat{C}'_\ell \end{bmatrix} ; \tag{133}$$

where $A^\#$ is the pseudo inverse of matrix $A$, hence for $\ell = 0$:

$$\hat{R} = \hat{C}'_0 - H(PH^T) \quad ; \tag{134}$$

and for $\ell > 0$:

$$\sum_{j=0}^{k-1} H\phi^j \Gamma \hat{Q} \Gamma^T (\phi^{j-\ell})^T H^T$$

$$= HP(\phi^{-\ell})^T H^T - H\phi^\ell PH^T$$

$$- \sum_{j=0}^{\ell-1} H\phi^j \hat{r} (\phi^{j-\ell})^T H^T \quad , \tag{135}$$

where

$$\hat{r} = \phi[- KHP - PH^T K^T + K\hat{C}'_0 K^T]\phi^T \quad , \tag{136}$$

and $K$ is the suboptimal filter gain. Equations (134) and (135) are the main results from which $Q$ and $R$ can be estimated and denoted by $\hat{Q}$ and $\hat{R}$.

Reference [43] solved the same problem using autocorrelation functions of the innovation sequence of the suboptimal filter to determine the optimum filter steady state gain directly, without the intermediate determination of the unknown covariances for the system and measurement noises. This approach is to identify an output equivalent representation of the original system that does not directly involve the unknown covariances. The difference between this solution and that given in ref. [21] is that this solution requires only matrix multiplication, whereas the other solution requires solving the Lyapunov-type matrix equation. Additionally, the stability of the algorithm presented in ref. [43] appears to be less dependent on the assumed values of the unknown noise covariance. However, the major disadvantage of both approaches is their implicit assumption that the sample correlation function can be considered as equivalent to the true correlation function for the purpose of developing the estimation algorithm. This assumption simplifies the algebraic manipulations, but at the same time rejects the probabilistic interpretation of the problem. In order to account for the errors in the measured correlation function, Belanger [64] has shown that if the covariance matrices are linear in a set of parameters, then the correlation function of the innovation sequence is also a linear function of this set of parameters. This fact was used to perform a weighted least-squares fit to a sequence of measured correlation product, and an algorithm was developed to provide recursive estimates of $Q$ and $R$. Neethling and Young [65] performed simulations for various innovation correlation techniques; results indicated that the weighted least-squares approach yields considerable improvements over the approaches discussed in refs. [21] and [43]. For this reason, the weighted least-squares technique will be used to solve a practical problem discussed in Section IV.

Adaptive schemes discussed thus far are emphasized on the identification of the second-order statistics of the white Gaussian noise that influenced the dynamics and measurements of a linear stochastic system. In other words, adaptive problems are formulated with *a priori* knowledge that the first-order statistics are known exactly. This is, of course, not the case in practical situations. Therefore the problem concerning the adaptation of both the second- and first-order statistics of a practical system will be discussed in Section IV.

### 2. *System Parameter Adaptation*

The topic of system parameter adaptation involves discussions of identifying uncertain elements in the $\phi$ and $H$ matrices. It is well known that this problem could be solved using the Bayesian approach, as previously mentioned. The goal here is to provide a different method that is closer to statistics than probability, i.e., the system output sequence $z_k$ is being treated as a mixed autoregressive moving average process. In developing the adaptive filter for on-line identification of unknown system parameters, the output correlation method is applicable and the following assumptions are necessary:

(i) Computation for parameter adaptation starts after the Kalman filter has reached a steady state.

(ii) $\phi$ is stable and nonsingular.

(iii) The system is completely controllable and completely observable.

(iv) The system is minimum phase with a scalar-input and scalar-output.

Recall that in the discussion of output correlation method for noise statistics identification it was shown that the output correlation function $C_\ell$ is given by Eq. (114). Using Eqs. (114) - (119), the following relation is obtained for $\ell = n+1$:

$$C_{n+1} = H\phi^{n+1} A^{-1} \begin{bmatrix} C_1 \\ \cdot \\ \cdot \\ \cdot \\ C_n \end{bmatrix} , \tag{137}$$

which can be simplified using the Caley-Hamilton theorem, i.e., a matrix is a root of its characteristic equation:

$$\phi^n + b_n\phi^{n-1} + \ldots + b_1 = 0 , \tag{138}$$

where $b_1, b_2, \ldots, b_n$ are the coefficients in the polynomial of $\phi$. Using Eq. (118), it can be seen that

$$H\phi^{n+1} A^{-1} = - [b_1, \ldots, b_n] . \tag{139}$$

Hence Eq. (137) can be simplified

$$C_{n+1} = - \sum_{i=1}^{n} b_i C_i , \tag{140}$$

which can be written in a more general form:

$$C_{n+j} = - \sum_{i=1}^{n} b_i C_{i+j-1} \quad \text{for} \quad j \geq 1 . \tag{141}$$

Equation (141) is equivalent to the Yule-Welker equation in the statistical time series analysis. Let $j = 1, 2, \ldots, n$, then Eq. (141) can be written as

$$\begin{bmatrix} C_{n+1} \\ \vdots \\ C_{2n} \end{bmatrix} = - \begin{bmatrix} C_1 & \cdots & C_n \\ \vdots & & \vdots \\ C_n & \cdots & C_{2n-1} \end{bmatrix} \begin{bmatrix} b_1 \\ \vdots \\ b_n \end{bmatrix} . \tag{142}$$

therefore,

$$\begin{bmatrix} b_1 \\ \vdots \\ b_n \end{bmatrix} = - \begin{bmatrix} C_1 & \cdots & C_n \\ \vdots & & \vdots \\ C_n & \cdots & C_{2n-1} \end{bmatrix}^{-1} \begin{bmatrix} C_{n+1} \\ \vdots \\ C_{2n} \end{bmatrix} . \tag{143}$$

The square matrix is known as the Hankel matrix and is always nonsingular provided Eq. (141) holds for all $j \geq 1$.

In real-time operation, the estimate of $C_k$ denoted by $\hat{C}_k$, is

$$\hat{C}_k = \frac{1}{N} \sum_{i=k}^{N} z_i \, z_{i-k} , \tag{144}$$

in which $N$ is the number of observations used in computing the estimation. By virtue of the ergodic property of $z_i$, it can be shown that $\hat{C}_k$ is a normal, unbiased, and consistent estimate of $C_k$. Therefore, the best estimate of $b_k$, denoted by $\hat{b}_k$, can be obtained by replacing each $C_k$ in Eq. (143) by $\hat{C}_k$, yielding

$$\begin{bmatrix} \hat{b}_1 \\ \vdots \\ \hat{b}_n \end{bmatrix} = - \begin{bmatrix} \hat{C}_1 & \cdots & \hat{C}_n \\ \vdots & & \vdots \\ \hat{C}_n & \cdots & \hat{C}_{2n-1} \end{bmatrix}^{-1} \begin{bmatrix} \hat{C}_{n+1} \\ \vdots \\ \hat{C}_{2n} \end{bmatrix} . \tag{145}$$

For real-time operation, a recursive algorithm for computing $\hat{b}_k$ is required, and it was derived in ref. [66]:

$$\begin{bmatrix} \hat{b}_1^{N+1} \\ \cdot \\ \cdot \\ \cdot \\ \hat{b}_n^{N+1} \end{bmatrix} = \begin{bmatrix} \hat{b}_1^{N} \\ \cdot \\ \cdot \\ \cdot \\ \hat{b}_n^{N} \end{bmatrix} + \frac{(L^{N+1})^{-1}}{N+1} \times \left\{ \begin{bmatrix} z_{N+1} \; z_{N-n} \\ \cdot \\ \cdot \\ \cdot \\ z_{N+1} \; z_{N+1-2n} \end{bmatrix} - \begin{bmatrix} \hat{C}_{n+1}^{N} \\ \cdot \\ \cdot \\ \cdot \\ \hat{C}_{2n}^{N} \end{bmatrix} \right\}, \tag{146}$$

where

$$\hat{C}_k^N = \frac{1}{N} \sum_{i=k}^{N} z_i \; z_{i-k} \tag{147}$$

$$L^N = \begin{bmatrix} \hat{C}_1^N & \cdot & \cdot & \cdot & \hat{C}_n^N \\ \cdot & & & & \cdot \\ \cdot & & & & \cdot \\ \cdot & & & & \cdot \\ \hat{C}_n^N & \cdot & \cdot & \cdot & \hat{C}_{2n-1}^N \end{bmatrix}. \tag{148}$$

Since $b_1$, ..., $b_n$ are directly related to $\phi$ and $H$, the estimate of $\phi$ and $H$ can be directly computed using Eq. (139). However, the efficiency (ratio of minimum variance to the variance of the estimator) of this estimator is good only for the case of pure autoregressive time series, i.e., $R = 0$ and

$$\begin{bmatrix} H \\ H\phi \\ \cdot \\ \cdot \\ \cdot \\ H\phi^{n-1} \end{bmatrix} \Gamma = \begin{bmatrix} 0 \\ \cdot \\ \cdot \\ \cdot \\ \cdot \\ 1 \end{bmatrix} . \tag{149}$$

In order to improve the efficiency, Eq. (143) is being reexamined based on the fact that for $k \geq 2n$, $C_k$ contains information that can be used for this purpose. Let $\rho_k$ be the normalized autocorrelation of $z_k$, i.e.,

$$\rho_k = \frac{C_k}{C_0} , \tag{150}$$

and

$$\hat{\rho}_k = \frac{\hat{C}_k}{\hat{C}_0} = \rho_k + \delta\rho_k \quad ; \tag{151}$$

for $j = 1, \ldots, m-n$, Eq. (140) can be written as

$$\hat{\rho}_{n+j} = - \sum_{j=1}^{n} b_i \, \hat{\rho}_{i+j-1} + e_{n+j} \quad , \tag{152}$$

where $e$ is the error defined by

$$e_{n+j} = - \sum_{i=1}^{n+1} b_i \, \delta\rho_{i+j-1} \quad , \tag{153}$$

which is normally distributed with zero mean and covariance:

$$\mathcal{E} = \sum_{i=1}^{n+1} \sum_{k=1}^{n+1} b_i b_k \operatorname{cov}(\delta\rho_{i+j-1}, \delta\rho_{k+\ell-1}) \quad . \tag{154}$$

It can be shown that for large $N$ (ref. [68]):

$$\operatorname{cov}(\delta\rho_{i+j-1}, \delta\rho_{k+\ell-1}) \approx \frac{1}{N} \sum_{s=-\infty}^{\infty} \rho_s \, \rho_{s+i+j-k-\ell} \quad . \tag{155}$$

Since both $b_k$ and $\rho_k$ are unknown, it is not possible to determine the actual values of $\mathcal{E}$ from Eqs. (154) and (155); however, it is possible (as it was shown in ref. [66]) to obtain its estimate if $b_k$ is replaced by $\hat{b}_k$, and $\operatorname{cov}(\delta\rho_{i+j-1}, \delta\rho_{k+\ell-1})$ is replaced by

$$\frac{1}{\left(1 + \frac{2n}{N} - \frac{n^2}{N^2}\right)} \sum_{s=-n}^{n} \hat{\rho}_s \, \hat{\rho}_{s+i+j-k-\ell} \quad .$$

Having estimated the value of $\mathcal{E}$, denoted by $\hat{\mathcal{E}}$, an algorithm for computing $b_k$ can be developed as follows:

Rewrite Eq. (152) in matrix form:

$$\rho = \Omega b + e \quad , \tag{156}$$

where

$$\rho = \begin{bmatrix} \hat{\rho}_{m+1} \\ \cdot \\ \cdot \\ \cdot \\ \hat{\rho}_n \end{bmatrix}, \quad b = \begin{bmatrix} b_1 \\ \cdot \\ \cdot \\ \cdot \\ b_n \end{bmatrix}, \quad e = \begin{bmatrix} e_{m+1} \\ \cdot \\ \cdot \\ \cdot \\ e_n \end{bmatrix}$$

and

$$\Omega = -\begin{bmatrix} \hat{\rho}_1 & \cdots & \hat{\rho}_n \\ \cdot & & \cdot \\ \cdot & & \cdot \\ \cdot & & \cdot \\ \hat{\rho}_n & \cdots & \hat{\rho}_{m+n-1} \end{bmatrix} .$$

Let $b^*$ be an *a priori* estimate of $b$, such that $(b - b^*)$ is Gaussian with zero mean and covariance $M$, then it can be shown (ref. [69]) that the minimum variance estimate of $b$ is given by

$$\hat{b} = b^* + M\Omega^T(\Omega M\Omega^T + \hat{\mathcal{E}})^{-1}(\rho - \Omega b^*) \tag{157}$$

and the estimation error is given by

$$E[(b - b^*)(b - b^*)^T] = M - M\Omega^T(\Omega M\Omega^T + \hat{\mathcal{E}})^{-1}\ \Omega M \tag{158}$$

It is interesting to know that Eq. (145) is a special case of Eqs. (157) and (158). Note that improvement in efficiency costs extra computations, thus making the on-line identification more difficult.

### 3. *System Structure Adaptation*

In the discussion of adaptive filtering up to this point, the order of the system $n$ is assumed to be fixed and known. In practical situations this may not be the case. Therefore the identification of $n$, sometimes called structure adaptation, is necessary. This discussion is treated from two viewpoints: the use of innovation sequence property and the application of Bayes' theory.

a. Innovation sequence approach. The basic concept of this approach is to fit the actual system with mathematical model of various system orders, then followed by goodness-of-fit tests. Earlier works in this area were done by Ho and Lee [70] and Box and Jenkins [71]. The present discussion utilized ideas developed in previous sections on innovation sequence and correlation techniques, i.e., the same test used to verify the optimality of the Kalman filter will be used here. Specifically, the following steps will be taken:

(1) Based on *a priori* knowledge, let the order of the system be $n$ and perform noise parameter and system parameter identifications using appropriate techniques.

(2) Use results obtained from step (1) to compute $K$, the filter gain.

(3) Test the innovation sequence for whiteness (Section II,C.2).

(4) If the innovation sequence is not white, repeat steps (1) - (3) with $n-1$, $n+1$, etc., for the order of the system, until convergence is obtained.

Step 1, the determination of noise parameters, can be achieved by various methods previously discussed. The determination of vector $b$ (consequently $\phi$ and $H$), in a recursive manner for different system orders can be done as follows: Let $b_1'$, $b_2'$, ..., $b_n'$, $b_{n+1}'$ be the values of a system of order $n+1$, using Eq. (143):

$$\begin{bmatrix} b_1' \\ \vdots \\ b_n' \\ \hdashline b_{n+1} \end{bmatrix} = - \left[\begin{array}{ccc:c} C_1 & \cdots & C_n & C_{n+1} \\ \vdots & & \vdots & \vdots \\ C_n & & C_{2n-1} & C_{2n} \\ \hdashline C_{n+1} & & C_{2n} & C_{2n+1} \end{array}\right]^{-1} \begin{bmatrix} C_{n+2} \\ \vdots \\ C_{2n+1} \\ \hdashline C_{2n+2} \end{bmatrix} . \tag{159}$$

Let

$$F = \begin{bmatrix} C_1 & \cdots & C_n \\ \vdots & & \vdots \\ C_n & & C_{2n-1} \end{bmatrix}$$

$$a = \begin{bmatrix} C_{n+1} \\ \vdots \\ C_{2n} \end{bmatrix} .$$

$$d = \begin{bmatrix} C_{n+2} \\ \vdots \\ C_{2n+1} \end{bmatrix}$$

Then, Eq. (159) can be written as

$$\begin{bmatrix} b_1' \\ \vdots \\ b_n' \\ \hdashline b_{n+1}' \end{bmatrix} = - \left[\begin{array}{c:c} F & a \\ \hdashline a^{\mathrm{T}} & C_{n+1} \end{array}\right]^{-1} \begin{bmatrix} d \\ \hdashline C_{2n+2} \end{bmatrix} . \tag{160}$$

Using the above notation, it can be seen that

$$\begin{bmatrix} b_1 \\ \vdots \\ b_n \end{bmatrix} = - F^{-1} a \quad , \tag{161}$$

$$\begin{bmatrix} F & \vdots & a \\ \cdots & \vdots & \cdots \\ a^T & \vdots & c_{n+1} \end{bmatrix}^{-1} = \begin{bmatrix} F^{-1} + F^{-1}aa^TF^{-1}f & -F^{-1}af \\ -fa^TF^{-1} & f \end{bmatrix}, \quad (162)$$

where

$$f = \left( \sum_{i=1}^{n+1} c_{n+i}\, b_i \right)^{-1} \quad ; \quad b_{n+1} = 1 \quad . \quad (163)$$

From Eqs. (160) - (162):

$$\begin{bmatrix} b_1' \\ \cdot \\ \cdot \\ \cdot \\ b_n' \end{bmatrix} = -\left\{ F^{-1} + \begin{bmatrix} b_1^2 & \cdots & b_1 b_n \\ \cdot & & \\ \cdot & & \\ \cdot & & \\ b_n b_1 & \cdots & b_n^2 \end{bmatrix} f \right\} d$$

$$- f\, c_{2n+2} \begin{bmatrix} b_1 \\ \cdot \\ \cdot \\ \cdot \\ b_n \end{bmatrix}, \quad (164)$$

$$b_{n+1}' = - f \sum_{i=1}^{n+1} b_i\, c_{n+i+1} \quad . \quad (165)$$

This concludes the discussion of step 1, the recursive computation of $b$ for increasing system order fitting. Algorithms for steps 2 and 3 are well known and available in many literatures, therefore, these discussions will not be included.

b. Bayesian approach. The development of combined system parameter and structure identifications from the Bayesian approach is found in ref. [51], in which unknown system parameters and structure are represented by a vector $\alpha$ and a scalar $\sigma$, respectively. Note that $\alpha$ has the same meaning as defined in Section 3.B.1.a and $\sigma$ is the dimensionality or order of the system. The problem is to obtain optimal estimates of the state when either or both $\alpha$ and $\sigma$ are unknown. The solution to this problem is based on the assumption that stochastic processes are generated at random from a finite set of structure models having $\sigma = m, m+1, \ldots, n$ where $m \leq n$, and parameter vector $\alpha_{ik}$, where $k = 1, 2, \ldots, L_i$, corresponding to the $i$th structure model, $i = m, m+1, \ldots, n$, with known *a priori* probabilities $p(\alpha_{ik}, \sigma = i)$. The optimality criterion will be a minimization of the mean square error, i.e.,

$$\min_{\hat{w}} E\{[w - \hat{w}]^T B[w - \hat{w}]\} \quad , \tag{166}$$

where $\hat{w}$ is the best estimate for a given set of measurement data $z$. $B$ is a symmetric positive definite matrix and

$$w = Ax \quad . \tag{167}$$

The dimensionality of $w$ is $r \times 1$, $x$ is $i \times 1$ and matrix $A$ is $r \times i$, where $i = m, m+1, \ldots, n$, $r \leq m$.

The mathematical solution to this adaptive problem is hinged on the generalized form of the partition theorem, which states:

$$\hat{w}(t|t) = \sum_{i=m}^{n} \sum_{k=1}^{L_i} p(\alpha_{ik}, \sigma = i|t) A_i(t) \hat{x}(t|t, \alpha_{ik}, \sigma = i) \quad , \tag{168}$$

where

$$\hat{x}(t|t, \alpha_{ik}, \sigma = i) = E[x_i(t)|Z, \alpha_i = \alpha_{ik}, \sigma = i] \quad , \tag{169}$$

$$p(\alpha_{ik}, \sigma = i|t) = \text{prob}\{\alpha_i = \alpha_{ik}, \sigma = i|Z\} \quad , \tag{170}$$

which is given by (Bayes rule):

$$p(\alpha_{ik}, \sigma = i|t) = \frac{1}{1 + \sum_{j=m}^{n} \sum_{\ell=1}^{L_i} \lambda_{ik,j\ell}(t) \dfrac{p(\alpha_{j\ell}, \sigma = j)}{p(\alpha_{ik}, \sigma = i)}} \quad , \tag{171}$$

where

$$\lambda_{ik,j\ell}(t) = \frac{p(Z|\alpha_{j\ell}, \sigma = j)}{p(Z|\alpha_{ik}, \sigma = i)} \quad . \tag{172}$$

It is interesting to know that Eqs. (168) and (171) are generalizations of Eqs. (80) and (81), respectively. Similar to the adaptive filter for unknown parameters only, the solution to the combined parameter and structure adaptation problem also consists of a sum of estimators, and each calculates the value of $A_i(t)\hat{x}(t|t, \alpha_{ik}, \sigma = i)$, which is being weighted by $p(\alpha_{ik}, \sigma = i|t)$.

Using the partition theorem plus the assumption that the stochastic processes are Gauss-Markov, an algorithm for the adaptive optimal estimator can be derived for the following linear discrete system:

$$\begin{aligned} x(t+1;\alpha_{ik}, \sigma=i) &= \Phi(t|\alpha_{ik}, \sigma=i)x(t;\alpha_{ik}, \sigma=i) \\ &\quad + \Gamma(t|\alpha_{ik}, \sigma=i)u(t|\alpha_{ik}, \sigma=i) \quad ; \end{aligned} \tag{173}$$

$$\begin{aligned} z(t; \alpha_{ik}, \sigma=i) &= H(t|\alpha_{ik}, \sigma=i)x(t;\alpha_{ik}, \sigma=i) \\ &\quad + v(t|\alpha_{ik}, \sigma=i) \quad ; \end{aligned} \tag{174}$$

accordingly, the following notations will be used in the Kalman filter: $Q(t|\alpha_{ik}, \sigma=i)$, $R(t|\alpha_{ik}, \sigma=i)$, $\hat{x}(t;\alpha_{ik}, \sigma=i)$, $K(t;\alpha_{ik}, \sigma=i)$, and $P(t;\alpha_{ik}, \sigma=i)$. Assuming $u$ and $v$ are Gaussian white noises with zero means and uncorrelated, let $\hat{x}(t|t,\alpha_{ik}, \sigma=i)$ be the best filter estimate, and then the following results can be obtained for the nonadaptive part of the estimator:

$$\begin{aligned} \hat{x}(t|t,\alpha_{ik}, \sigma=i) &= \Phi(t,t-1|\alpha_{ik}, \sigma=i)\hat{x}(t-1|t-1,\alpha_{ik}, \sigma=i) \\ &\quad + K(t,t|\alpha_{ik}, \sigma=i)[z(t) - H(t|\alpha_{ik}, \sigma=i) \\ &\quad \times \Phi(t,t-1|\alpha_{ik}, \sigma=i)\hat{x}(t-1\ t-1,\alpha_{ik}, \sigma=i)] \quad , \end{aligned} \tag{175}$$

$$K(t,t|\alpha_{ik},\ \sigma=i) = P(t|t-1,\alpha_{ik},\ \sigma=i)H^{\mathrm{T}}(t|\alpha_{ik},\ \sigma=i) \times P_z^{-1}(t|t-1,\alpha_{ik},\ \sigma=i) \quad , \tag{176}$$

$$P(t|t,\alpha_{ik},\ \sigma=i) = P(t|t-1,\alpha_{ik},\ \sigma=i) - K(t,t|\alpha_{ik},\ \sigma=i)P_z(t|t-1,\alpha_{ik},\ \sigma=i) \times K^{\mathrm{T}}(t,t|\alpha_{ik},\ \sigma=i) \quad , \tag{177}$$

$$P(t|t-1,\alpha_{ik},\ \sigma=i) = \Phi(t,t-1|\alpha_{ik},\ \sigma=i) \times P(t-1|t-1,\alpha_{ik},\ \sigma=i)\Phi^{\mathrm{T}}(t,t-1|\alpha_{ik},\ \sigma=i) + \Gamma(t-1|\alpha_{ik},\ \sigma=i)Q(t-1|\alpha_{ik},\ \sigma=i) \times \Gamma^{\mathrm{T}}(t-1|\alpha_{ik},\ \sigma=i) \quad , \tag{178}$$

$$P_z(t|t-1,\alpha_{ik},\ \sigma=i) = H(t|\alpha_{ik},\ \sigma=i) \times P(t|t-1,\alpha_{ik},\ \sigma=i)H^{\mathrm{T}}(t|t-1,\alpha_{ik},\ \sigma=i) + R(t|\alpha_{ik},\ \sigma=i) \quad . \tag{179}$$

Equation (179) is the conditional error covariance matrix for $\tilde{z}(t|t-1,\alpha_{ik},\ \sigma=i)$, i.e.,

$$P_z(\cdot|\cdot) = \mathrm{E}\{\tilde{z}(\cdot|\cdot)\tilde{z}^{\mathrm{T}}(\cdot|\cdot)\} \quad , \tag{180}$$

where by definition:

$$\tilde{z}(t|t-1,\alpha_{ik},\ \sigma=i) = z(t) - \hat{z}(t|t-1,\alpha_{ik},\ \sigma=i) \tag{181}$$

in which

$$\hat{z}(t|t-1,\alpha_{ik},\ \sigma=i) = H(t|\alpha_{ik},\ \sigma=i)\Phi(t,t-1|\alpha_{ik},\ \sigma=i) \times \hat{x}(t-1|t-1,\alpha_{ik},\ \sigma=i) \quad . \tag{182}$$

The adaptive part of the estimator is given in Eq. (171), in which the *a posteriori* probability can be computed recursively:

$$p(\alpha_{ik},\sigma=i|t) = \frac{|P_z(t|t-1,\alpha_{ik},\sigma=i)|^{-1/2} \times \exp[-\frac{1}{2} Q(t|\alpha_{ik},\sigma=i)]p(\alpha_{ik},\sigma=i|t-1)}{\sum_{j=m}^{n} \sum_{\ell=1}^{L_i} |P_z(t|t-1,\alpha_{j\ell},\sigma=j)|^{-1/2} \times \exp[-\frac{1}{2} Q(t|\alpha_{j\ell},\sigma=j)]p(\alpha_{j\ell},\sigma=j|t-1)} , \quad (183)$$

where

$$p(\alpha_{ik}, \sigma=i|t_0) = p(\alpha_{ik}, \sigma=i) , \quad (184)$$

$$Q(t|\alpha_{ik}, \sigma=i) = \tilde{z}^T(t|t-1,\alpha_{ik}, \sigma=i)P_z^{-1}(t|t-1,\alpha_{ik}, \sigma=i) \times \tilde{z}(t|t-1,\alpha_{ik}, \sigma=i) . \quad (185)$$

A goodness-of-fit test for the Bayesian approach may be performed by examining the error covariance $P_w$ of the combined nonadaptive and adaptive algorithms.

Defining

$$P_w(t_1|t_2) = E\{[W(t_1) - \hat{W}(t_1|t_2)] \times [W(t_1) - \hat{W}(t_1|t_2)]^T|Z\} . \quad (186)$$

Applying the total probability theorem, it can be shown that for $t_1 = t_2 = t$:

$$P_w(t|t) = \sum_{i=m}^{n} \sum_{k=1}^{L_i} p(\alpha_{ik}, \sigma=i|t) \times [B_i(t)P(t|t,\alpha_{ik}, \sigma=i)B_i^T(t) + P_{ik}(t|t)] , \quad (187)$$

where

$$P(t|t) = E\{[x_i(t) - \hat{x}(t|t,\alpha_{ik}, \sigma=i)] \times [x_i(t) - \hat{x}(t|t,\alpha_{ik}, \sigma=i)]^T|Z,\alpha_{ik}, \sigma=i\} , \quad (188)$$

$$P_{ik}(t|t) = [\hat{W}(t|t,\alpha_{ik}, \sigma=i) - \hat{W}(t|t)] \times [\hat{W}(t|t,\alpha_{ik}, \sigma=i) - \hat{W}(t|t)]^T . \quad (189)$$

It is apparent that the quantity $p_{ik}(t|t)$ is the performance degradation due to processing data from a known model by an adaptive filter instead of the optimal Kalman filter that matches the model. This information as well as $P_w(t|t)$ can be used for sensitivity studies and performance evaluations.

### *C. ERROR BOUNDING TECHNIQUES*

Another approach to filter design, entirely different from the adaptive techniques just discussed, is the so-called error bounding techniques, which are developed to deal with most practical situations in which the digital computer used to implement the Kalman filter equation may not be adequate in terms of speed and memory requirements or effects of either intended or inadvertent discrepancies between the true system and the system assumed by the Kalman filter, which creates the divergence problem. This and the other (computer constraints) problem will be discussed next.

#### *1. Computer Constraints*

Constraints imposed by the computer are usually relaxed by reducing the number of calculations performed in the implementation of the filter. Depending on the particular problem at hand, one or more of the following methods can be employed: state reduction, decoupling, and prefiltering.

a. State reduction. Reducing the number of states in the system model is one way to reduce the computer burden. This decision cannot be made in an arbitrary manner. However, in the absence of an organized method, state reduction becomes an art rather than a science; it usually requires good engineering judgments and experiences in knowing how the filter actually works. Since the design is trial-and-error in nature, a great deal of computer simulations are required to evaluate the sensitivity of states to be deleted. In general, the purpose of the sensitivity analysis is to examine the effects of filter performance caused by deleting various states. In order to determine what state to be deleted, the following criteria can be used as a guideline: states that cannot be estimated accurately; states have no practical interest; states that can be combined into fewer equivalent states; states have little effect on other state measurements of interest; and states with small rms values.

b. Decoupling. When the number of states has been reduced to its minimum (based on sensitivity analysis and engineering judgment), other approaches may be taken to further reduce the burden of the computer. For example, consider the following system:

$$\begin{pmatrix} x_{n+1} \\ x^*_{n+1} \end{pmatrix} = \begin{pmatrix} \phi_{11} & \phi_{12} \\ \phi_{21} & \phi_{22} \end{pmatrix} \begin{pmatrix} x_n \\ x^*_n \end{pmatrix} + \begin{pmatrix} u_n \\ u^*_n \end{pmatrix} . \tag{190}$$

If the elements of $\phi_{12}$ and $\phi_{21}$ are small compared to those elements in $\phi_{11}$ and $\phi_{22}$, also if $u_n$ and $u^*_n$ are uncorrelated, then it may be possible to separate $x_n$ from $x^*_n$ and compute only the uncoupled filter equations for

$$x_{n+1} = \phi_{11} x_n + u_n , \tag{191}$$

$$x^*_{n+1} = \phi_{22} x^*_n + u^*_n . \tag{192}$$

The validity of separating weakly-coupled states has to be proven by examining the covariances of the coupled system [Eq. (190)] and the decoupled system [Eqs. (191) and (192)] to see if they are in good approximations. Once the decoupling decision is made, a relatively high-order filter can be processed as several low-order disjoint filters--each can be computed separately. The advantage is apparent by the fact that the number of multiplications required to calculate the covariance matrix varies as the third power of the state size. For example, if it is possible to separate an $n$-state filter into three disjoint filters, each with $n/3$ states, then the number of calculations required to propagate the $P$ matrix of three $(n/3)$-state filter is $kn^3/9$, while the number of calculations for the $n$-state filter is $kn^3$. Thus the computer burden is reduced by 90%!

c. Prefiltering. Prefiltering is an averaging method performed on measurement data in cases where they are continuously or almost continuously available. Sometimes this method is called "data compression," which is applicable only if $\Delta T >> \Delta t$, where $\Delta t$ is the time interval in which measurement data are available and $\Delta T$ is the Kalman cycle time, i.e., the time that it takes to process the Kalman filter equations. Under this condition, it is not possible to use every measurement individually to update the filter. However, they may be used collectively in some averaging manner. Thus this method is also referred to as "measurement averaging." Let

$$\Delta T = n\ \Delta t \quad ; \tag{193}$$

considering the scalar case in which

$$z_k = x_k + v_k \quad , \tag{194}$$

the average measurement over the $\Delta T$ interval can be computed by

$$\overline{z} = \frac{1}{N} \sum_{i=1}^{N} z_i \tag{195}$$

and Eq. (195) can be written as

$$\bar{z} = \frac{1}{N} \sum_{i=1}^{N} x_i + \frac{1}{N} \sum_{i=1}^{N} v_i \tag{196}$$

in which $N$ is the number of measurements taken over the $\Delta T$ interval. In order to use $\bar{z}$ in the standard Kalman filtering equations, $\bar{z}$ must be put into a standard form which is similar to Eq. (194). Since by definition:

$$\bar{v} = \bar{z} - \bar{x} \quad , \tag{197}$$

using Eq. (196), the average measurement noise $\bar{v}$ can be written as

$$\bar{v} = \left( \frac{1}{N} \sum_{i=1}^{N} v_i \right) + \left( \frac{1}{N} \sum_{i=1}^{N} x_i - \bar{x} \right) \quad . \tag{198}$$

The first term on the right side of Eq. (198) shows that the original measurement noise has been reduced through the averaging technique, but at the same time noise due to the averaging has been added; this can be identified as the second term on the right side of Eq. (198).

In order to use $\bar{z}$ in the standard Kalman filter form, the variance of $\bar{v}$ must be expressed in terms of $\sigma_v^2$ and $\sigma_x^2$, which are defined by

$$E[v_k{}^2] = \sigma_v{}^2 \quad , \tag{199}$$

$$E[v_k{}^2] = \sigma_x{}^2 \quad . \tag{200}$$

It should be recognized that $\bar{v}$ is not a white noise sequence. Furthermore it is correlated with $x$. These phenomena are caused by the averaging noise, which must be accounted for in the implementation of the filter algorithm. An approximate method of accomplishing this is given in ref. [67].

*2. Divergence*

Practical experiences indicated that several existing methods are quite effective in preventing filter divergence. Two of these methods, process noise addition and finite memory filter, are selected for discussion here.

a. Process noise addition. If it is known that certain elements in the $Q$ matrix (process noise) are zero due to inexact modeling, the designer may purposely change the zero elements to have nonzero values. The action taken by the designer is referred to as "process noise addition." The reason for the addition of fictitious noise to the system can be seen by examining the error covariance equation:

$$\dot{P} = FP + PF^{\mathrm{T}} - PH^{\mathrm{T}}R^{-1}HP + GQG^{\mathrm{T}} \quad . \tag{201}$$

Under steady-state condition, Eq. (201) becomes

$$FP + PF^{\mathrm{T}} - PH^{\mathrm{T}}R^{-1}\ HP = -\ GQG^{\mathrm{T}} \quad . \tag{202}$$

If some elements in $Q$ are zero, then corresponding elements in $P$, and consequently in $K$, are zero. Thus, for those states originally assumed not being driven by white noise, the filter does not improve its performance although new measurements are available. However, if these elements in $Q$ are not zero, then the corresponding elements of $K$ will not be zero, and the filter will always try to improve its performance as long as measurement data are available.

b. Finite memory filter. The finite memory filter, sometimes called a "moving window," employs various techniques to truncate old data that are no longer considered important. For this reason it is necessary to cast the filter in the following form:

$$\hat{x}_{k+1}^{(N)} = \hat{\psi}_k \; x_k^{(N)} + K_k \; z_k \quad , \tag{203}$$

where $\hat{x}_k^{(N)}$ is the best estimate of state $x$ at time $t_k$ using the last $N$ data points; $\psi_k$ is the transition-matrix, and $K_k$ is the filter gain. Development of the finite memory filter in the form of Eq. (203) has not been completely satisfactory, due to the fact that the finite memory property of the filter must be embodied in $K_k$ and $\psi_k$, also additional information involving $z_{k-N}$ is needed to compensate for the effect of omitting old measurement data. Three different approaches to the solution of this problem will be discussed. They are the $\varepsilon$ technique, direct limited memory filter, and fading memory filter and age-weighting.

*(1) The $\varepsilon$ technique.* It was pointed out earlier that as the filter gain approaches zero, the normal filtering process stops and divergence occurs. One method to prevent this type of divergence is to have an effective gain consisting of two parts: the regular Kalman gain and a gain acts as a weighting term, i.e.,

$$K_{\text{eff}} = K_{\text{reg}} + K_{\text{wei}} \quad , \tag{204}$$

in which $K_{\text{wei}}$, the weighting gain is computed based on the most recent measurement. The motivation for this development as well as the fundamental concepts are given in ref. [22], which may be summarized as follows:

Consider the equation

$$\hat{x}_{k+1}(+) - \phi_k \hat{x}_k(+) = [\hat{x}_{k+1}(+) - \phi_k \hat{x}_k]_{\text{reg}} + \varepsilon' \; \Delta\hat{x}_{k+1}(+) \quad , \tag{205}$$

where

$$[\hat{x}_{k+1}(+) - \phi_k \hat{x}_k]_{\text{reg}} = K_{k+1}[z_{k+1} - H_{k+1}\phi_k \hat{x}_k(+)] \quad , \tag{206}$$

$$\Delta\hat{x}_{k+1}(+) = H_{k+1}^{\mathrm{T}}(H_{k+1}\ H_{k+1}^{\mathrm{T}})^{-1}\ [z_{k+1} - H_{k+1}\phi_k\hat{x}_k(+)] \quad . \tag{207}$$

$K_k$ is the regular Kalman gain and $\varepsilon'$ is a scalar factor to be determined. Equation (207) represents the best estimate of $[x_{k+1} - \phi_k\hat{x}_k(+)]$ based on the most recent measurement residuals. Note that the best estimate is zero when $\hat{z}_{k+1} = H_k\phi_k\hat{x}_k(+)$.

Next define a new scalar $\varepsilon$ such that

$$\varepsilon' = \frac{\varepsilon\ R_{k+1}^{\mathrm{S}}}{H_{k+1}P_{k+1}(-)H_{k+1}^{\mathrm{T}} + R_{k+1}^{\mathrm{S}}} \quad , \tag{208}$$

where $R_k^{\mathrm{S}}$ is a scalar value of the $R_k$ matrix. Substitute Eqs. (208) and (207) into Eq. (205):

$$\hat{x}_{k+1}(+) = \phi_k\hat{x}_k(+) + (K_{\mathrm{reg}} + K_{\mathrm{wei}}) \times [z_{k+1} - H_{k+1}\phi_k\hat{x}_k(+)] \quad , \tag{209}$$

where

$$K_{\mathrm{reg}} = \frac{P_{k+1}(-)\ H_{k+1}^{\mathrm{T}}}{H_{k+1}P_{k+1}(-)H_{k+1}^{\mathrm{T}} + R_{k+1}^{\mathrm{S}}} \quad , \tag{210}$$

$$K_{\mathrm{wei}} = \varepsilon\ \frac{R_{k+1}^{\mathrm{S}}H_{k+1}^{\mathrm{T}}\,\big|\,H_{k+1}H_{k+1}}{H_{k+1}P_{k+1}(-)H_{k+1}^{\mathrm{T}} + R_{k+1}^{\mathrm{S}}} \quad . \tag{211}$$

Note that $K_{\mathrm{wei}}$ is proportional to $\varepsilon$, which has values ranging from 0 to 1, e.g., for $\varepsilon = 0$, the modified filter becomes a regular Kalman filter; for $\varepsilon = 1$, $z_{k+1} = H_{k+1}\hat{x}_{k+1}(+)$ implying that the estimate of the measurement equals the measurement itself.

Corresponding to $K_{\mathrm{reg}}$ and $K_{\mathrm{wei}}$, the covariance matrix also consists of two terms, $P_{\mathrm{reg}}$ and $P_{\mathrm{wei}}$, i.e.,

$$P_{\mathrm{eff}} = P_{\mathrm{reg}} + P_{\mathrm{wei}} \tag{212}$$

and it can be shown that $P_{\mathrm{wei}}$ is a function of $\varepsilon^2$:

$$P_{\text{wei}} = \varepsilon^2 \frac{(R_k^{\text{S}})^2 \; H_k^{\text{T}} \; H_k}{[H_k P(-) H_k^{\text{T}} + R_k^{\text{S}}][H_k H_k^{\text{T}}]^2} \quad . \tag{213}$$

It is apparent now that the $\varepsilon$ technique is designed to prevent both the gain and covariance from becoming too small, and it is done by weighting the most recent measurement data.

*(2) Direct limited memory filter.* Various algorithms for the limited memory filter are given in available literatures, the following concept is discussed in ref. [57], in which the filter equations, for the case when $Q = 0$, are given:

$$\hat{x}_k^{(N)} = P_k^{(N)} (P_{k/k}^{-1} \hat{x}_k - P_{k/i}^{-1} \hat{x}_{k/i}) \quad \text{for} \quad i < k \quad , \tag{214}$$

$$(P_k^{(N)})^{-1} = P_{k/k}^{-1} - P_{k/i}^{-1} \quad , \tag{215}$$

where $\hat{x}_k^{(N)}$ is the best estimate of $x_k$, given $N = k - i$ measurements. $P_k^{(N)}$, $P_{k/k}$, and $P_{k/i}$ are covariances of $(\hat{x}_k^{(N)} - x_k)$, $(\hat{x}_k - x_k)$ and $(\hat{x}_{k/i} - x_k)$, respectively.

Note that the limited memory filter estimate is a linear combination of two weighted regular (infinite memory) filter estimates. Therefore, direct computation of Eqs. (214) and (215) is not desirable since the original intent is to avoid infinite memory filter computations. For this reason, approximation methods have been devised to circumvent this and other (e.g., inversions of matrices $P_{k/k}$ and $P_{k/i}$) difficulties. Basically it is required to select $N$ first and followed by processing the measurement data to obtain $\hat{x}_N$ and $P_{N/N}$, then computing $\hat{x}_{2N}$ and $P_{2N/2N}$, from which $\hat{x}_{2N/N}$ and $P_{2N/N}$ can be obtained, i.e.,

$$x_{2N/N} = \phi(2N, N) \hat{x}_N \quad , \tag{216}$$

$$P_{2N/N} = \phi(2N, N) P_N \, \phi^{\text{T}}(2N, N) \quad . \tag{217}$$

Finally, calculate $\hat{x}_{2N}^{(N)}$ and $P_{2N}^{(N)}$ using Eqs. (214) and (215). To continue the processing of data for the next time interval, $\hat{x}_{2N}^{(N)}$ and $\hat{P}_{2N}^{(N)}$ will be used as initial conditions. Note that for $k > N$, $\hat{x}_k$ is conditioned on the last group of data measured between $N$ and $2N$.

*(3) Fading memory filter and age-weighting.* Another method to achieve finite memory filtering is to discard old data by weighting them according to the time of their occurrence, i.e., to place more emphasis on new data and less emphasis on old data. In other words, this method requires $R$, covariance of measurement noise, to be increased for old data. This concept can be implemented easily by introducing an age-weighting factor, $s$ $(s \geq 1)$, such that

$$R_k^* = s^{i-k} R_k, \quad k = i, \quad i-1, \ i-2, \ \ldots , \tag{218}$$

where $R_k$ is the regular measurement noise covariance and $R_k^*$ is the new covariance. A convenient choice of the age-weighting factor would be

$$s = \exp\left( \frac{\Delta t}{T} \right) , \tag{219}$$

where $\Delta t$ is the measurement interval and $T$ the time constant. For a constant $R_K = R$,

$$R_{k-m}^* = \left(\exp \frac{m\Delta t}{T} \right) R \quad , \quad m = 0, 1, 2, \ldots . \tag{220}$$

The finite memory filter constructed under the above conditions is

$$\hat{x}_k(+) = \phi_{k-1}\hat{x}_{k-1}(+) + K_k[z_k - H_k \ \phi_{k-1} \ \hat{x}_{k-1}(+)] \quad , \tag{221}$$

$$K_k = P_k^*(-) \ H_k^{\mathrm{T}} \ [H_k P_k^*(-) \ H_k^{\mathrm{T}} + R_k]^{-1} \quad , \tag{222}$$

$$P_k^*(+) = P_k^*(-) - P_k^*(-) \ H_k^{\mathrm{T}} \ [H_k P_k^*(-) H_k^{\mathrm{T}} + R_k]^{-1} \ H_k P_k^*(-) \ , \tag{223}$$

$$P_k^*(-) = s\phi_{k-1}P_{k-1}^*(+)\ \phi_{k-1}^{\mathrm{T}} + Q_{k-1} \quad . \tag{224}$$

Note that for $s = 1$, the above set of equations becomes the regular Kalman filter provided proper interpretations are given to $P_k^*(-)$ and $P_k^*(+)$.

### D. *CONCLUDING REMARKS*

In conclusion, it should be pointed out that there are no general techniques currently available that can be applied with absolute confidence in designing optimal Kalman filters for practical systems. Nonetheless, physical intuition and off-line simulations are considered to be effective tools used to obtain good designs. To this end, the following section, on computer simulation, is written to demonstrate the usefulness of some ideas developed in this section.

## IV. APPLICATIONS

### A. *PROBLEM STATEMENT*

Consider an augmented inertial navigation system (INS), which employs a Kalman filter to estimate and reset system errors, using external position measurements. Assume that the error dynamics of the system can be represented by a set of linear differential equations, which are suitable for state space formulations. That is, the state vector is composed of elements including errors in latitude, longitude, gyro drift rates, etc., and the measurement vector is viewed as the difference between INS position and the position indicated by the reference source. Assume also that only additive random errors appear as driving forces in both the state and measurement vector equations. The driving forces are series of random sequences that can be generated from appropriate shaping filters with white noise inputs. However, the first- and second-order statistics of the white noises are unknown. Specifically, this sample problem considers an error dynamic system that can be described by the following set of differential equations:

$$\dot{\psi}_x = -\Omega_V \psi_y + \varepsilon_x \quad , \tag{225}$$

$$\dot{\psi}_y = \Omega_V \psi_x + \Omega_H \psi_z + \varepsilon_y \quad , \tag{226}$$

$$\dot{\psi}_z = -\Omega_H \psi_y + \varepsilon_z \quad , \tag{227}$$

$$\dot{\varepsilon}_x = u_x \quad , \tag{228}$$

$$\dot{\varepsilon}_y = u_y \quad , \tag{229}$$

$$\dot{\varepsilon}_z = R_z + u_z \quad , \tag{230}$$

$$\dot{R}_z = 0 \quad , \tag{231}$$

where $\psi_x$, $\psi_y$, and $\psi_z$ are inertial platform tilts; $\varepsilon_x$, $\varepsilon_y$, and $\varepsilon_z$ are gyro drift rates; $R_z$ is the $z$-gyro ramp state; and $\Omega_V$ and $\Omega_H$ are vertical and horizontal components of the earth rate.

The discrete time form of the state space formulation of Eqs. (225)-(231) is

$$x_{i+1} = \phi_i x_i + \Gamma_i u_i \quad , \tag{232}$$

where $x$ is a seven-state vector, dimensions of matrices $\phi$ and $\Gamma$ as well as vector $u$ can be identified correspondingly.

Position measurements are described by

$$z_i = H_i x_i + v_i \quad , \tag{233}$$

where

$$z = \begin{pmatrix} \delta_L \\ \delta_\lambda \end{pmatrix} \; ; \quad H = \begin{bmatrix} 0 & -1 & 0 & 0 & 0 & 0 & 0 \\ 1 & 0 & 0 & 0 & 0 & 0 & 0 \end{bmatrix} \; ;$$

$$v = \begin{pmatrix} v_L \\ v_\lambda \end{pmatrix}$$

$\delta_L$ and $\delta_\lambda$ are latitude and longitude measurement errors, respectively; $v_L$ and $v_\lambda$ are latitude and longitude measurement noises, which are Gaussian and white. The first- and second-order noise characteristics are given below:

$$E\{u_i\} = \mu_u \quad ;$$

$$E\{v_i\} = \mu_v \quad ;$$

$$E\{(u_i - \mu_u)(u_j - \mu_u)^T\} = Q\ \delta_{ij} \quad ;$$

$$E\{(v_i - \mu_v)(v_j - \mu_v)^T\} = R\ \delta_{ij} \quad ;$$

$$E\{(u_i - \mu_u-(v_j - \mu_v)^T\} = S\ \delta_{ij} \quad ;$$

where $\mu_u$, $\mu_v$, $Q$, $R$, and $S$ are unknown except that $Q$ and $R$ are positive definite and $S$ is consistent with $Q$ and $R$. The problem is to develop an algorithm which will improve the filter optimality by identifying the second-order ($Q$, $R$, and $S$) and first-order ($\mu_u$ and $\mu_v$) noise statistics using external position measurement data.

## B. *SOLUTION*

The filter to be considered is

$$\hat{x}^*_{i+1/i} = \phi_i \hat{x}^*_{i/i} + \Gamma_i \mu^*_u + K^*_i [z_i - H_i \hat{x}^*_{i/i} - \mu^*_v] \quad ; \tag{234}$$

$$\hat{x}^*_{i/i} = \hat{x}^*_{i/i-1} + B^*_i\ \nu^*_i \quad ; \tag{235}$$

$$K^*_i = \Gamma_i S^*_i (R^*_i)^{-1} \quad ; \tag{236}$$

$$B^*_i = P^*_{i/i-1} H^T_i [H_i P^*_{i/i-1} H^T_i + R^*_i]^{-1} \quad ; \tag{237}$$

$$P^*_{i/i-1} = [\phi_i - K^*_{i-1}H_{i-1}]P^*_{i-1}[\phi_i - K^*_{i-1}H_{i-1}]^T + \Gamma_{i-1}Q^*_{i-1}\Gamma^T_{i-1} - K^*_{i-1}R^*_{i-1}K^{*T}_{i-1} \quad ; \tag{238}$$

$$P^*_i = [I - B^*_i H_i]P^*_{i/i-1} \quad ; \tag{239}$$

$$\nu^*_i = z_i - H_i\hat{x}^*_{i/i-1} - \mu^*_v \quad ; \tag{240}$$

$$\hat{x}^*_0 = \hat{x}^*_{0/0} = E\{x_0\} \quad ; \tag{241}$$

$$P^*_0 = P^*_{0/0} = \text{var}\{x_0\} \quad ; \tag{242}$$

in which the starred notations indicate that the uncertain descriptions for the noise statistics are being used to operate the Kalman filter. That is, $\hat{x}^*_{i+1/i}$ denotes the optimal estimate of $x_{i+1}$ using measurements up to and including the $i$th interval, based on the assumed values $\mu^*_u$, $\mu^*_v$, $Q^*$, $R^*$, and $S^*$ for $\mu_u$, $\mu_v$, $Q$, $R$, and $S$. $B^*_i$ is the usual Kalman filter gain and $\nu^*_i$ is the innovation sequence.

The solution will be given in two parts. The algorithm for adapting the second-order noise statistics is discussed in part one and the algorithm for adapting the first-order noise statistics is discussed in part two. The reason for this approach is based on the fact that estimation of the second-order noise statistics does not require the explicit knowledge of the first-order noise statistics.

Part one: Adaptation of second-order noise statistics

Substituting Eq. (235) into (234):

$$\hat{x}^*_{i+1/i} = \phi_i\hat{x}^*_{i/i-1} + \Gamma_i\mu^*_u + G^*_i\nu^*_i \quad , \tag{243}$$

where

$$G^*_i = \phi_i B^*_i + K^*_i - K^*_i H_i B^*_i \quad . \tag{244}$$

Defining,

$$e^*_{i+1} = x_{i+1} - \hat{x}^*_{i+1/i} \quad . \tag{245}$$

Using Eqs. (232), (240), and (243) we obtain a recursive expression for

$$e^*_{i+1} = (\phi_i - G^*_i H_i) e^*_i + \Gamma_i (u_i - \mu^*_u) - G^*_i (v_i - \mu^*_v) \quad . \tag{246}$$

Using the definition for $e^*_{i+1}$, Eq. (240) becomes

$$\nu^* = H_i e^*_i + v_i - \mu^*_v \quad . \tag{247}$$

Equations (246) and (247) contain all necessary information for generating an adaptive algorithm. Prior to developing the algorithm, however, it is desired to transform these equations into the standard form, in which the mean values do not appear explicitly. To this end, take the expectation of Eqs. (246) and (247), then subtract (246) and (247) from their expectations. The results are

$$\varepsilon^*_{i+1} = (\phi_i - G^*_i H_i) \varepsilon^*_i + \Gamma_i \tau_i - G^*_i \xi_i \quad , \tag{248}$$

$$\sigma^*_i = H_i \varepsilon^*_i + \xi_i \quad , \tag{249}$$

where

$$\varepsilon^*_i = e^*_i - \mathrm{E}\{e^*_i\} \quad , \tag{250}$$

$$\sigma^*_i = \nu^*_i - \mathrm{E}\{\nu^*_i\} \quad , \tag{251}$$

$$\tau_i = u_i - \mu_i \quad , \tag{252}$$

$$\xi_i = v_i - \mu_i \quad . \tag{253}$$

Note that $\varepsilon^*_i$, $\sigma^*_i$, $\tau_i$, and $\xi_i$ have the same statistical properties as $e^*_i$, $\nu^*_i$, $u_i$, and $v_i$, respectively, except for the mean values, which are zero instead of $\mathrm{E}\{e^*_i\}$, $\mathrm{E}\{\nu^*_i\}$, $\mu_u$, and $\mu_v$. Note that $\varepsilon^*_i$ and $\sigma^*_i$ are now viewed as the predictor error and innovation sequence, respectively, for the case where the noise means are zero.

In order to obtain the observation vector $\sigma_i^*$, which contains all information needed to adapt the second-order noise statistics, Eq. (251) will be used. The mean values for the innovation sequence can be computed using the formula

$$E\{\nu_i^*\} = \frac{1}{N+1} \sum_{i=1}^{N} \nu_i^* \quad , \tag{254}$$

which is unbiased, consistent, and efficient if the steady-state condition of the filter and the ergodicity of the sequence are satisfied. Having ascertained $\sigma_i^*$, the next step is to form the correlation function for $\sigma_i^*$, called the observed correlation function, and the weighted least-squares minimization is performed using the difference (error) between the observed correlation function and the theoretical correlation function, which can be derived as follows:

$$E\{\sigma_i^* \sigma_{i-\ell}^{*T}\} = H_i E\{\varepsilon_i^* \varepsilon_{i-\ell}^{*T}\} H_{i-\ell}^T + H_i E\{\varepsilon_i^* \xi_{i-\ell}\} + R^* \delta_\ell \quad . \tag{255}$$

In order to express Eq. (255) in a more desirable form, an explicit function for $\mathcal{E}_i^*$ is required. From Eq. (248):

$$\varepsilon_i^* = \psi(i,0)\varepsilon_0^* + \sum_{k=0}^{i-1} \psi(i,k+1)[\Gamma_k \tau_k - G_k^* \xi_k] \quad , \tag{256}$$

where $\psi(i_1,i_2)$ is the state transition matrix

$$\psi(i_1,i_2) = A(i_1-1)A(i_1-2),\ldots,A(i_2) \quad \text{for } i_1 > i_2 \tag{257}$$

and $A_i$ is defined as

$$A_i = \phi_i - G_i^* H_i \quad . \tag{258}$$

For an asymptotically uniform stable filter, the following inequality holds:

$$\|\psi(k,0)\| \leq d_1 e^{-d_2 k} \quad , \quad d_1 \text{ and } d_2 > 0 \quad . \tag{259}$$

Hence, the term $\psi(i,0)\varepsilon_0^*$ in Eq. (256) may be dropped if a sufficiently long time is allowed prior to the processing of the measurement data. Under this condition, Eq. (256) becomes

$$\varepsilon_i^* = \sum_{k=0}^{i-1} \psi(i, k+1)[\Gamma_k \tau_k - G_k^* \xi_k] \quad . \tag{260}$$

Let

$$Q^* = \sum_{n=1}^{N} Q_n \alpha_n \quad ; \quad R^* = \sum_{n=1}^{N} R_n \alpha_n \quad ; \quad S^* = \sum_{n=1}^{N} S_n \alpha_n \quad .$$

Substitute Eq. (260) into (255):

$$\begin{aligned} E\{\sigma_i^* \sigma_{i-\ell}^{*T}\} = {} & H_i E\{(\sum_{k=0}^{i-\ell} \psi(i, k+1)[\Gamma_k \tau_k - G_k^* \xi_k]) \\ & \times (\sum_{k=0}^{i-\ell-1} \psi(i-\ell, k+1)[\Gamma_k \tau_k - G_k^* \xi_k])^T\} H_{i-\ell}^T \\ & + H_i E\{(\sum_{k=0}^{i-\ell} \psi(i, k+1)[\Gamma_k \tau_k - G_k^* \xi_k]) \xi_{i-\ell}^T\} \\ & + R^* \delta_\ell \quad . \end{aligned} \tag{261}$$

After simplification, Eq. (261) becomes

$$E\{\sigma_i^* \sigma_{i-\ell}^{*T}\} = \sum_{n=1}^{N} F_n(i, \ell) \alpha_n \quad , \tag{262}$$

where

$$\begin{aligned} F_n(i, \ell) = {} & H_i \{ \sum_{k=0}^{i-\ell-1} \psi(i, k+1)[\Gamma_k Q_n \Gamma_k^T - G_k^* S_n \Gamma_k^T \\ & - \Gamma_k S_n G_k^{*T} + G_k^* R_n G_k^{*T}] \psi^T(i-\ell, k+1)\} H_{i-\ell}^T \\ & + [H_i \psi(i, i-\ell+1) \Gamma_{i-\ell}] S_n \\ & - [H_i \psi(i, i-\ell+1) G_{i-\ell}^*] R_n + R_n \delta_\ell \quad . \end{aligned} \tag{263}$$

In order to apply the conventional least-squares minimization technique, both the observed ($\sigma_i \sigma_{i-\ell}^T$) and the theoretical ($\sum_{n=1}^{N} F_n(i, \ell) \alpha_n$) correlation matrices will be transformed into vectors. First, vector $\beta(i, \ell)$ is defined as

$$\beta(i, \ell) = \mathrm{vec}[\sigma_i \sigma_{i-\ell}^T] \quad . \tag{264}$$

For example, if $[\sigma_i \sigma_{i-\ell}^T]$ is a 2 × 2 matrix:

$$[\sigma_i \sigma_{i-\ell}^T] = \begin{bmatrix} S_{11} & S_{12} \\ S_{21} & S_{22} \end{bmatrix} ,$$

then,

$$[\sigma_i \sigma_{i-\ell}^T] = \begin{pmatrix} S_{11} \\ S_{21} \\ S_{12} \\ S_{22} \end{pmatrix}$$

Similarly, define a set of vectors, $\text{vec}[F_1(i,\ell)]$, $\text{vec}[F_2(i,\ell)]$, ..., $\text{vec}[F_n(i,\ell)]$ using matrices $F_1(i,\ell)$, $F_2(i,\ell)$, ..., $F_n(i,\ell)$ for $\ell > 0$. In the case when $\ell = 0$, both vectors $\text{vec}[F_n(i,0)]$ and $\text{vec}[\sigma_i \sigma_i^T]$ have repeated entries. That is, in the above example, $S_{12} = S_{21}$ for $\ell = 0$. Repetition of this type will be removed when forming the vectors.

Next, define a matrix

$$\mathcal{F}(i, \ell) = [\text{vec } F_1(i, \ell)$$
$$\text{vec } F_2(i, \ell), \ldots, \text{vec } F_N(i, \ell)] \quad . \tag{265}$$

Clearly, the observation vector defined in Eq. (264) is related to the matrix defined in (265):

$$\beta(i, \ell) = \mathcal{F}(i, \ell)\alpha + \eta(i, \ell) \quad , \tag{266}$$

in which $\eta(i, \ell)$ is the observation noise vector. Equation (266) represents the measurement equation, in which $\alpha$ is the state vector to be estimated.

The double variables $(i, \ell)$ appearing in Eq. (266) may be converted into a single variable using the transformation:

$$t = (i-1)(L+1) + \ell \quad , \tag{267}$$

where $L$ is the maximum value of the shift integer. For example, if $L = 3$, then

$$i = 1,\ \ell = 0 \rightarrow t = 0 \ ;\quad \beta(1,0) \rightarrow \beta(0) \ ;$$
$$i = 1,\ \ell = 1 \rightarrow t = 1 \ ;\quad \beta(1,1) \rightarrow \beta(1) \ ;$$
$$i = 1,\ \ell = 2 \rightarrow t = 2 \ ;\quad \beta(1,2) \rightarrow \beta(2) \ ;$$
$$i = 1,\ \ell = 3 \rightarrow t = 3 \ ;\quad \beta(1,3) \rightarrow \beta(3) \ ;$$
$$i = 2.\ \ell = 0 \rightarrow t = 4 \ ;\quad \beta(2,0) \rightarrow \beta(4) \ ;$$
$$i = 2,\ \ell = 1 \rightarrow t = 5 \ ;\quad \beta(2,1) \rightarrow \beta(5) \ ;$$
$$\vdots \qquad\qquad \vdots$$
etc. etc.

Hence Eq. (266) can be written as

$$\beta(t) = \mathcal{G}(t)\alpha + \eta(t) \tag{268}$$

or

$$\beta_t = \mathcal{G}_t\alpha + \eta_t \ , \tag{269}$$

with the understanding that the observations are ordered as follows: $\beta(1,0)$, $\beta(1,1),\ldots,\beta(1,L)$, $\beta(2,0)$, $\beta(2,1),\ldots,\beta(2,L)$, $\ldots,\beta(i,0)$, $\beta(i,1),\ldots,\beta(i,L)\ldots$, $(L,L)$.

The problem is to find the best estimate of $\alpha_t$, denoted by $\hat{\alpha}_t$, that minimizes the weighted square function:

$$J_t = \eta_t^T W_t^{-1} \eta_t \tag{270}$$

in which $W_t^{-1}$ is the weighted function. Solution to this problem is well documented in Kalman filtering literatures:

$$\hat{\alpha}_{t+1} = \hat{\alpha}_t + P_{t+1}\mathcal{G}_{t+1}^T W_{t+1}^{-1}[\beta_{t+1} - \mathcal{G}_{t+1}\hat{\alpha}_t] \ , \tag{271}$$

where

$$P_{t+1} = P_t - P_t\mathcal{G}_{t+1}^T[W_{t+1} + \mathcal{G}_{t+1}P_t\mathcal{G}_{t+1}^T]^{-1}\mathcal{G}_{t+1}P_t \ . \tag{272}$$

Recursive computation begins with initial values for $\alpha_0$ and $P_0$:

$$\alpha_0 \equiv E\{\alpha\} \quad (\textit{a priori} \text{ estimate}) \ , \tag{273}$$

$$P_0 \equiv E\{(\alpha - \alpha_0)(\alpha - \alpha_0)^T\} \quad . \tag{274}$$

For minimum variance estimation errors, $W_t$ must be chosen to be

$$W_t = E\{\eta_t \, \eta_t^T\} \quad . \tag{275}$$

Part two: Adaptation of first-order noise statistics

Consider Eqs. (232) and (233) again, in which the mean values $\mu_u$ and $\mu_v$, are now the only unknowns to be estimated. Thus, all quantities that do not depend on the noise means are assumed optimal, e.g., $G_i^*$ will be replaced with $G_i$, $Q^*$ will be replaced with $Q$, etc. The cost function to be minimized is

$$J(\nu^*) = \sum_{i=1}^{m} \nu_i^{*T} \, V_i^{-1} \, \nu_i^* \quad , \tag{276}$$

where $\nu_i^*$ is the innovation sequence, $m$ is the terminal time index in the interval of interest, and $V_i$ is the covariance of $\nu_i$ (the innovation sequence of optimal filter). Defining,

$$\Delta\mu_v = \mu_v - \mu_v^* \quad , \quad \text{and} \quad \Delta\mu_u = \mu_u - \mu_u^* \quad . \tag{277}$$

From the above definitions, Eqs. (247) and (246) can be written as

$$\nu_i^* = H_i e_i^* + \Delta\mu_v + (v_i - \mu_v) \quad , \tag{278}$$

and

$$e_{i+1}^* = (\phi_i - G_i H_i) e_i^* + \Gamma_i \Delta\mu_u - G_i \Delta\mu_v + \Gamma_i (u_i - \mu_u) - G_i (v_i - \mu_v) \quad , \tag{279}$$

respectively. In the optimal case ($\Delta\mu = \Delta\mu = 0$), Eqs. (278) and (279) become

$$\nu_i = H_i e_i + v_i - \mu_v \quad , \tag{280}$$

$$e_{i+1} = (\phi_i - G_i H_i) e_i + \Gamma_i (u_i - \mu_u) - G_i (v_i - \mu_v) \quad . \tag{281}$$

Defining,

$$\Delta\nu_i^* = \nu_i^* - \nu_i \quad \text{and} \quad \Delta e_{i+1}^* = e_{i+1}^* - e_{i+1} \quad . \tag{282}$$

Combining Eqs. (278) and (280):

$$\Delta\nu_i^* = H_i \, \Delta e_i^* + \Delta\mu_v \quad . \tag{283}$$

Combining Eqs. (279) and (281):

$$\Delta e_{i+1}^* = (\phi_i - G_i H_i) \Delta e_i^* + \Gamma_i \, \Delta\mu_u - G_i \, \Delta\mu_v \quad . \tag{284}$$

Let

$$\Delta b = b - b^* = \begin{pmatrix} \Delta\mu_u \\ \Delta\mu_v \end{pmatrix} ; \quad b = \begin{pmatrix} \mu_u \\ \mu_v \end{pmatrix} ; \; b^* = \begin{pmatrix} \mu_u \\ \mu_v \end{pmatrix} ; \tag{285}$$

and

$$D_{i+1} = (\phi_i - G_i H_i) D_i + [\Gamma_i \; \vdots \; -G_i] \quad . \tag{286}$$

Combining Eqs. (284), (285), and (286):

$$\Delta e_i^* = D_i \, \Delta b \quad . \tag{287}$$

Hence, Eq. (283) can be written as

$$\Delta\nu_i^* = H_i D_i \, \Delta b + \Delta\mu_v \quad . \tag{288}$$

But from Eq. (285),

$$\Delta\mu_v = [0 \; \vdots \; I] \Delta b \quad , \tag{289}$$

therefore

$$\Delta\nu_i^* = (H_i D_i + [0 \; \vdots \; I]) \Delta b \quad . \tag{290}$$

Let

$$\Omega_i = (H_i D_i + [0 \; \vdots \; I]) \quad , \tag{291}$$

then Eq. (290) can be written as

$$\nu_i^* = \Omega_i (b - b^*) + \nu_i \quad . \tag{292}$$

The reason for expressing $\nu_i^*$ in terms of $b^*$ is that Eq. (276) may be formulated such that some gradient techniques can be applied to estimate the vector $b$, which is the solution to the problem. Therefore an expression for the gradient of $J(\nabla_{b^*} J)$ shall be formed using Eq. (276). Let $b^*$ be changed by an amount $\delta b^*$, then

$$\nu_i + \delta\nu_i^* = \Omega_i(b - b^* - \delta b^*) + \nu_i \tag{293}$$

or

$$\nu_i^* + \delta\nu_i^* = \nu_i^* - \Omega_i\ \delta b^* \quad . \tag{294}$$

Returning to Eq. (276) and replacing $\nu^*$ by $(\nu^* + \delta\nu^*)$

$$J(\nu^* + \delta\nu^*) = \sum_{i=1}^{m} (\nu_i^* - \Omega_i\ \delta b^*)^{\mathrm{T}}\ V_i^{-1}(\nu_i^* - \Omega_i\ \delta b^*) \tag{295}$$

or

$$J(\nu^* + \delta\nu^*) = \sum_{i=1}^{m} \nu_i^{*\mathrm{T}}\ V_i^{-1}\ \nu_i^* - 2\sum_{i=1}^{m} \delta b^{*\mathrm{T}}\ \Omega_i^{\mathrm{T}}\ V_i^{-1}\ \nu_i^*$$

$$+ \sum_{i=1}^{m} \delta b^{*\mathrm{T}}\ \Omega_i^{\mathrm{T}}\ V_i^{-1}\ \Omega_i\ \delta b^* \quad . \tag{296}$$

Defining

$$dJ = J(\nu^* + \delta\nu^*) - J(\nu^*) \quad , \tag{297}$$

then from Eqs. (276) and (296):

$$dJ = -2\sum_{i=1}^{m} \delta b^{*\mathrm{T}}\ \Omega_i^{\mathrm{T}}\ V_i^{-1}\ \nu_i^*$$

$$+ \sum_{i=1}^{m} \delta b^{*\mathrm{T}}\ \Omega_i^{\mathrm{T}}\ V_i^{-1}\ \Omega_i\ \delta b^* \tag{298}$$

If $\delta b^*$ is small, the second term can be neglected, hence

$$\nabla_{b^*} J = -2\sum_{i=1}^{m} \Omega_i^{\mathrm{T}}\ V_i^{-1}\ \nu_i^* \tag{299}$$

Equation (299) is used to implement the estimator. In order to verify that the estimated value for $b$ is corresponding to a minimum point of $J$, it is necessary to examine the second derivative of $J$ with respect to $b^*$, i.e.,

$$\nabla_{b^*} J(\nu^* + \delta b^*) = -2 \sum_{i=1}^{m} \Omega_i^T V_i^{-1}(\nu_i^* - \Omega_i \delta b^*) \quad . \qquad (300)$$

Defining:

$$\Delta(\nabla_{b^*} J) = \nabla_{b^*} J(\nu^* + \delta b^*) - \nabla_{b^*} J(\nu^*) \quad . \qquad (301)$$

Using Eqs. (300) and (301), the derivative of the gradient is

$$\Delta(\nabla_{b^*} J) = 2 \sum_{i=1}^{m} \Omega_i^T V_i^{-1} \Omega_i \quad . \qquad (302)$$

The estimation of vector $b$ may be accomplished through the following steps:

(1) Initialize the estimator with a starting guess of $b_0^*$, $V_0$, and a predetermined value of $m$.

(2) Use the observed data $z_i$, the suboptimal means $b_i^*$ and Eq. (262) for updating $V_i$ to compute $\nu_i^*$, $J_i$, and $\Delta J_i/\Delta b_i^*$. Since $V_i = E\{\nu_i \nu_i^T\}$ and $\nu_i$ and $\sigma_i$ have the same statistical properties, $V_i = E\{\sigma_i^* \sigma_{i-\ell}^{*T}\}$ with the asterisks removed and set $\ell = 0$.

(3) Apply a gradient technique to adjust $b_i^*$ to $b_{i+\ell}^*$ such that $J_{i+1} < J_i$.

(4) Repeat steps 2 and 3 until convergence is obtained.

## *C. COMPUTER SIMULATIONS*

An adaptive Kalman filtering computer program has been developed using algorithms presented in the previous subsection. The program is coded in Fortran for the CDC 6600 system. Conditions and results of a two-part simulation are given below:

Part I. Conditions for adaptation of second-order statistics

(1) Initial values of $Q$ and $R$ are 2 times the true value:

$$Q_0 = 10^{-10} \begin{bmatrix} 3600 & 0 & 0 \\ 0 & 3600 & 0 \\ 0 & 0 & 81 \end{bmatrix} , \tag{303}$$

$$R_0 = \begin{bmatrix} (0.028)^2 & 0 \\ 0 & (0.028)^2 \end{bmatrix} \tag{304}$$

Value of $S$ is zero at all times (assume no correlation), therefore

$$S_0 = [0] \quad . \tag{305}$$

(2) Assume the initial estimate of $\alpha$ to be

$$\alpha_0 = (\ 1 \quad 1 \quad 1 \quad 1 \quad 1\ ) \quad . \tag{306}$$

Based on the assumptions stated in (1), the true final value of $\alpha$ is

$$\alpha_F = (\ \tfrac{1}{2} \quad \tfrac{1}{2} \quad \tfrac{1}{2} \quad \tfrac{1}{2} \quad \tfrac{1}{2}\ ) \quad . \tag{307}$$

(3) The initial covariance matrix is assumed to be

$$P_0 = 100\ I \tag{308}$$

where $I$ is a $5 \times 5$ identity matrix.

(4) It is desired to observe and compare the rms latitude and longitude errors for two cases:

Case 1. The true values of $Q$ and $R$ in a nonadaptive optimal Kalman filter.

Case 2. Use two times the true value of $Q$ and $R$ as

initial values in the adaptive Kalman filter.

Results are given in Fig. 4. Figure 4a gives rms latitude error, $\delta L$ as a function of number of measurements in time, and Fig. 4b gives rms longitudinal error $\delta\lambda$. Vertical scales have been normalized; measurements are 1-hr apart. Curves A and B on each figure indicate rms error obtained in a nonadaptive Kalman filter. In curve C, rms errors for both $\delta L$ and $\delta\lambda$ converge on their true values after approximately 250 measurements.

Part II. Conditions for adaptation of first-order statistics

(1) The true values of vector $b$ are zero. The initial values are assumed to be

$$b_0^T = (\ 1\quad 1\quad 1\quad 1\quad 1\ )\ . \tag{309}$$

(2) Steady-state values of $Q$ and $R$ obtained from Part I, together with the known $S = 0$, are used as known parameters.

(3) The gradient technique given in ref. [72] is used,

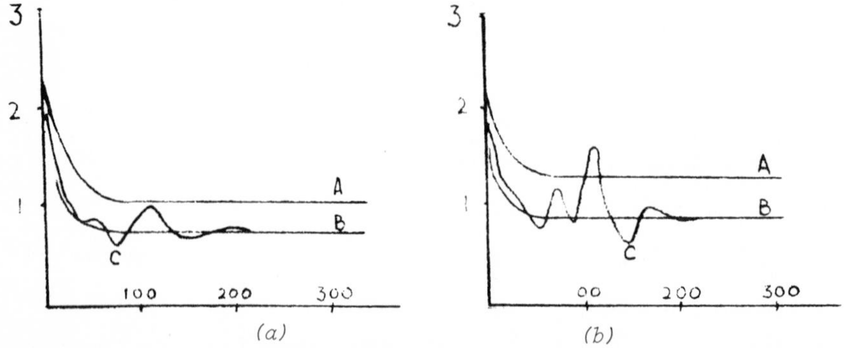


*FIG. 4. (a) rms latitude error versus number of measurements; (b) rms longitude error versus number of measurements. In both diagrams, A represents nonadaptive suboptimal Kalman filter with noise covariances twice the true value; B represents nonadaptive optimal Kalman filter with true noise covariances; and C is adaptive Kalman filter with unknown noise covariances.*

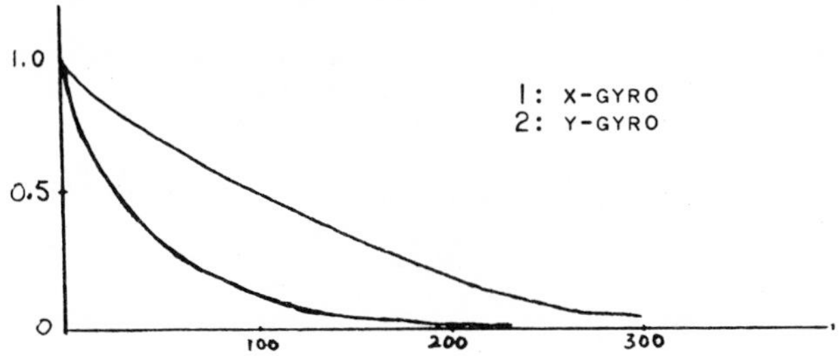


*FIG. 5. Estimated noise means versus number of measurements.*

i.e., choose $\delta b^*$ to be in opposite direction to $\nabla_{b^*} J$:

$$\delta b^* = -\beta \nabla_{b^*} J \quad , \tag{310}$$

where $\beta = 0.0001$, a value selected for rapid convergence.

(4) Other parameters are also chosen for the convenience of making computations, e.g., $K_f = 10$ and $N = 5$ are used.

Results: Two of the five estimated values of the $b$ vector are given in Fig. 5, in which the mean values of the $x$ and $y$ gyro noise are shown converging toward their true values after 200 measurements.

## V. SUMMARY

This chapter has presented an overview of recent advances in linear adaptive filtering theory. An application of the theory to solve a practical problem was also given. The overview of the theory was divided into two parts: analysis and synthesis. In the analysis section, three main topics were reviewed: covariance analysis, model testing, and implementation errors. In the synthesis section, two major areas were examined: adaptive techniques and error bounding techniques. The adaptive techniques discussed encompass noise parameter adaptation, system parameter adaptation, and system structure adaptation. Error bounding techniques included computer constraints and the filter divergence problem.

An application of the noise parameter adaptation technique was considered to demonstrate the feasibility of performing on-line identification of unknown error model parameters of an inertial navigation system, and several interesting results were obtained and shown in Figs. 4 and 5.

This chapter gives an extensive overview of the topic, but by no means has the discussion been exhausted. It is hoped that materials presented in this chapter will provide a proper frame for continuous research in the advancement of adaptive filtering theory and applications.

## *REFERENCES*


1. O. NEUGEBAUER, "The Exact Sciences in Antiquity," Princeton University Press, New Jersey, 1952.
2. H. W. SORENSON, *Proc. SW IEEE Conf.* (1972).
3. R. A. FISHER, "Contributions to Mathematical Statistics," Wiley, New York, 1950.
4. A. N. KOLMOGOROV, *Bull. Acad. Sci. USSR, Math. Ser. 5*, 3 (1941).
5. N. WIENER, "The Extrapolation, Interpolation, and Smoothing of Stationary Time Series," Wiley, New York, 1949.
6. P. SWERLING, *J. Astron. Sci. 6*, 46 (1959).
7. A. G. CARLTON and J. W. FOLLIN, *AGARDograph*, 21 (1956).
8. R. E. KALMAN, *J. Basic Eng. Trans., ASME, 82*, 35 (1960).
9. R. E. KALMAN and R. S. BUCY, *J. Basic Eng. Trans., ASME, 83*, 95 (1961).
10. R. E. KALMAN, T. S. ENGLAR, and R. S. BUCY, "Fundamental Study of Adaptive Control Systems," RIAS Report No. ASD-TR-61-27, 1962.
11. H. HEFFES, *IEEE Trans. Auto. Cont., AC-11*, 541 (1966).
12. T. NISHIMURA, *IEEE Trans. Auto. Cont., AC-12* (1967).
13. S. R. NEAL, *IEEE Trans. Auto. Cont., AC-12* (1967).
14. R. E. GRIFFIN and A. P. SAGE, *IEEE Trans. Auto. Cont., AC-13*, 320 (1968).
15. C. F. PRICE, *IEEE Trans. Auto. Cont., AC-13*, 699 (1968).
16. J. R. HUDDLE and D. A. WISMER, *IEEE Trans. Auto. Cont., AC-13*, 421 (1968)
17. H. W. SORENSON, *IEEE Trans. Auto. Cont., AC-12*, 557 (1967).
18. R. J. FITZGERALD, *Proc. 2nd IFAC Symp. Auto. Cont. in Space*, Vienna, Austria, 1967.
19. F. H. SCHLEE, C. J. STANDISH, and N. F. TODA, *AIAA Journal*,

*5*, 1114 (1967).
20. J. W. BERKOVEC, *JACC Proc. 488* (1969).
21. R. K. MEHRA, *IEEE Trans. Auto. Cont.*, *AC-15*, 175 (1970).
22. S. F. SCHMIDT, "Theory and Applications of Kalman Filtering," *AGARDograph*, *139* (1970).
23. J. R. HUDDLE, "Theory and Applications of Kalman Filtering, *AGARDograph*, *139* (1970).
24. R. H. BATTIN and G. M. LEVINE, "Theory and Applications of Kalman Filtering," *AGARDograph*, *139* (1970).
25. A. GELB, "Applied Optimal Estimation," The MIT Press, Massachusetts, 1974.
26. R. A. NASH, J. A. D'APPOLITO, and K. J. ROY, *AIAA Guid. Cont. Conf.*, Stanford, California, 1972.
27. R. S. BUCY and P. D. JOSEPH, "Filtering for Stochastic Processes with Applications to Guidance," Interscience Publisher, New York, 1968.
28. A. ANDREWS, *AIAA Journal*, 1165 (1968).
29. J. F. BELLANTONI and K. W. DODGE, *AIAA Journal*, 1309 (1967).
30. P. DYER and McREYNOLDS, *J. Opt. Theory Appl.*, *3*, 444 (1969).
31. P. G. KAMINSKI, A. E. BRYSON, and J. F. SCHMIDT, *IEEE Trans. Auto. Cont.*, *AC-16*, 727 (1971).
32. N. A. CARLSON, *AIAA Journal*, *11*, 1259 (1973).
33. T. KAILATH, *IEEE Trans. Inf. Theory*, *IT-19*, 750 (1973).
34. A. LINDQUIST, *SIAM J. Cont.*, *12* (1974).
35. M. MORF, G. SIDHU, and T. KAILATH, *IEEE Trans. Auto. Cont.*, *AC-19*, 315 (1974).
36. S. CHANDRASEKHAR, *Astrophys. J.*, *106*, 152 (1947); *107*, 48 (1948).
37. A. A. DESALU, L. A. GOULD, and F. C. SCHWEPPE, *IEEE Trans. Auto. Cont.*, *AC-19*, 904 (1974).
38. D. T. MAGILL, *IEEE Trans. Auto. Cont.*, *AC-10*, 434 (1965).
39. C. G. HILBORN and D. G. LAINIOTIS, *IEEE Trans. Sys. Sci. Cybernetics*, *SSC-5*, 38 (1969).
40. J. C. SHELLENBARGER, *Proc. Nat. Elect. Conf.*, *22*, 698 (1966).
41. T. KAILATH, *IEEE Trans. Auto. Cont.*, *AC-13* (1968)
42. R. K. MEHRA, *IEEE Trans. Auto. Cont.*, *AC-17*, 693 (1972).
43. B. CAREW and P. R. BELANGER, *IEEE Trans. Auto. Cont.*, *AC-18*, 582 (1973).
44. B. FRIEDLAND, *IEEE Trans. Auto. Cont.*, *AC-14*, 359 (1969).
45. J. T. LIN and A. P. SAGE, *IEEE Trans. Sys. Mgt. Cybernetics*, *SMC-1*, 314 (1971).
46. A. P. SAGE and G. W. HUSA, *Proc. 8th IEEE Symp. Adaptive Proc.* (1969).
47. S. S. GODBOLE, *IEEE Proc. Decision Cont. Conf.* (1973).
48. L. CHIN, *Proc. JACC*, Philadelphia, Pennsylvania (1978).
49. D. M. BOROSON and S. C. SCHWARTZ, *Proc. JACC*, San Francisco, California (1977).
50. D. J. KRAUSE and D. GRAUPE, *Proc. 3rd Symp. Nonlinear Estimation Theory*, San Diego, California (1972).
51. D. G. LAINIOTIS, *IEEE Trans. Auto. Cont.*, *AC-16*, 160 (1971).
52. D. G. LAINIOTIS, *Proc. IEEE*, *64*, 1126 (1976).
53. O. M. MOSTAFA, *Proc. JACC*, San Francisco, California (1977).
54. S. F. SCHMIDT, "Advances in Control Systems," Vol. 3 (C. T. LEONDES, ed.), Academic Press, New York, 1966.

55. R. COSAERT and E. GOTTZEIN, *IFAC Symp.*, Vienna (1967).
56. A. H. JAZWINSKI, *IFAC Symp.*, Dusseldorf (1968).
57. A. H. JAZWINSKI, "Stochastic Processes and Filtering Theory," Academic Press, New York, 1966.
58. T. S. TARN and J. ZABORSKY, *AIAA Journal*, *8*, 1127 (1970).
59. R. W. MILLER, *AIAA Journal*, *9*, 537 (1971).
60. J. E. SACKS and H. W. SORENSON, *AIAA Journal*, *9*, 767 (1971).
61. N. E. NAHI and B. M. SCHAEFER, *IEEE Trans. Auto. Cont.*, *AC-17*, 61 (1972).
62. A. H. JAZWINSKI, *Automatica*, *5*, 975 (1969).
63. P. D. ABRAMSON, "Simultaneous Estimation of the State and Noise Statistics in Linear Dynamic Systems," M.I.T. Report TE-25, 1968.
64. P. R. BELANGER, *Automatica*, *10*, 267 (1974).
65. C. NEETHLING and P. YOUNG, *IEEE Trans. Auto. Cont.*, *AC-19*, 623 (1974).
66. R. K. MEHRA, *IEEE Trans. Auto. Cont.*, *AC-16*, 12 (1971).
67. R. G. BROWN and G. L. HARTMAN, *Proc. Nat. Elect. Conf.*, *24*, 67 (1966).
68. M. S. BARTLETT, "An Introduction to Stochastic Processes," Cambridge University Press, London, 1962.
69. A. E. BRYSON and Y. C. HO, "Applied Optimal Control," Waltham, Massachusetts, Blaisdell, 1969.
70. Y. C. HO and R. C. K. LEE, *J. Inf. Cont.*, *8*, 93 (1965).
71. E. E. BOX and G. M. JENKINS, "Statistical Models for Forecasting and Control," Holden-Day, California, 1970.
72. H. J. KELLEY, *J. Am. Rocket Soc.* (1960).

# INDEX

## D



## E



## F



## G



## H



## I



## K

## S



## T



## V



## W



## Y



## Z